Information Systems

Chapman & Hall/CRC
Textbooks in Computing

Series Editors

John Impagliazzo
Andrew McGettrick

Pascal Hitzler, Markus Krötzsch, and Sebastian Rudolph, Foundations of Semantic Web Technologies

Henrik Bærbak Christensen, Flexible, Reliable Software: Using Patterns and Agile Development

John S. Conery, Explorations in Computing: An Introduction to Computer Science

Lisa C. Kaczmarczyk, Computers and Society: Computing for Good

Mark Johnson, A Concise Introduction to Programming in Python

Paul Anderson, Web 2.0 and Beyond: Principles and Technologies

Henry Walker, The Tao of Computing, *Second Edition*

Ted Herman, A Functional Start to Computing with Python

Mark Johnson, A Concise Introduction to Data Structures Using Java

David D. Riley and Kenny A. Hunt, Computational Thinking for the Modern Problem Solver

Bill Manaris and Andrew R. Brown, Making Music with Computers: Creative Programming in Python

John S. Conery, Explorations in Computing: An Introduction to Computer Science and Python Programming

Jessen Havill, Discovering Computer Science: Interdisciplinary Problems, Principles, and Python Programming

Efrem G. Mallach, Information Systems: What Every Business Student Needs to Know

Iztok Fajfar, Start Programming Using HTML, CSS, and JavaScript

Mark C. Lewis and Lisa L. Lacher, Introduction to Programming and Problem-Solving Using Scala, *Second Edition*

Aharon Yadin, Computer Systems Architecture

Mark C. Lewis and Lisa L. Lacher, Object-Orientation, Abstraction, and Data Structures Using Scala, *Second Edition*

Henry M. Walker, Teaching Computing: A Practitioner's Perspective

For more information about this series please visit:

https://www.crcpress.com/Chapman--HallCRC-Textbooks-in-Computing/book-series/CANDHTEXCOMSER?page=2&order=pubdate&size=12&view=list&status=published,forthcoming

Information Systems

What Every Business Student Needs to Know

Second Edition

by

Efrem G. Mallach

CRC Press is an imprint of the
Taylor & Francis Group, an **informa** business
A CHAPMAN & HALL BOOK

CRC Press
Taylor & Francis Group
6000 Broken Sound Parkway NW, Suite 300
Boca Raton, FL 33487-2742

Printed on acid-free paper

International Standard Book Number-13: 978-0-367-18353-0 (Paperback)
International Standard Book Number-13: 978-0-367-18354-7 (Hardback)

Library of Congress Cataloging-in-Publication Data

Names: Mallach, Efrem, 1942-author.
Title: Information systems : what every business students needs to know /
by Efrem G. Mallach.
Description: Second edition. | Boca Raton, FL : CRC Press, [2020] | Series:
Chapman & Hall/CRC textbooks in computing | Includes bibliographical
references and index.
Identifiers: LCCN 2019049011 | ISBN 9780367183530 (paperback ; alk. paper)
| ISBN 9780367183547 (hardback ; alk. paper) | ISBN 9780429061011
(ebook)
Subjects: LCSH: Management information systems. | Business–Data
processing. | Business–Information technology. | Information
technology–Management.
Classification: LCC HD30.213.M342 2020 | DDC 658.4/038011–dc23
LC record available at https://lccn.loc.gov/2019049011

Visit the Taylor & Francis Web site at
http://www.taylorandfrancis.com

and the CRC Press Web site at
http://www.crcpress.com

Contents

1 Why Information Systems Matter in Business—And to You

CHAPTER OUTLINE

WHY THIS CHAPTER MATTERS

In 2020, computer literacy is a given. Every middle-school student is computer-literate, just as they all know how to read. This book won't try to repeat what you've known for years.

Knowing how to read isn't the same as understanding literature or poetry. Computer literacy isn't the same as information literacy either. Middle-school students are computer-literate, but they are not information-literate. They don't understand how information can benefit an organization, what affects its usefulness for that purpose, and how it should be managed to be as useful as possible. Middle-school students don't have to understand that any more than they have to understand the structure of a play or the use of a metaphor—but college students who major in English do have to understand those things, so they study them. As a business student, and later in your career as a knowledge worker and as a manager, you have to be *information-literate*. That's what this chapter is about.

Another way of looking at it: it's the difference between "How do I use this?," which you already know, and "How do I use this to make a difference?"

CHAPTER TAKE-AWAYS

As you read this chapter, focus on these key concepts to use on the job:

1. Intelligent use of information can help any type of organization.
2. The value of information depends on its quality. Information quality can be described by a few specific factors.
3. Computers are basic to using information intelligently. A company can't use information intelligently without using computer-based information systems intelligently.
4. You will benefit personally in your career if you understand information systems.

1.1 THE VALUE OF INFORMATION

Consider these three business scenarios.

Scenario 1

A toy manufacturer gets an order for 5,000 wagons. It needs 20,000 wheels to produce them. It has only 4,000 wheels in stock. It must order at least 16,000 more wheels.

Possible Outcome A

The production control manager phones the purchasing agent who handles wheels. She's out for the rest of the day, so the production control manager leaves a voice mail. When the purchasing agent returns the call the next morning, the production control manager is on the factory floor, so she leaves a voice mail. They talk the day after that. The purchasing agent then calls three suppliers from whom the toy manufacturer has bought wheels in the past. A is no longer in business. B's sales representative has just left on a three-week vacation. She leaves a voice mail for C's salesperson. The salesperson senses desperation in the purchasing agent's message and quotes a price that is 25% higher than usual. C gets the order anyhow.

Possible Outcome B

The toy company's production planning system calculates the need and sends an electronic message to the purchasing department. Workflow software in the purchasing department routes the request to the purchasing agent, but will reroute it to her manager if she doesn't process it within 24 hours. She processes it the next morning, sending electronic Request for Bid messages to three firms that have supplied wheels in the past. She gets bids back from the two that are still in business. C quotes standard prices. B, whose sales rep is eager to close some business before leaving on a three-week vacation, offers a 20% discount. This time, B gets the order.

Scenario 2

A college professor enjoys solving puzzles and often buys puzzle books online. Two bookstores get a new puzzle book edited by Will Shortz, who is known for high-quality puzzles.

Possible Outcome A

Seller A prices the books at a discount to attract business. It puts them on a shelf and its web site, and waits for people to walk in and buy them or to order them online. Some do, but the professor doesn't happen to visit the store or its site, so she doesn't see it.

Possible Outcome B

Seller B keeps a list of people who have ordered books online in the past, indexed by category. It sends people who bought puzzle books an email to inform them of the new book. The professor orders the book from the store that sent her the email. She doesn't feel like taking time to check other sources, so she doesn't know the other store would sell it for $2 less.

Scenario 3

An athletic equipment store sells light-weight, impact-absorbing running socks for $10 per pair, of which $3.50 is its profit margin. The store's owner figures that many people who buy running shoes could also use a pair of those socks, so he plans a sock promotion to tie in with shoe sales.

Possible Outcome A

The store offers running shoe buyers a coupon for $3 off a pair of these socks. Sock sales go up a lot. The manager decides to include a $3 sock coupon with every pair of running shoes they sell.

Possible Outcome B

The store tries three one-week experiments: coupons for $1, $2, and $3 off a pair of these socks. They plot sock sales as in Figure 1.1(a). They know sales volume is not their objective: they sell more socks with higher discounts, but they also make less profit per pair. They graph profit on socks as in Figure 1.1(b). Based on this information, the manager decides to include a $2 discount coupon with every pair of running shoes.

FIGURE 1.1 (a) Sock sales and (b) profit analysis.

What This Means

Think about these outcomes. In each scenario, which business will do better? *The person or organization that uses information effectively is more likely to come out ahead.* Information can be more important than product features: a product with less capability or a higher price can outsell a better one, if prospective buyers know what it can do and how it can be used.

These three scenarios illustrate the three main ways that information and information systems help companies succeed: linking parts of an organization, connecting organizations to customers and suppliers, and helping make better decisions. A chapter is devoted to each of these key concepts later in this book.

EFFECTIVENESS VERSUS EFFICIENCY

You just read that businesses should use information *effectively.* You've also heard people talk about *efficiency.* A businessperson must know the difference.

Effectiveness describes how well you achieve your objective. To measure effectiveness, you must know what that objective is.

Efficiency describes how much output you get from each unit of input. To measure efficiency, you must be able to measure input and output.

Example: You must choose between two software packages to prepare your income tax return. One is difficult to use, with complicated data entry procedures that must be followed precisely, and it can't prepare your state return. However, it is very efficient: it runs quickly on your old, slow computer. The other has a much more modern interface, and can prepare returns for any U.S. state, so it gets the job done more effectively—but it runs slowly on your computer. It is inefficient. You would need to upgrade your computer or spend a lot of time waiting for it to complete its calculations. You must know your objectives and tradeoffs before choosing.

Why does this matter? With information systems, you are concerned with both effectiveness and efficiency. Ideally, you want both. However, there are often tradeoffs between them. Their relative importance can vary from one situation to the next. You must understand the difference between the two concepts to make those tradeoffs properly.

Where you fit in: These tradeoffs are business decisions. They shouldn't be made on technical grounds. Technical factors are involved, but the final decision must be made by businesspeople such as you'll be after you graduate.

1.2 SYSTEMS AND INFORMATION SYSTEMS

The scenarios in Section 1.1, "The Value of Information," used computers to improve a business outcome. Using computers is incidental. Computers are a technology that enables certain things to happen. What's important in business is what happens. That's where systems come in.

A *system* is *a group of components that interact for a purpose*. A transportation system consists of vehicles, signals, roads, and more. Each of these components is itself a system. A bus, which is a *subsystem* of the transportation system, has a propulsion system, a braking system, a steering system, and so on. These are made up of still smaller subsystems, down to subatomic particles.

A system has a *boundary*. The boundary defines what is inside the system, what is outside. It is important to know the boundary of any system. Transportation planners may take one approach to improving a system if it goes only as far as the city limits and they cannot control what happens outside them. They may take a different approach if their plan can include the city's suburbs.

Most systems are *open systems*: that is, they communicate with their environment across their boundary.* A system that can't communicate with its environment is a *closed system*.

Many systems incorporate *feedback*. Feedback means that a system output becomes an input to it. For example, as a driver, you maintain speed through the pressure of your foot on the accelerator. That is a system input. Output is the car's speed. Speed is *fed back* into the system via your eyes and the speedometer. If it's too high, you lift your foot and the car slows down. If it's too low, you press harder and the car speeds up. Feedback keeps the system operating in a desirable range.

Feedback can be *internal* or *external*. If we define the system to include the car but not its driver, the feedback in the previous paragraph is *external*. A cruise control feature would, with that same system boundary, use *internal* feedback.

Figure 1.2 shows how input, output, and feedback combine. This system uses external feedback: it selects an advertising program, and the feedback comes via customer behavior after the ads run. Since customers are not part of the system under discussion, this feedback is external.

An *information system* (IS) is a system whose purpose is processing information. Processing information includes the five activities shown in Figure 1.3:

1. Entering data into the system.
2. Processing that data. Processing usually involves *calculating* and *comparing*.
3. Storing data, either processed or not, for future use.

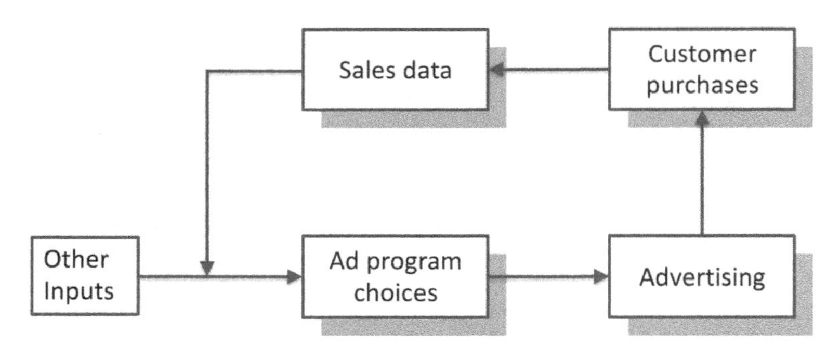

FIGURE 1.2 Conceptual diagram of system showing external feedback.

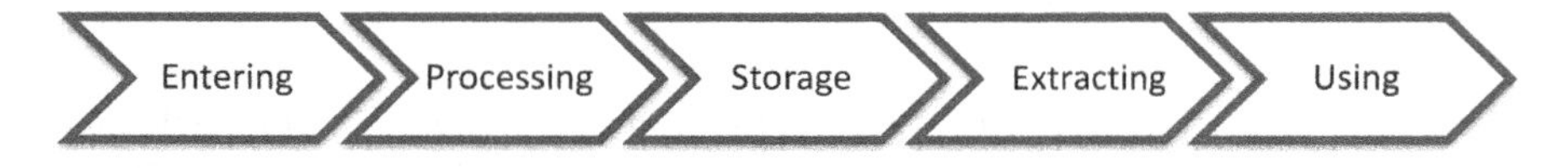

FIGURE 1.3 Conceptual information system processing flow.

* The term *open system* also has other meanings. Context will tell you which is meant in a given situation.

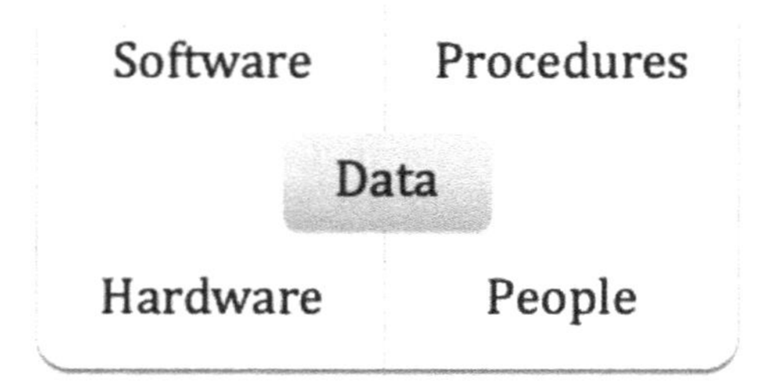

FIGURE 1.4 Conceptual diagram of information system.

4. Extracting information from storage.
5. Using extracted information to benefit the business.

In a well-designed system, each of these activities is performed by a device or person chosen for that purpose. Choosing them is an important part of system design.

Computers are central to information systems today. When most people hear *information system*, they think of computers. However, people are also an important part of almost every information system. We can visualize the components of an information system as in Figure 1.4.

In this diagram:

- The elements in the lower half exist physically. We can point to a person or to a computer, and say "That part of the system carries out Activity X."
- The elements in the upper half tell the lower elements on each side what to do. Those two elements must be designed to work with each other.
- The element in the middle, data, holds the system together. The physical elements of the system access the data. The instructions tell them what to do with it.

The term *information technology* (IT) refers to parts of the system on the left and in the middle. It is important not to confuse IS with IT. Information systems include people, but IT doesn't.

> ***Where you fit in:*** People with a technical background and orientation tend to focus on the left side of the diagram. Businesspeople must sometimes remind them that the elements on the right side are just as important to the overall system, and more important to the organization's satisfaction with what the system does.

1.3 WHAT *IS* INFORMATION, REALLY?

It is impossible to discuss information systems without discussing information. We all have an informal concept of information, but an informal concept isn't enough for serious study.

Definitions

There are several types of what we call "information." Knowing the difference is important to avoid misunderstandings in discussing information systems: what might seem to be a difference of opinion can reflect just using a word differently. Here's how we'll use these terms:

Data is* raw facts in isolation from other facts. The name of this book's publisher is data.

---

* Strict grammarians consider *data* to be a plural noun that refers to many data items. One data item is, in Latin and in formal English, a *datum*. However, most people use *data* as a singular noun in 2020.

Information is data that has been organized and processed to be meaningful to a person (or other information system) who (or which) will use it. A list of publishers who offer textbooks for this course could be information. An instructor selecting a book for it would find such a list useful.

Another definition of information is *anything that reduces uncertainty.* This definition lets us assess the amount of information we have: "The temperature is 85°F" (30°C) conveys more information than "it's in the 80s," which in turn conveys more information than "it's warm." "A new 2019 BMW 230i starts at $36,395 delivered" conveys more information than "you can get a new BMW in the mid-thirties."

Knowledge is the ability to apply information in a business situation. Your instructor may have an effective and efficient process for going through the list of publishers. That knowledge could be shared with other instructors. Some of them might suggest improvements to it, based on their own processes. Knowledge sharing can benefit everyone.

Some people discuss *wisdom* above those three levels, but that's seldom necessary in discussing information systems.

How These Work

Information system activity begins with data. Data comes from somewhere and is entered somehow. Knowing where data comes from and how it gets into an information system is important for evaluating its credibility, as you'll see below.

Data is processed into information via two types of activities: computation and comparison.

Computation creates information by carrying out a predefined process on two or more data items. For example, by taking the start time of a movie and subtracting the time it takes you to get to the theater (both data) you can learn how late you can leave for the theater without missing the opening credits. That's information.

Comparison selects processing steps on the basis of data. Suppose dinner takes you 30 minutes. By subtracting 30 minutes from the time you must leave for the movie, and comparing the result with the opening time of the campus dining hall, you can decide to eat dinner there—or not to.

Comparison can be used to select what data to output (students with grades below 70, accounts with balances over $10 that are more than 60 days old), organize data (sorting into alphabetical order involves comparing two items to see which comes before the other), format text ("if the end of this letter is beyond the line length, move the word it is in to a new line"), control a computer ("if the coordinates of the mouse pointer are within this rectangle, highlight this item") and more.

Information Quality

The degree to which information can contribute to a value-adding business process, and hence the value of that information, depends on its *quality.*

Differences in information value matter when information is used for management decisions. High-quality information enables managers to make good decisions quickly. Lower-quality information leads to poor decisions and wastes decision makers' time.

Information can also link members of an organization, or an organization with its customers. Low-quality information cannot perform these tasks effectively. If the manufacturing department gets a low-quality sales forecast, it might produce the wrong items or the wrong quantities.

Over time, poor information quality can damage a good relationship. When information is used for record keeping, low-quality information impairs the records' value. In extreme cases, low-quality information can lead to financial or legal difficulties.

You understand this informally. Your next step is to use these concepts in working with business information. Whenever you use computer output, you should be aware of the factors that affect its quality and what that quality is likely to be. The key information quality factors are:

Correctness

Information is *correct* if it is derived from the proper data values through the proper processing steps. In this example, an information system fell short on this score:

> *In early computers, where storage cost much more than it does now, storing only the last two digits of years saved space. One woman's birth year was* stored as 95. In 1999, she was invited to enroll in kindergarten. However, she was born in 1895. The school's computer wasn't programmed to allow for 104-year-olds.*

Some correctness situations are absolute. A person is four years old, or she isn't. An email address is correct, or it isn't. A stock number refers to Oikos vanilla Greek yogurt, or it doesn't.

A data element can be correct even if it's a bit off from its actual value. Approximate values are good enough for much numerical data. In fact, the question is often not whether numeric data are exact, because they can't be, but whether they are close enough. Writing $\pi = 3.14159$ is correct to six significant digits. Approximate data are not incorrect, though we must take their possible error into account when we use them. Whether approximate data are of sufficiently high quality is the subject of the next quality factor, *accuracy*.

Non-numeric data can also be "nearly correct." My name is often misspelled. Most misspellings do not prevent the postal service from delivering my mail. Informally, one might say that the spelling "Efram Mallack" is incorrect. As we use the terms here, it is correct—it does, after all, identify me—but inaccurate. U.S. airport security screeners accept boarding passes that differ from names on identification documents by up to three letters, since the Transportation Security Administration (TSA) knows that correct data can be inaccurate.

Most incorrect input data is due to human error. An information system can reduce human data entry errors through *source data automation*, where data is entered automatically. Bar code readers are source data automation devices. A system can also use *data validation* to minimize incorrect data. Checks can never completely eliminate input errors, and they do not address errors that might arise in processing correct inputs.

Case 1 at the end of this chapter discusses an example of incorrect information.

> ***Where you fit in:*** You will often depend on information from computers even though you have no direct knowledge of where the raw data came from or how it was processed. You must make an effort to confirm that the information is based on correct data.

Accuracy

Information is *accurate* if its value is acceptably close to the value that a perfect calculation using perfect inputs would produce. The meaning of "acceptably close" depends on the use of the information. It varies from one situation to another. In the example near the end of the previous section, the TSA defines "acceptably close" as "names differ by no more than three letters."

> *Pre-election poll results may be* ***inaccurate*** *for several reasons:*
>
> - *People might not be honest about their real intentions.*
> - *People's intentions, in terms of how they plan to vote and whether they will vote at all, may change between the poll and the election.*
> - *The sample may not be representative of the population. Reluctance to answer surveys might not be spread equally over supporters of all parties.*
>
> *Some of these can be addressed. If history shows a consistent pattern of poll results differing from election outcomes, it might be possible to compensate for dishonesty and response rate differences. Changes in intent, though, are often driven by external events that no polling organization can foresee.*

---

* Or so it is said. This story may not be true. At least one source, Crandall et al. (2013), says it isn't.

A data item may be inaccurate because it is inherently an estimate. Nobody knows what interest rates will be in two years, though many economists will gladly exchange their guesses for your money. A business can often improve data accuracy by taking more time, spending more money, or both. It's a business decision as to whether the increased accuracy is worth it.

Sensitivity analysis is one way to gauge the need for accuracy. Suppose the cost to develop a new product is estimated at $15,000,000 and its sales are estimated at $100,000,000 over five years. On that basis, the product is expected to be profitable. But what if the estimates are off?

To carry out a sensitivity analysis, we ask if the product would be profitable if development costs and sales estimates were off by, say, 20%. We repeat our analysis for a development cost of $18,000,000 and see that it still would be profitable. We then check for sales of $80,000,000 and learn, to our dismay, that it wouldn't be. This means we should try to improve the accuracy of our sales forecast, but we don't need a more accurate development cost estimate. The sensitivity analysis helps us focus on areas where better accuracy can affect our decision.

Many people use the terms *correctness* and *accuracy* interchangeably, especially informally. This is understandable, since errors in either lead to information that isn't what it should be. However, they are different. The important thing is to be aware of potential sources of error, so you can look out for them. Using different names for different error sources helps do that.

Precision

The *precision* of a data element is the smallest difference that can be represented by the way a data element is stored in a computer or presented to its users. It applies only to numeric data.

Question 1: What is your bank's interest rate for a 30-year home mortgage loan?
"About 4%." Reply: "Come on. Surely you know the exact rate."
"4.125%." Reply: "Thank you, that's what I wanted to know."
Question 2: What will next year's inflation rate be?
"About 4%." Reply: "Fine. Now I can plan my investment approach."
"4.125%." Reply: "Get serious. Nobody can predict it that precisely."
The same pair of answers, which differ only in precision, got opposite reactions in the two scenarios. Why was this? Because different uses of data have different needs for precision, and because precision beyond the available accuracy is useless.

We can visualize precision as a series of steps, in an information storage system or in presenting data to users. Figure 1.5 shows precision this way. Pairs of data elements can be shown at any point where two lines intersect, such as point A, but not at any other points. Data at points such as B and C must be approximated. For example, 7% sales tax on a $1.45 item should be 10.15¢, but in the U.S. consumers pay in whole cents, so this tax is rounded to 10¢.

Excess precision may not reflect the underlying data properly. (It also makes data harder to use.)

The precision of data presentation sends a message about its accuracy. Precision beyond accuracy is misleading. If we are told that one advertising plan has a predicted brand awareness increase of 36.717% and a second 35.941%, we may be tempted to choose the first. If we know both figures are accurate to within 5% or so, we will consider the plans equivalent on this score. Showing the figures as 37% and 36% would be less misleading. Giving them as 32–42% and 31–41%, or as $37 \pm 5\%$ and $36 \pm 5\%$, would make the situation crystal-clear.

Never give more precision than the accuracy justifies. If you plan to buy a car for about $15,000 and a burger for $3.59, it would be misleading to budget $15,003.59 for both. It would be better to say "about fifteen thousand dollars." Creating an appearance of accuracy which isn't there can be unethical, if done to obtain an advantage that the quality of the information doesn't justify.

Issues of excess precision seldom arise in school. If an accounting assignment says depreciation is $10,229, you know it's not a penny more or less. You don't have to worry about where the figure came from or its possible errors. If you round it to $10,230, let alone $10,000, you'll get a poor grade. Business situations are seldom like that.

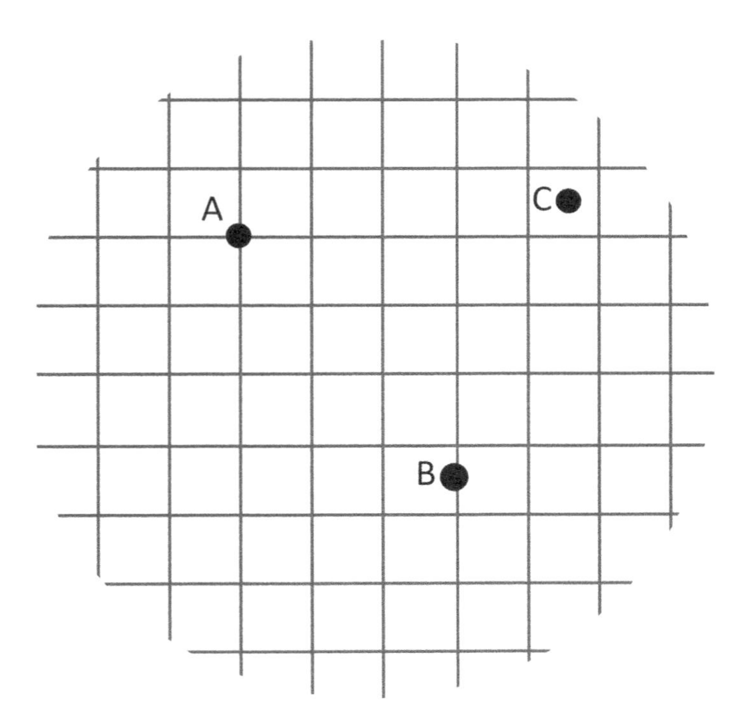

FIGURE 1.5 Circle diagram showing precision.

It may be appropriate to provide *less* precision than the accuracy of the data could justify:

1 Corporate earnings per share are usually stated to the nearest cent. That's precise enough to evaluate profitability trends and to compare different firms' stock. More precision, which can be obtained from data in annual reports, would be pointless.
2 Graphs are often the best way to convey data relationships, but are seldom precise. The graph in Figure 1.6 conveys the history of Disney's earnings per share more quickly than the table, with sufficient precision for most purposes. Where high precision must be combined with the convenience of graphs, graphs can be supported by tables.
3 Color coding classifies data into ranges such as green for "OK," yellow for "marginal," and red for "problem." Like graphs, color can be backed up by numeric data, or used to highlight specific numbers which a user can then read.

Since computers make it easy to calculate with high precision, it is tempting to give results extra digits. Printouts with six digits after the decimal point look impressive, but examine the data first. Displaying more digits than are justified gives a false sense of precision. Suppose we read that the United States (U.S.) covers 3,618,770 square miles, presumably accurate to 5+ significant digits, and we know its population is about 300 million. A computer can divide one by the other for a density of 82.90109623 people per square mile. However, our population figure was a rough guess. It is accurate to one significant digit (the 2020 U.S. population is between 250 and 350 million) but not two (it isn't between 295 and 305 million). All we really know about its population density is that it's about 80 people per square mile. A computer can't tell that one of the inputs was a rough approximation, but an information-literate user must know this or find it out.

When you provide information, you're responsible for presenting it to reflect its accuracy. If a number with six-digit accuracy happens to be exactly 80, showing it as 80.0000 will convey that message. But what about 20,000? Is it accurate to one significant digit, to five, or in between? In

DISNEY (DIS) EARNINGS	
FY	*EPS*
2008	$2.28
2009	$1.76
2010	$2.03
2011	$2.52
2012	$3.13
2013	$3.38

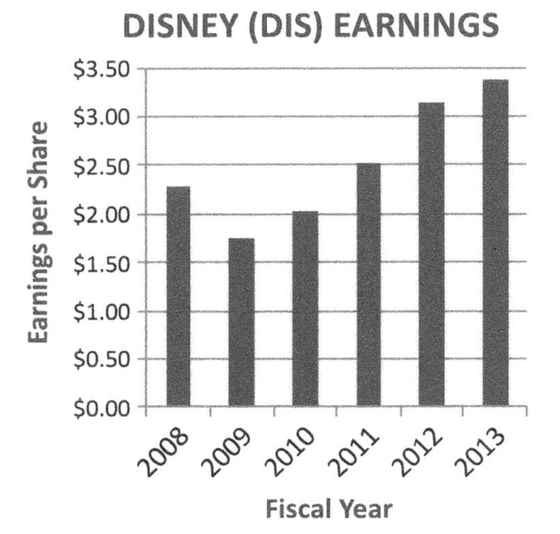

FIGURE 1.6 Earnings per share shown (a) to the nearest cent and (b) on a graph.

Sig. Digits:	*1*	*3*	*5*
	20000	20000	20000
	30000	36100	36122
	+40000	+42700	+42749
	90000	98800	98871

FIGURE 1.7 Three columns of numbers with different accuracies.

science and engineering, its accuracy could be conveyed by writing 2×10^4 if it has one significant digit of accuracy, 2.000×10^4 if it has four. In business, where scientific notation is seldom used, it's hard to convey accuracy when a number happens to end with several zeroes. If the number appears with other numbers, readers can infer its accuracy from the others as in Figure 1.7.

Where you fit in: Precise computer output may not be as accurate as it looks. You must always ask "Where did this number come from, and how accurate were its inputs?"

Timeliness

The *timeliness* of an information element refers to the relationships among when information is needed, when it becomes available, and when its underlying data was obtained.

The Treaty of Ghent, ending the War of 1812 between the U.S. and Great Britain, was signed on December 24, 1814. Not knowing this, their armies fought the Battle of New Orleans on January 8, 1815. News of the treaty arrived February 8—too late to prevent about 300 soldiers from being killed and another 1,300 wounded.

Information timeliness has two characteristics.

- *Information must be available in time for its intended use.* This reflects the relationship between the time information is needed and when it becomes available. Information value often deteriorates with time. Many business decisions are worth more if made earlier. As

General Patton said, "a good plan today is better than a perfect plan tomorrow." The benefit of having better information must be weighed against the harm of a delayed decision.

- *Information must reflect up-to-date data.* This reflects the difference between the time information is made available and when its underlying data was obtained. The information available to New Orleans combatants before the battle was six weeks old. A chart of stock prices up to a month ago is of little use to day traders who work with short-term fluctuations. An ATM that dispenses cash on the basis of day-old account balances is useless.

Where you fit in: Instant information delivery does not mean that information is timely. It can only be if the entire process, starting with gathering source data, is fast enough. You must be aware of the entire business process, including human steps, when you evaluate the timeliness of an information system's output.

Consistency

Consistency means that all data elements that contribute to an information item, or to a set of related items, are based on the same assumptions, definitions, time periods, etc. Consistency applies to information, or processed data—not to individual data elements.

A sales manager wants to calculate the productivity of a firm's sales force. This manager divided last year's sales by the size of its sales force—and was severely disappointed because productivity was far below the industry average. What's going on here?
Chances are the inputs were ***inconsistent****. When asked for the size of the sales force, an information retrieval system would probably give its size at the time of the request. If a firm is growing, it has more salespeople now than it averaged during the previous year. Meaningful productivity calculations must use sales force size for a period that is consistent with the sales data.*

Globalization increases opportunities for information inconsistency. Terms can have different meanings in different countries due to differences in culture, business practices, legislation, or regulations. The definition of a depreciable asset, for example, can differ from one country to another. The sum of financial figures from different countries might not be meaningful in terms of any one country's practice. Small differences are an accuracy issue. Some inaccuracy may be acceptable for purposes such as strategic planning. Large differences may make figures useless.

Where you fit in: The sales manager in the above example was sufficiently computer-literate to put the question directly to a computer, but not sufficiently information-literate to understand the pitfalls. Nobody can prevent people from misusing computers. You must protect yourself from such errors by being aware of consistency issues. Later on, as a manager, you'll need to make sure that everyone in your organization is aware of these issues, and give them the training they need to achieve this.

Conformity to Needs and Expectations

Information must conform to the needs and expectations of whoever will use it. *Conformity to needs and expectations* is how well its processing, timeliness, etc., match what they expect.

Consider a request for "last year's dollar volume."

- *Does "dollar volume" mean what customers ordered, what was shipped, what they accepted, or what they paid for? What if they paid for part of their order in this year, part in another?*
- *Is "last year" the most recently ended calendar year, the most recent calendar year for which we have complete data, the most recent fiscal year, the most recent four fiscal quarters, or the most recent twelve months?*

- *Should this count items shipped during the year but returned after it? Shipped before the year, returned during it? Ordered during, canceled after? Ordered before, canceled during?*
- *Should it include the full value of multi-year agreements (such as subscriptions) sold during the year, the value of the fraction of their term within the year, or something else?*

These may have different answers for taxation, managerial accounting, commission payments, customer preference analysis, and other possible uses.

Computers make problems such as this worse. Before them, this request would have gone to an assistant. The assistant would ask questions like these. More questions would come up in talking to the people who have the data. Computers don't ask questions. They output what a programmer thought "last year's dollar volume" ought to mean. By the time anyone raises these issues, that programmer will probably have left the company.

Conformity to expectations is only meaningful in the context of a specific user or group of users. If several people will use information, it may have to be provided in different ways.

Organizations with international operations or customers must also consider language differences. Information in a language that a person does not understand does not meet that person's needs. If it's just labels on a graph, it may be possible to provide multiple versions in all necessary languages. Other situations can be more complex:

- Consider a customer list in alphabetical order. *Haakon* comes after *Huldre* in Norwegian because *aa* is an alternate way to write the letter å, which comes at the end of the alphabet after *z*. Other languages offer other examples.
- Many programs provide on-screen menus. Users can often choose a menu item by keying in its first letter. It may be difficult to find translations that start with the letters that were used in English. (It may be impossible. Native Italian words don't use *j*, *k*, *w*, *x*, or *y*, though Italian keyboards include them for names and foreign words.) Changing a program to accept different letters can be more work than merely replacing text.

Where you fit in: Computer-based information systems are developed by systems analysts, programmers, database designers, and more. This affects conformity to expectations in two ways:

- The more people involved in anything, the more misunderstandings can and will happen. Computer systems are full of details. Misunderstandings about them can impact the result.
- People who go into programming differ from those who gravitate toward business positions. When a pre-computer executive asked an assistant to collect data, that assistant's educational background and business perspective resembled the executive's. A programmer may not have the same grasp of business concepts, or even any interest in them.

Awareness of potential problems in this area can go a long way toward avoiding them. If you're not certain that your understanding of what computer output means matches the understanding of the people who generated it, *ask*. Keep asking until you're sure.

Completeness

Information is *complete* when it is based on all the relevant factors, omitting none.

Anne: What section of Accounting 202 should I take?
Beth: How about 8 o'clock Tuesday/Thursday with Prof. LaFond?
Anne: No, that's too early. I don't even eat breakfast until 8.

Beth: You didn't say it can't be early morning. How about 2 pm Monday/Wednesday with Prof. Higgins?
Anne: No, I work Mondays.
Beth: You didn't say that. How about 1 pm Tuesday/Thursday with Prof. O'Malley?
Anne: No, I had him for Accounting 201 and didn't like him.
Beth: You didn't say to avoid O'Malley. How about Wednesday evening with Prof. Rai?
Anne: No, I have orchestra rehearsals on Wednesday evenings.
Beth: You didn't say Wednesday evenings are out. How about taking it next fall?
Anne: I guess I'll have to do that.

How can Anne expect good advice if her questions are incomplete? Beth needs complete information about Anne's needs to provide meaningful answers. No wonder she gave up!

If a company has 12 sales offices and needs to calculate overall sales, it must include sales data from all 12. If one is missing, the data will be incomplete. As with conformity to expectations, this would have been obvious decades ago when a person collected the information by phoning the sales offices—but a computer will probably include zero for Chicago sales if a zero is in the database, not knowing that it means Chicago hasn't reported yet.

Question 11 at the end of this chapter is at least partly about information completeness.

Where you fit in: Computers can be programmed to flag incomplete information, but often aren't. Businesspeople such as yourself must check for potential incompleteness. If you trust computer output when you shouldn't, "that's what the computer said" won't get you off the hook.

Cost

Cost is a measure of the resources an organization uses, expressed in financial terms. Information has a cost.

Information may seem to be free or nearly so. Yet even when an organization doesn't pay for data, there are costs to collect it, store it, process it, and send results to their destinations. These costs must be minimized.

The cost of information is important because we must often justify an information system by comparing its costs and benefits. Costs are usually easy to estimate. Benefits are more difficult. Sometimes it isn't possible or necessary to calculate benefits precisely, but even then, cost is a consideration.

The value of information quality must be traded off against its cost. Source data automation can improve accuracy. Is that worthwhile? We can only answer by considering costs and benefits. Companies are increasingly investing in information quality. The need for up-to-date, accurate information increases as business speeds up. The cost of obtaining it drops every year.

Planning and developing information systems also involve cost tradeoffs. It may be possible to trade off higher initial cost for lower operating costs, or a more expensive software license for lower incremental cost per additional user.

Where you fit in: Cost tradeoffs are business decisions, not technical decisions. Businesspeople must understand their technical implications to make them properly.

In a Nutshell

Where you fit in: The biggest problem you'll face with information processing on the job is not computer literacy. Knowing how to use a computer is the easy part.

The real issue is *information* literacy. That means understanding what information is, where it comes from, how to assess its quality, and how computer storage and processing influence it. These have a far-reaching impact on organizations and their profitability. Twenty-first-century managers must understand them and be able to deal with them.

1.4 LEGAL AND ETHICAL INFORMATION USE

When you learned to write, you also learned that you should not write ransom notes. Writing is not a crime, but using it to commit a crime is punishable.

A person who has access to information and information systems must not use them for unethical or illegal ends either. There are several types of unethical or illegal use:

Using a Computer, or Information, as a Tool to Commit a Crime

Information in computers can be worth money. That comes about in several ways:

- Companies may expect to make money by selling that information. It may be in a form that consumers buy, such as music or videos. It may be of value to businesses, such as economic forecasts by industry and country.
- Companies may hope to make money by having information that others do not, such as oil exploration findings, market research data, or circuit designs. Criminals may try to bypass protections to obtain information that is a company's *intellectual property.*
- Having information may enable a person to steal money or other items of value. This applies to credit card numbers, bank account numbers, and similar information that grants access to valuable assets. Lists of email addresses enable criminals to send *phishing* emails, enticing people to provide personal information. Some people who get these emails provide it.
- Some people pursue political agendas at any cost. Modifying information in computers may be worth a great deal to them, though perhaps not in direct monetary terms.

If information is in a computer, that computer can be compromised and the information in it extracted. Copying a song from a friend's phone, rather than paying the iTunes Store 99¢, is a small example of theft—but millions of 99¢ payments add up to substantial losses to creators, performers, and publishers of music. Other information involves potentially greater loss per theft. There is no way to put a monetary value on what a *cyberterrorist* could do with access codes to a nuclear power plant or air traffic control computer.

Using a Computer, or Information, as the Object of a Crime

People try to damage computers for a variety of reasons. They may want to steal information, as discussed above; harm the organization whose computers they attack; or just seek intellectual challenge in proving that they can. This topic is discussed more in Section 6.6, "Network Security."

Using Information Unethically

Illegal use of computers and information is usually obvious. You don't need to be told not to do it. If you want to do it anyhow, nothing you read here will stop you.

Illegal and unethical use of computers and information overlap, but laws can't forbid all unethical activity, and some illegal activities are not unethical. (There have been laws in many countries that most people find unethical. In the U.S., it was illegal for Rosa Parks to sit in the front of an Alabama bus in 1955.) Ethics can be defined, informally, as the application of a moral code to one's behavior. This requires both having a moral code and applying it to one's actions.

An example of unethical information use: A bank uses an information system to calculate interest rates for home mortgage loans. Bank managers know that people in certain groups have difficulty getting mortgages, so they will pay higher rates. They also know that members of those groups tend to live in certain areas. The system could add 1% to the interest rate of loans on homes in those areas, after calculating what the rate should be on the basis of other criteria. The effect, the bank knows, is to raise the interest rate it charges to members of those groups. A few people who are not in those groups might be affected, and a few members of those groups who live elsewhere might not be, but most of the time this system would work as intended.

This use of information systems would be unethical because it has the effect of discriminating against specific ethnic groups. It might be illegal or not, depending on applicable laws, but the ethical question is (to this author) not in doubt. If one's moral code says that people should not be penalized for their ethnicity, applying that moral code to the bank's behavior would prevent it from developing this information system.

A complete discussion of ethical behavior is beyond the scope of this book. What is important is that you have a moral code, that you apply it to your behavior, and that this behavior includes your use of information systems.

> ***Where you fit in:*** As a working professional, you will have access to information and information systems. Your use of that information, and of those systems, ***must*** conform to ethical principles. You don't have to deal with the technical aspects of protecting your employer's information. Specialists can, and should, do that. You should be aware of the need for such protection and support it.
>
> You may be pressured to act unethically. Your future managers may not all be ethical. You may be asked to access information which you should not, to give others information you should not pass on, to modify data you are not authorized to modify, to withhold information that could lead to a lost sale if disclosed, or to alter financial records. Try to resist those pressures. Actions that seem expedient in the short run tend to work out poorly in the long run. You can't afford that.

KEY POINT RECAP

Intelligent use of information can help any type of organization.

As a businessperson, you will be in a position to use information intelligently, to help those who work for you use information intelligently, or not to do either of those.

The value of information depends on its quality. Information quality can be described by a small number of specific factors.

As a businessperson, you must always be alert to information quality considerations:

- in evaluating the worth of information that others supply to you,
- in maximizing the value of information that you supply to others, and
- in ensuring proper design of information systems that your company uses.

Computers are basic to using information intelligently. A company can't use information intelligently without using computer-based information systems intelligently.

As a businessperson, you will have no choice about using computer-based information systems. Your only choice will be whether you use them by rote, following instructions without thinking about them, or intelligently. You and your company will benefit if you use them intelligently.

You will benefit personally in your career if you understand information systems.
As a businessperson, the value you get from information systems at work will add to the value you provide to your employer. Providing value to your employer leads to raises and promotions.

KEY TERMS

Where you fit in: This section of each chapter will list key terms that were defined in the chapter, with their definitions (perhaps in slightly different words). These terms were italicized the first time they were used. As a businessperson, you should understand these terms in context and be able to use them in ways that reflect this understanding.

Accuracy (measure of information quality): Difference between an information value and the value that a perfect calculation using perfect data would produce.
Boundary (of a system): A conceptual line separating components of a system from its surroundings.
Comparison: Selecting one or more alternative processing paths on the basis of a data value.
Completeness (measure of information quality): The degree to which information is based on all the relevant data, omitting none of it.
Computation: Creating information by carrying out a predefined process on two or more data items.
Conformity to needs and expectations (measure of information quality): The degree to which the way an information item was obtained matches the needs and expectations of its users.
Consistency (measure of information quality): The degree to which all the data items that contribute to information are based on the same time period, assumptions, etc.
Correctness (measure of information quality): The extent to which information is derived from the proper data values through the proper processing steps.
Cost: A measure of the resources an organization expends to obtain information.
Cyberterrorist: A person who tries to advance political or social objectives by attacking computers, networks, and/or information in them.
Data: Raw facts in isolation from other facts.
Effectiveness: The degree to which a person or organization achieves an objective.
Efficiency: The degree to which an activity is carried out using the least possible resources.
External feedback (in a system): Passing a system output back to its inputs through something that is outside the system.
Feedback (in a system): Using a system output as a system input.
Information: (a) Data that has been organized and processed so as to have meaning; (b) anything that reduces uncertainty.
Information system: System whose purpose is to process information.
Information technology: The electronically based elements of an information system: hardware, software, data, and communications.
Intellectual property: Something that comes from a person's mind and in which the law recognizes property rights.
Internal feedback (in a system): Passing a system output back to its inputs within the system.
Knowledge: The ability to apply information in a business situation.
Open system: System that communicates with its environment across its boundary.
Precision (measure of information quality): The maximum accuracy with which a data element is stored or presented.
System: A group of components that interact to achieve a stated purpose.
Timeliness (measure of information quality): the relationships between the time that information is available to users and (a) the time it is needed and (b) the period to which its underlying data refers.

REVIEW QUESTIONS

1. Describe, in your own words, the difference between *efficiency* and *effectiveness*.
2. Can information make an organization more effective? What makes you think so?
3. Informally, what is the connection between *uncertainty* and *information*?
4. Define *data* and *information*, giving an example of the difference between them.
5. What are the two fundamental operations involved in turning data into information?
6. What are the three general ways in which information systems provide the greatest value to organizations?
7. List the eight factors that define information quality.
8. What are the two ways in which information can have poor timeliness?
9. Give an example, not mentioned in the book, of illegal use of information.
10. Give an example, not mentioned in the book, of unethical use of information.
11. What is a system? An information system?
12. Define *feedback* in a system. What are the two types of feedback?
13. What are the five types of activities that take place in an information system?

DISCUSSION QUESTIONS

1. Think of a real situation in which having information helped a business by reducing its costs, increasing sales, or in some other way that increased its profits. Identify the business (by type if you don't want to give its name), state what the information was, describe what happened, and describe what would have happened if the business had not had that information.
2. Consider the shoe store giving customers discount coupons for socks in Section 1.1. With no coupon, 10% of shoe purchasers bought socks. With a $3 coupon, 65% bought socks. Which outcome is better for the store?
3. Assume that running burns about as many calories for a given distance as walking, but a runner covers twice the distance that a walker covers in the same amount of time.
   a. Which is more efficient for weight loss, if time is the input unit?
   b. Suppose a person is unable to run more than two miles (3.2 km) but can walk four miles (6.4 km). If that person must choose between running two miles or walking four miles per day, assuming he or she has time for either, which would be more effective?
   c. If the objective were maximum weight loss for 20 minutes of activity per day, which would be more effective?
   d. How would the possibility that a runner needs to shower after exercise, but a walker does not, affect this decision? Make *and state* any necessary assumptions.
4. You work as a supermarket cashier. How would each of the following actions on your part affect the quality of information in the store's inventory or sales databases?
   a. You see six yogurt containers. You scan one, press the "Multiple" key, and enter 6 for the quantity. You do not notice, or do not care, that three are strawberry yogurt and three are peach. (The price of both flavors is the same.)
   b. You see six yogurt containers. You scan one, press the "Multiple" key, and enter 6 for the quantity. You do not notice, or do not care, that three are store brand yogurt @59¢ and three are Chobani Greek yogurt @$1.29.
   c. You see a bag of apples with product code 4664. You key in 6446 by mistake. That is the product code for garlic, which is more expensive per pound/kg.
   d. The customer is your best friend. You bag some of his purchases without scanning them.
5. The value of information depends on its quality. For each of the following decisions, (i) give an example of low information quality in terms of each information quality factor,

(ii) state how that low information quality could affect the decision, and (iii) state how you would avoid that type of low information quality. Example: incorrect information (the second quality factor) on the first decision could be an overstated GPA. That could lead the firm to choose a less qualified applicant over another who did not overstate his or her GPA. This could be avoided by checking the applicant's college transcript. (Some quality factors may have no impact on some decisions.)

You will have to choose a specific information item to use with each information quality factor. In the example, the information item chosen was the applicant's GPA. You will therefore need eight items for each of parts 3(a), 3(b), and 3(c). You can use the same item eight times, use eight different items once each, or anywhere in between.

a. Choosing a financial analyst from six new college graduate applicants.
b. Deciding how to allocate a $250,000 advertising budget among various media.
c. Selecting a warehouse location to serve the western U.S. Assume that all customers have been served until now from its Vermont headquarters, and that business in the West has grown enough to justify a second distribution center.

6. Four students from New Bedford, Mass., want to see a show in Providence, R.I. Each gets into his or her own SUV, which gets 10 miles per gallon (23.5 liters/100 km). They drive west on I-195, arriving half an hour later, and get good seats. A second group piles into one Toyota Prius, which gets 50 miles per gallon (4.7 liters/100 km). They head north on routes 140 and 24, then west on I-495, and finally south on I-95. They arrive over an hour later, missing the start of the show and standing for the rest of it.
   a. Which group was more effective? Why?
   b. Which group was more efficient? Why?
   c. What is the ideal combination of efficiency and effectiveness for this situation?
7. Consider the five fundamental activities of an information system (Section 1.3). Give an example of each activity as it applies to each of the following information systems. (The example system functions listed may not correspond to these fundamental activities.)
   a. A restaurant system. Some of the functions it supports are servers entering orders in the dining area, cooks preparing dishes in the kitchen, servers printing checks in the dining room, and managers getting a variety of reports in their offices.
   b. A student information system. Some of the functions it supports are students registering for courses, instructors assigning grades, students checking their grades, and preparing transcripts.
   c. A social networking system. Some of the functions it supports are photographers posting photos, others commenting on those photos, and all participants discussing cameras and photography.
   d. A cruise line information system. Some of the functions it supports are the line posting information about cruises, travelers booking cabins and optional extras such as shore excursions, and management getting reports that enable them to optimize pricing according to demand.
8. To see the effect of data representation on precision, open a spreadsheet program. Then:
   a. In cell B2, enter the formula *=1000000/3*. Display the result to four decimal places.
   b. In cell C2 (directly under it), enter *333333.3333*. Both cells should look the same.
   c. In cell D2, enter the formula *=B2 – C2*, for their difference. You will see a small number. Depending on your computer, software, and settings, you may see several leading zeroes or *E–05* at the end, indicating that the decimal point should be five places to the left of where you see it. The significant digits will probably all be 3s.
   d. In cell E2, enter the formula =D2*10000000 to multiply the content of cell D2 by ten million. The result will no longer consist of all 3s. Explain why.
9. A NASA press release described Comet ISON as having "a tail about 186,400 miles (300,000 kilometers) long." A comet's tail is fuzzy. Its length cannot be measured

accurately. Assuming the writer of this press release was given the length in metric units (km), what can you say about the conversion to miles here?

10. For each of the four information systems a–d in Question 7, give examples of five different information quality problems that could arise. State the business impact of each problem.
11. In 1984, James A. Cummings Inc. of Fort Lauderdale, Fla., sued Lotus Development Corp., claiming that a defect in its Symphony spreadsheet program caused Cummings to underbid a construction contract by $254,000. Cummings dropped its suit when it found that one of its employees had failed to include a cost item in a sum, but only after wasted time, expense, and aggravation on both sides. Explain the information quality issues here. If you had worked for Cummings at the time, could you have avoided this error? How?
12. There are laws in the U.S. against *insider trading*: buying or selling stocks on the basis of information that is not publicly known. (Style guru Martha Stewart was imprisoned for six months for selling stock the day before an announcement reduced the value of that stock considerably, after being tipped off to a probable drop by her broker.) In terms of ethical use of information, do you agree with the concept behind those laws? Explain your position.

KHOURY CANDY DISTRIBUTORS, INC.

Each chapter of this book will show how its topic applies to Khoury Candy Distributors (KCD) Inc., based in Springfield. Jake and Isabella, senior MIS students at nearby Standish College, met KCD co-founder Jason Khoury when he gave a talk to the Standish business student society. He invited them to do a term project on KCD's information systems. This chapter provides the background on KCD and those systems.

KCD, Standish College, and all people mentioned in this case are fictional.

Khoury Candy Distributors Background

Jason Khoury's grandparents came to the U.S. from Lebanon in the 1950s with their then-young children. Lebanon was peaceful at the time, but they were concerned (prophetically, as things turned out) with future turmoil. The Khourys settled in Springfield and opened a candy store to serve students at Standish College. They soon expanded their offerings to include pastries, soft drinks, school supplies, and other items that students bought.

Jason's father Anthony took over the business in 1973, giving his parents a well-earned rest. Jason, then in high school, learned the business from the ground up by working after school and on weekends. Even when he left Springfield to earn a business degree at the state university, he came back to work during school vacations. He saw that as a good way to contribute to his own tuition and living expenses at State.

Jason graduated with honors in 1980, but instead of pursuing a corporate career like many of his classmates, he returned to the family business. He supervised its evolution from a candy store selling a few other items to a full convenience store stocked with what students wanted in the 1980s. At the same time, he enrolled in the Standish part-time MBA program.

In 1982, one of the store's suppliers approached it with a problem and a proposition. The problem: the supplier had overestimated how many chocolate bars it would sell before Hallowe'en. The proposition: if the Khourys would take about ten times their usual order, they'd get a 40% discount from their usual wholesale price.

While Anthony Khoury was lukewarm to the idea, Jason persuaded his father to let him see what he could do with it. The Khourys bought the chocolate. Jason used his contacts in the area and State's alumni network to find buyers at 15% below the usual wholesale price. The difference between 40% and 15% covered shipping costs and left the Khourys with a handsome profit.

That experience got Jason thinking. Most stores that buy chocolate don't sell candy in volume, so they don't get huge discounts from suppliers. On the other hand, the supplier didn't lose money on this deal, or they wouldn't have proposed it. He thought he'd be able to negotiate similar discounts for large quantities in the future. Perhaps candy distribution had more profit potential than a convenience store? It certainly had more opportunities to use what he was then studying in school.

Jason also considered the impact of information systems here. He didn't know if that supplier used computers to forecast sales. If they didn't, perhaps computers could improve forecasting. If they did, it might be a good example of how a poorly programmed computer or poor input data can mess things up!

KCD Today

Nearly forty years after Jason Khoury resold those chocolate bars, KCD is the second-largest candy distributor in North America and the largest to focus solely on candy. Their annual revenue is about $1 billion, and they have about 450 employees. Their headquarters is still in Springfield, as is one of their distribution centers, but they have six additional centers in the U.S. and one in Canada. From those, they serve about 10,000 retail locations in 20 U.S. states and four Canadian provinces. Jason Khoury is still in charge, though one of his daughters manages KCD's Atlanta, Ga., distribution center and is scheduled to take over all of its distribution operations in a few months.

KCD's Information Systems

Medium and large businesses all used computers in the early 1980s. Personal computers were still in their infancy, though. Stores like the Khourys' seldom had them. Information systems were part of all university business programs, so Jason Khoury knew about their potential to improve operations and gain competitive advantage. Soon after his candy distributorship began operating, he purchased an IBM PC/XT. He used the Lotus 1-2-3 spreadsheet program to organize data and make sales projections.

KCD has kept up with technology ever since. It uses an IBM Z computer in Springfield for corporate information processing. In addition, each distribution center has a server for that center's data. Each server is the hub of a local network that links computers in the office and warehouse areas. Order picking optimization is vital to warehouse efficiency, so KCD tries to stay at the forefront of this technology.

You'll learn more about KCD's information systems in the following chapters.

KCD People

You'll meet these KCD employees as you read the case episodes:

- Jason Khoury, President and CEO (Ch. 1)
- Sandra Steere, Vice President of Business Development (Ch. 2)
- Chris Evans, Chief Information Officer (mentioned in Ch. 2, meet in Ch. 3)
- Lakshmi Agarwal, Manager of Application Software (Ch. 4)
- Visal Phan, Database Administrator (mentioned in Ch. 4, meet in Ch. 5)
- Armand Rocher, Network Manager (Ch. 6)
- Harvey Leonard, Vice President of Marketing (Ch. 8)
- Brian Greenwood, Systems Planner (Ch. 9)
- Jennifer Khoury, Manager, Atlanta Distribution Center and soon to be COO (Ch. 10)
- Nikau Taumata, Senior Programmer (Ch. 11)

Questions

1. Consider the information quality factors in this chapter. For each one, say how poor information quality can, in the last words of the background section, "mess things up."
2. The last part of the case states "Order picking optimization is vital to warehouse efficiency." Can it also affect warehouse effectiveness? Discuss. (Suggestion: Start by saying what you think warehouse effectiveness is. You may also want to do a little research on order picking optimization.)
3. Draw a diagram of KCD as a system. Show six types of entities that it interacts with outside its boundary. ("Types of entities" means you can't get six by saying "customer 1, customer 2," and so on. Customers are one type of entity.) On arrows from each entity type to or from KCD, indicate what goes between that entity type and KCD. Most entity types will have one arrow in each direction. Some of what goes between KCD and the entities will be information. Some will be physical items.

CASE 1: TRUSTING A SPREADSHEET

In 2010, Harvard professors Carmen Reinhart and Kenneth Rogoff published "Growth in a Time of Debt." This paper was widely cited to justify cutting government spending during a recession. It was the basis for some economic proposals during the 2012 U.S. presidential election, as it appeared to show that stimulus spending (which the Democratic Party supported and the Republican Party opposed) tends not to have long-term benefits.

On April 15, 2013, UMass Amherst graduate student Thomas Herndon and two faculty members posted a paper which documented multiple errors in Reinhart and Rogoff's spreadsheets. For example, they found that the growth rate for countries with high debt was +2.2%, not −0.1% as Reinhart and Rogoff had reported.

Profs. Reinhart and Rogoff did the responsible thing: they went over their work and conceded every error that Herndon identified. They also wrote, however, that those errors do not affect their overall conclusions. Other economists feel that they do.

The point is not about government spending or the specific errors. Economists can debate those. The point is that *anyone* can make errors that lead to incorrect information. Reinhart and Rogoff are experienced researchers. Their paper was reviewed by other experienced researchers before publication. It was read by thousands of economists and studied by hundreds before Herndon tried to repeat its analysis, could not, and asked why. If errors can persist for years under those conditions, you cannot assume that *any* information you get at work is error-free. *Anyone* can make a mistake. The question is not *if* it will happen, but *when*. The answer may be "When you least expect it and it can do the most harm."

Questions

1. Your English instructor uses a spreadsheet to calculate class grades. Its formulas consider the weight of each item, policies such as dropping the lowest homework grade, and adjustments to create a desired grade distribution. You are concerned that this instructor, who is not a spreadsheet expert, is at least as likely to make errors as were Drs. Reinhart and Rogoff. You do not want such an error to impact your grade. How do you proceed? Does your answer depend on whether you are or are not a spreadsheet expert yourself?
2. Suppose Herndon had released his findings before the U.S. presidential election, rather than five months after it. How do you think that might have affected the authors' response? What does this suggest about reactions to finding errors in a spreadsheet that a business uses? How could this affect what you do if you find errors in a spreadsheet you are given at work?

3. You think that finding errors in a paper published by senior scholars in your field could boost your professional reputation, much as it (justifiably) did Herndon's. How do you proceed?

CASE 2: LYNN MCCALLUM, MD

Dr. Lynn McCallum operates a family practice in Redding, California. In addition to herself, Dr. McCallum's practice employs a nurse practitioner, two back office personnel, a receptionist, and a biller. They see all ages of patients and handle general family practice visits as well as minor outpatient surgical procedures.

Dr. McCallum started using Practice Fusion in 2009. One factor in choosing it was cost: the software was free, with costs covered by ads at the bottom of the screen. (An ad-free version was available for a fee. From mid-2018 on, all users pay $100 per physician per month.)

Another factor was that Dr. McCallum didn't need a technology specialist to implement the software, since it is accessed via a web site and Practice Fusion manages the data on its servers. Her office just needs computers with Internet access. As she says, "I have a laptop on a small desk on rollers in each exam room. I review meds and chart on the patient as I sit in front of them. This way I can chart on my laptop and still maintain eye contact with my patient. If a referral for a consultation with a specialist is required, or a referral for an imaging study, I send a message on Practice Fusion right in front of the patient. Often the entire note is completed before we leave the room. This minimizes my charting time at the end of the day."

Dr. McCallum describes the benefits of Practice Fusion: "Using Practice Fusion has benefited the office in so many ways! I have pulled up a patient file on Practice Fusion on the computer at the hospital to assist me in completing [forms] for an admission, and since Practice Fusion is web based, I can access it anywhere! I have used Practice Fusion from home when I get an after-hours patient call and I can instantly access their med list and their visit history, etc., to help me help them. I also chart that note right at the time they are on the phone with me; no more sticky notes and random pieces of paper that get lost!"

In the past, Dr. McCallum used speech-to-text programs to dictate medical information for patient charts. However, she finds that she now has "very little in the way of dictation or typing" because of the templates that Practice Fusion provides. (See Figure 1.8.) Her practice has augmented those templates with additional ones that are customized for it.

FIGURE 1.8 Practice Fusion screen shot.

Dr. McCallum continues, "I love that the pediatric growth charts are automatically graphed when you input the vitals. I love that the Past Medical/Surgical history is completed once, then just added on to as time goes on, rather than the redundant way we had been charting in the past. We also have a current medication list that we can print for the patients if needed and can easily review for completeness. I could go on …."

Questions

1. What is the information system here? What are its main components?
2. Who are the users of this system? How does it improve their efficiency? Their effectiveness?
3. Give two information quality issues that could arise with this system. How would you try to prevent them or minimize their impact? (State what users would be affected by these issues and what "impact" means to them in this context.)
4. Practice Fusion has competitors. If you were advising a physician with a small practice like Dr. McCallum's, what are three additional factors (in addition to the two near the start of the case) that you would consider in choosing among them? You don't have to find competing software, let alone evaluate it. Just list three other factors to consider.

BIBLIOGRAPHY

1. Associated Press, "Agreement reached in Lotus suit," December 10, 1986, news.google.com/newspapers?id=TFlWAAAAIBAJ&sjid=bu8DAAAAIBAJ&pg=4089,6387832, accessed September 16, 2019.
2. Crandall, W.R., J.A. Parnell and J.E. Spillan, *Crisis Management* (2nd ed.), Sage, 2013.
3. Fitzgerald, J., "Uncovering flaws in an influential study," *The Boston Globe*, June 23, 2013, p. G1.
4. Herndon, T., M. Ash and R. Pollin, "Does high public debt consistently stifle economic growth? A critique of Reinhart and Rogoff," University of Massachusetts Amherst, Political Economy Research Institute, April 15, 2013, www.peri.umass.edu/images/WP322.pdf, accessed September 16, 2019.
5. Jet Propulsion Laboratory, California Institute of Technology, "NASA's Spitzer observes gas emission from Comet ISON," July 23, 2013, www.jpl.nasa.gov/news/news.php?release=2013-231, accessed September 16, 2019.
6. Practice Fusion web site, www.practicefusion.com, accessed September 16, 2019.
7. Reinhart, C. and K. Rogoff, "Growth in a time of debt," *American Economic Review*, vol. 100, no. 2 (May 2010), p. 573.
8. Reinhart, C. and K. Rogoff, "Errata: 'Growth in a time of debt,'" May 5, 2013, www.carmenreinhart.com/user_uploads/data/36_data.pdf, accessed September 16, 2019.

2 The Role of Information Systems in Business

CHAPTER OUTLINE

WHY THIS CHAPTER MATTERS

You've used computers since you could reach a keyboard. You can't imagine life without tweets, apps on your phone, Instagram, and Google. Do you really need to be convinced that information systems are useful?

Convinced? Perhaps not. But the ways information systems are useful to businesses are often not obvious from how we use them in our personal lives.

The ways in which information systems bring the greatest value to businesses are not the personal applications we use. Word processing (for example) is useful, but it's not what has the most business value. Neither are record-keeping applications such as payroll, which at best saves the cost of hiring payroll clerks. We have to look for uses with a broader impact on how a company does business, how it competes with others, how it relates to its customers, how it can reshape its industry. That, in turn, means that we must understand how businesses operate and compete.

This chapter will give you the perspective to put information systems into that business context.

CHAPTER TAKE-AWAYS

As you read this chapter, focus on these key concepts to use on the job:

1. Information systems are more than just computers.
2. Any organization can use information systems to improve its position in its industry.
3. There are many points in any business where information systems can be applied usefully.
4. Using information systems effectively in these ways requires understanding how a business wants to compete: its *competitive strategy*.

2.1 A PERSPECTIVE ON INFORMATION SYSTEMS

Every organization wants to succeed. Every organization has a measure for success. For a business, the traditional measure is profit. (Responsible businesses look beyond profit to their impacts on stakeholders and the environment.) For a startup, it may be sales or market share. For a university it may be moving up in student quality or research grants. For a sports team it's wins/losses or reaching league playoffs. Whatever the measures are, even if people in the organization don't think about them as measures of success, they exist.

The ultimate goal of using information systems is to improve an organization's success by those measures. The problem with this goal is that it doesn't provide practical guidance. We have to connect it to reality. That's where this chapter comes in.

The connection between the three topics of this chapter and success is shown in Figure 2.1:

- A company's industry determines, to a high degree, its *value chain*: the sequence of activities through which it produces an output that is worth more than the total cost of its inputs. Most organizations of a given type have similar value chains.
- A company's *competitive strategy,* the approach it takes toward competing in its industry, also influences its value chain.
- A company's competitive strategy determines how it will try to use five *competitive forces* and how it will respond when other firms try to leverage them.
- The way a company's value chain functions and the way it deals with competitive forces combine to determine its success.

There are other success factors. Product design and advertising must be executed well. Timing is important: a product that was a good idea when its development began may be useless three years later. Leveraging the three factors in italics above will still make a firm more successful than it would be otherwise.

Information systems can make a difference to a company's success in three fundamental ways:

1. Improving communication within an organization.
2. Connecting an organization with its customers and suppliers.
3. Helping people in the organization make better decisions.

Figure 2.2 shows them overlapping. An information system that helps choose a marketing plan helps make better decisions and connect with customers. An information system that orders parts for production links purchasing and production, and connects the organization with its suppliers.

To put these together to make an organization more successful, we look for opportunities to apply information systems to (1) the stages of the value chain and (2) the competitive forces, focusing (3) on making a difference in one of these three ways. In a given situation the combinations won't all apply, but some of them will. Those combinations can guide information system planning.

Where you fit in: This means thinking like a manager, not an employee who just does his or her own job. Businesses reward people who think like managers.

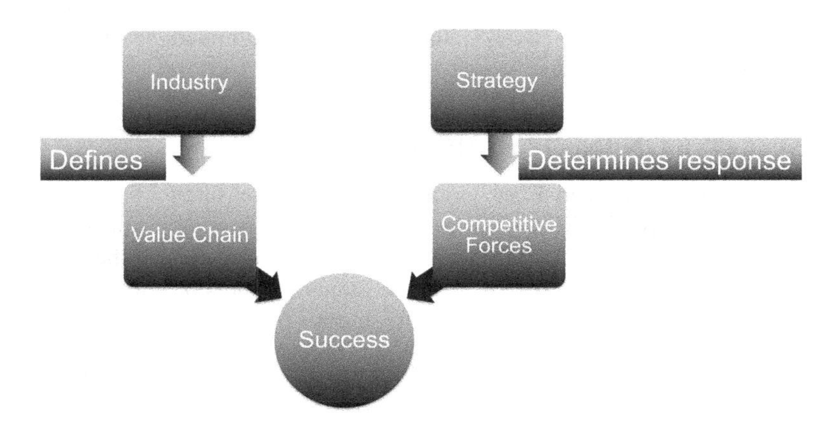

FIGURE 2.1 Relationships among strategic considerations.

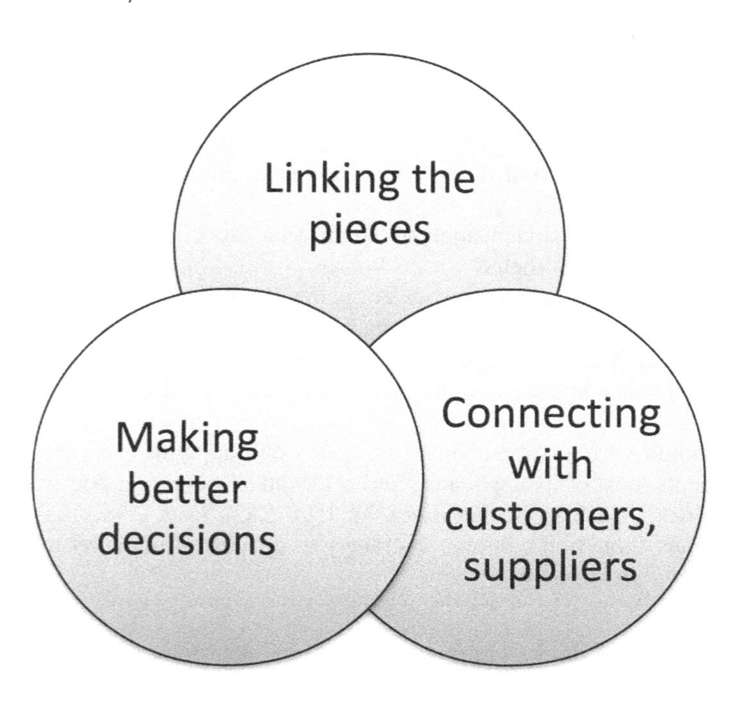

FIGURE 2.2 Three ways information systems make a difference.

Many information systems don't do these. Computerizing payroll doesn't improve competitive position, but that doesn't mean a company should do payroll by hand. It should go for the savings that automating payroll provides, but to make a big difference, we must look at these three areas.

WHAT WILL INFORMATION SYSTEMS DO IN THE FUTURE?

In December 2013 IBM predicted the top information systems innovations in the next five years. Their predictions are at *venturebeat.com/2013/12/16/ibm-reveals-its-top-five-predictions-for-the-next-five-years*

1. The classroom will learn you. No more needing half the semester for a new teacher to figure out who you are.
2. Innovations in physical stores will enable it to beat online shopping.
3. Doctors will use your DNA to keep you well. Your treatment won't be the same as your buddy's, even if you both have the same condition.
4. A digital guardian will protect you online. Crooks will get smarter, but technology will keep them from winning.
5. The city will help you live in it. It will respond to needs faster than people could.

This book is being written five years later. Did these happen, in whole or in part? If they didn't, why do you think that is? What factors influence the speed with which advances occur? You can find IBM's 2018 predictions at *dzone.com/articles/ibms-5-technology-predictions-for-the-next-5-years* or watch an IBM executive present them at *research.ibm.com/5-in-5*.

2.2 COMPETITIVE STRATEGIES

To be successful, an organization must be able to answer the question, "Why should a customer do business with us, not someone else?" (You may have heard it called its *unique selling proposition* in marketing courses.) This applies to different organizations in different ways:

- University: Why should a student apply to our university, not the one across town, or accept our offer of admission, not theirs?
- Hotel: Why should a visitor stay with us, not at the Sheraton two blocks away?
- Local hardware store: Why should a customer buy from us, not Home Depot?
- Car manufacturer: Why should someone buy our sedan, not a Toyota Camry?
- Dentist: Why should someone want me to fill their cavity, not Dr. Atkins?

These questions should have answers. Those answers won't apply to every possible customer. A university that accepts most of its applicants, and prides itself on giving poorly prepared students a good education, won't steal a valedictorian with 1590 SATs from Princeton. In knowing what it offers and to whom, though, it can plan a strategy to do well in its target market. *Information systems can help with that strategy.*

> ***Where you fit in:*** As you think of reasons why a potential customer should do business with your employer, also ask yourself how information systems can help make that happen.

Business strategies are of three main types (Figure 2.3):

1. *Low-cost strategy*, offering customers the lowest cost for a product that is fundamentally about the same as most other suppliers'. Example: Kia, when it first sold cars in the United States.
2. *Differentiation strategy*, offering customers a product or service that is different from most others in a way that applies to many, if not all, potential buyers. Example: BMW. Its cars are differentiated by performance and handling. Most car buyers want good performance and handling, though perhaps not enough to overcome other factors such as cost.
   There are many possible differentiation strategies. Some that apply in almost any market are:
   - Quality: "Our products have higher quality than the rest." *Information systems can support this strategy by monitoring and analyzing quality control testing.*
   - Innovation: "We're first with the latest ideas." *Information systems can support this strategy by improving internal communication in the design process.*
   - Customer service: "We take better care of you than our competitors do." *Information systems can support this strategy by giving customer support staff information about this customer, what products he or she has, and the history of their interactions.*

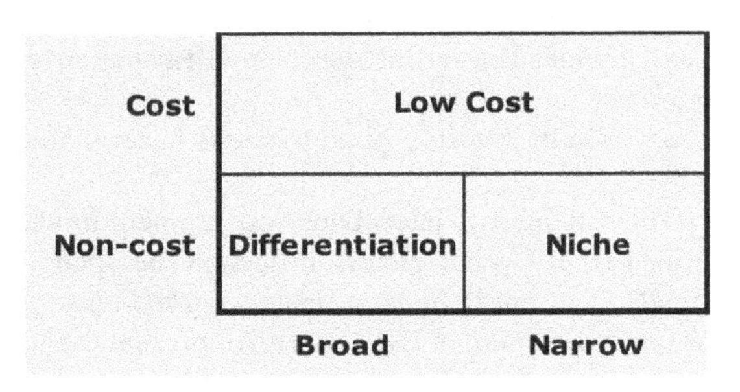

FIGURE 2.3 Three generic business strategies.

There are also differentiation strategies that apply in specific markets:

- Airlines: "Our planes have more legroom in economy than other airlines' planes." *Information systems can justify this strategy by analyzing whether this airline's target market will pay enough for more legroom to offset revenue loss from carrying fewer passengers per plane.*
- Dentistry: "Our implants take less time than other types of implants." *Information systems can support this strategy by customizing each patient's implant, enabling it to be placed more quickly and accurately.*
- Roofing materials: "Our roof shingles have a longer warranty than others." *Information systems can support this strategy by determining the added costs associated with a longer warranty, and evaluating whether the target market will support these costs via increased sales volume and/or accepting a higher price.*

These apply to every customer. Everyone wants high-quality products, innovative products, and good service. Every economy-class air traveler wants more legroom. Everyone who gets a dental implant wants it done quickly. Everyone who buys a new roof wants it to last a long time. These strategies are aimed at all customers in those markets, not just some of them.

3 *Niche strategy*, offering products for the unique needs of a market segment. Example: Jeep. Jeep cars are differentiated from most others by their ability to be driven off paved roads. If you don't do that, or don't at least want to know that you could, they may not be for you.

Which strategy is best? There is no single right answer. No product or service is right for all customers all the time, but most products and services are right for some customers some of the time. A firm should have a clear idea of its strategy and focus its efforts on making that strategy succeed. Information systems can help make the strategy succeed. Using them effectively requires understanding the strategy itself.

An organization's strategy is the most important determinant of how it responds to competitive pressures. An organization with a low-cost strategy, faced with a competitor that undercuts its prices or with a new approach to its market whose main advantage is lower cost, will respond in one way—and will need certain information systems capabilities to support that response. An organization with a niche strategy, faced with similar less expensive competition, will respond differently—and will require different information systems capabilities to support that response. Therefore, we'll look at the five basic competitive forces next.

Where you fit in: Whatever your job, try to understand your employer's competitive strategy. When you have a choice among alternative courses of action, one that fits that strategy better than others is almost always a better choice.

2.3 THE FIVE BASIC COMPETITIVE FORCES

Competition has existed for ages. When a cook in ancient Athens walked down the street of pot sellers to buy a clay pot, every seller wanted to sell one. Sellers who persuaded cooks to buy their pots prospered. Others did not.

In the last few decades, researchers have looked closely at competition and how it works. What factors is it based on? What forces make some firms more successful than others? How can firms influence these forces, and what can other firms do about it?

Information systems can improve a firm's position with respect to these forces and facilitate its response to other firms' moves. Understanding these forces is therefore basic to understanding how information systems help organizations succeed. The benefit of being more competitive is far greater than the benefit of preparing a forecast with a spreadsheet instead of a calculator.

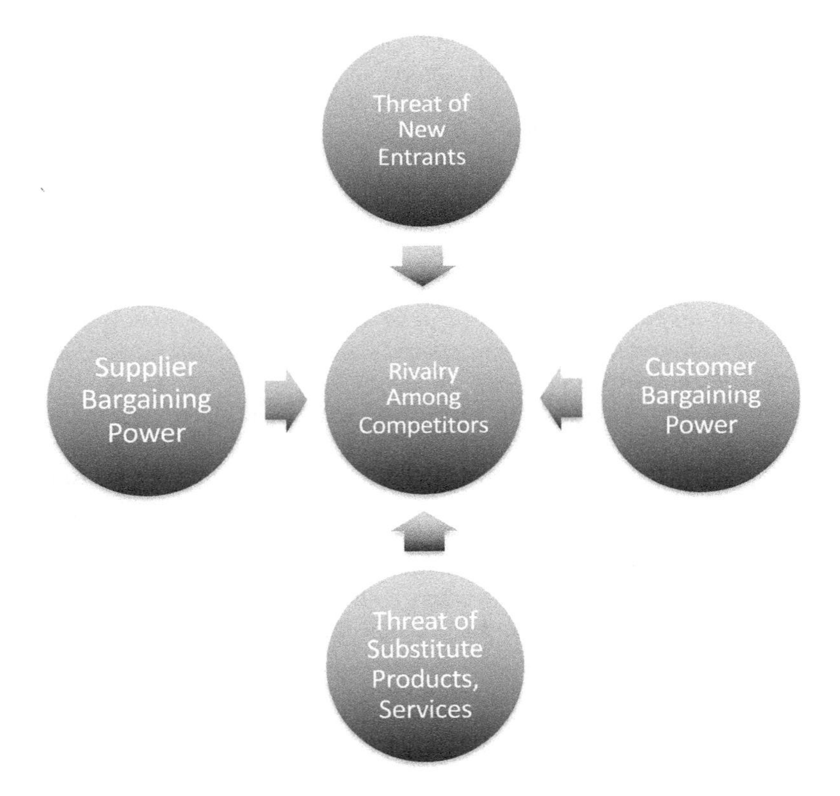

FIGURE 2.4 Porter's five competitive forces.

Today's approach to describing competitive forces is due to Michael Porter, so these forces are often referred to as "Porter's competitive forces." Porter found that the key factors determining the outcomes of competitive situations fit into five major categories. Figure 2.4 shows them.

Rivalry among Existing Competitors

This is what we tend to think of when we hear "competition" (Figure 2.5). TGI Friday's competes with Ruby Tuesday in casual restaurants; Old Navy competes with American Eagle in clothing. Pizza shops in one town compete with each other. Buyers' choices depend on how the offerings appeal to them at that time.

Information systems can affect that appeal in many ways. These begin with market research to identify the most compelling products or services and their characteristics and continue through all steps of the value chain (see the next section). All three ways in which information systems can improve a company's position—making better decisions, linking parts of the organization, and connecting with customers and suppliers—can affect a firm's position relative to competitors.

The Impact of the Internet and the Web on Rivalry among Existing Competitors

The web creates new ways for companies to compete. Customers may choose a supplier for a capability such as online customer support. Restaurants with good reviews on Trip Advisor, Urban Spoon, and yelp.com have an advantage over those that don't. If you Like a restaurant on Facebook, your friends are more likely to go there than to one across the street.

Companies must watch competitors to avoid being left behind. If BMW lets customers search dealer inventory online for a certified used car that meets their needs, Audi must do the same or lose

FIGURE 2.5 CVS and Walgreen's next to each other.

sales. Does Honda, which doesn't compete with BMW directly, also need this capability? Perhaps not, but they may see it as a way to get an edge in their segment of the auto market.

Suppose BMW develops this capability first. Audi sees it and responds.* Audi can't respond immediately. There will be a lag until they notice it. More time will pass as they decide what to do, develop this capability, and upload cars to its database. During this time, buyers of certified used cars will tend to buy BMWs, not Audis. The playing field will tilt in BMW's favor.

Once Audi has this capability, the playing field is again level, but at a new level. Buyers have a new, desirable capability. However, imagine a soccer (football in most of the world) field with dozens of balls. If it is tilted, the balls will roll toward one end. If it becomes level again, the balls will remain at that end until play eventually redistributes them. When a company gains market share because of an innovative information system, it tends to keep its share after the original reason for it has gone away. Here, if people buy certified used BMWs because, at the time, only BMW offers this search capability on their web site, they will (if satisfied) tend to buy BMWs for a while after that. Market inertia extends the value of being first beyond the early sales gains.

The web also enables companies to strengthen their competitive positions by creating *alliances*. Alliances can be among companies in different businesses, such as the agreement by which Shell fuel stations in parts of the U.S. give discounts to customers of certain supermarkets. They can also be among companies in different parts of the same business, such as airlines that serve different geographic regions or hardware stores in different cities.

Bargaining Power of Customers

A sale involves an offer by a seller and a decision by a customer to accept it. There is an implied negotiation, even if there isn't a real one. The relative power of the two parties to this negotiation plays a part in its outcome.

---

* These firms are simply examples. Feel free to reverse them or to substitute others.

In most retail situations, the seller has most of the power. If we don't like the price of apples at our local supermarket, we are free to shop elsewhere, but we know other stores probably charge about the same. Asking for a discount will be met with a polite "we don't do that."

Business purchasing is often less one-sided. When Japan Airlines purchased Boeing 787 aircraft, it didn't accept the first price Boeing asked. Both firms knew that Airbus stood ready to sell comparable aircraft if negotiations broke down. That strengthened JAL's hand, undoubtedly resulting in a lower price than they would have paid if they didn't have another potential supplier.

In any situation there is some power on both sides. If you think bananas should cost 59¢ a pound and a store's price is 69¢, you'll probably buy them anyhow. At 99¢ you'll buy fewer, perhaps none at all, or look elsewhere. Even in the unbalanced retail situation, there are limits to a seller's power over its customers.

Reducing customer power can overcome a moderate competitive advantage but not a strong one. Suppose Starmark gives coffee customers a card that earns them a free cup after buying nine. If a cup of coffee costs $2, this discount effectively reduces the price to repeat buyers to $1.80 ($18 for nine paid cups and a free one). If McD's also charges $2, regular buyers will prefer Starmark. The card will be effective as long as McD's charges at least $1.80. Some customers will stay with Starmark if McD's cuts its price to $1.69, from force of habit or because they like its coffee. But if McD's charges $1, the card won't keep customers from leaving. Sellers can reduce their power, but can't take it away. If a competitor's offer is a lot better, customers will take it.

Free products are of limited value in increasing customer loyalty, because they can be countered by lower prices. That is why airlines, masters of customer loyalty programs, offer more than free flights to their best customers. They offer upgrades to first class, which cost the airline little (if the seat would otherwise be empty) but are expensive to purchase. They offer access to phone lines with shorter waits and early boarding. These benefits cost the airline little or nothing—phone calls have to be handled sooner or later, all passengers will eventually board—but they make a big difference to flyers who get them (Figure 2.6). Information systems are essential to these programs. (There's more about loyalty programs in Chapter 8.)

Changing the price, quality, or some other aspect of the basic offering is not using this force. If a company cuts prices, it changes the basis of its competition with its competitors. It will probably get more customers, but as soon as competitors match its prices, or they return to their previous level, those customers will leave. Reducing the bargaining power of customers allows a company to keep its customers even though a competitor's offering is, at this time, somewhat better.

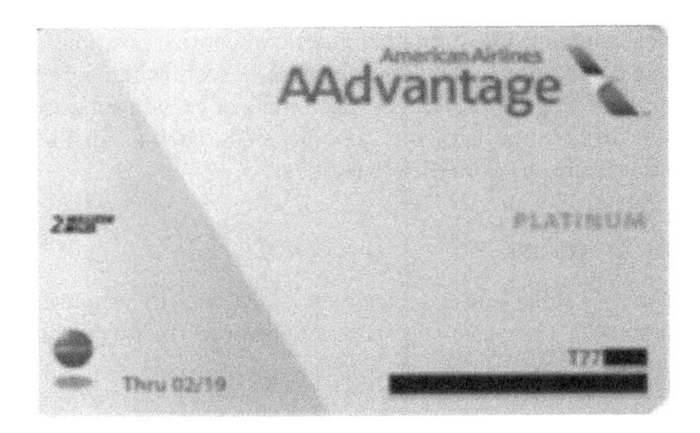

FIGURE 2.6 Airline elite loyalty card.

The Impact of the Internet and the Web on the Bargaining Power of Customers

The web increases the bargaining power of customers by making it easier to find suppliers. Want a hotel in Vancouver, Canada? Hotels.com shows 293. One is sure to meet your needs.

Without the web, travelers would choose a hotel that their travel agent knows, that advertises in their local newspaper, or that's in their guidebook; or perhaps they'd call the reservation line of a familiar chain to see if it has a hotel there. In theory they could check a Vancouver phone book or a complete hotel directory, but few travelers would.

Companies can reduce customer bargaining power by increasing *switching costs*. Consider shipping packages. If you log into FedEx's site and ship a package, the recipient is now in their database. Next time you send her something, you select her name from your address list on the FedEx site—or start to type it, and you'll see a list of previous recipients whose names match what you've typed so far. A company may have thousands of names in its FedEx address list. It will be reluctant to change shippers since that would mean re-entering all those names and addresses. This company may stay with FedEx even if another shipper is less expensive.

The cost of using a more expensive supplier can exceed the switching cost over time, but few companies make that calculation. If this company moves to that less expensive shipping firm as soon as it notices the cost difference, entering addresses as it uses them, it will probably save the data entry cost many times over—but the desire to avoid switching costs is strong.

Loyalty programs such as the Starmark card don't require information systems, but information systems and the web make sophisticated programs practical. Airlines' frequent flyer programs are more effective in motivating travelers than simpler programs would be. Information systems also keep program cost under control, since careful monitoring and analysis of demand ensure that upgrades and flights given as awards would seldom have been sold to paying passengers.

Bargaining Power of Suppliers

A company also has power over its suppliers. It can go elsewhere to obtain what it needs. Suppliers know this. Buyers can use this power to their advantage.

Suppose you supply a critical component to the air conditioning industry. Customer A tells you about its needs well in advance, doesn't change its orders unless absolutely necessary, and pays promptly. Customer B orders at the last minute, demands rush service, changes its orders on a whim, and pays when your lawyer threatens to sue. You put up with that because selling to them is profitable. Then a materials shortage limits your production. You can produce 5,000 units next year. Each of these customers wants 4,000. How do you think you will allocate them? Which one will get the 4,000 it wants? Which will hear "I'm sorry, but 1,000 are all we can sell you?"

All things being equal, large customers have more power over their suppliers than small ones. All things need not be equal, though. Information systems can even things out. A small company that can, thanks to a demand forecasting system, plan its needs well can obtain delivery and price commitments that would otherwise only go to a supplier's largest customers.

Information systems can link a customer to its suppliers, creating what is in effect a partnership in many respects. They can automate the ordering process, saving time and money on both sides and reducing errors. You'll read more about *supply chain management* in Chapter 8.

The Impact of the Internet and the Web on the Bargaining Power of Suppliers

The web reduces supplier bargaining power by making it easier for purchasers to find new ones. This is the other side of the earlier statement that the web increases customer bargaining power.

In the other direction, it can strengthen suppliers' hand by integrating the supplier's supply chain with the customer's. It is harder for a company to change suppliers when its operations are linked with those of its existing suppliers.

Threat of New Entrants

The best way to limit competition is to prevent others from competing with you. Sometimes this is done by creating legal barriers, such as limiting the number of taxi licenses in a city. If legal barriers are impractical or insufficient, companies can try to limit new entrants in other ways.

Information systems can create technological barriers to entry. If customers expect suppliers to have a certain capability, new companies can't go into that business without it. If the capability is expensive, that will deter new entrants. For example, a regional or national shipping business today has to let shippers and addressees track packages. That requires scanners everywhere, wireless links to a central database, and shipper/customer access to that database. This increases the cost of entering the shipping business. That increased cost can deter potential entrants.

The earlier example of automobile companies creating databases of certified used cars is also a technological barrier to entry. Once this capability becomes part of used car buyers' expectations, and the playing field has stabilized at its new, higher level, a firm hoping to enter the automobile business must have it. It's a small fraction of the cost of starting an automobile company, but every additional cost has an effect at the margin.

The Impact of the Internet and the Web on the Threat of New Entrants

The web has reduced *geographic* barriers to market entry. Unless location is a factor in providing a product or a service, the web has opened worldwide markets to businesses of any size.

By enabling online sales, the web has also reduced some financial barriers to market entry. A one-person business can have the online presence of a global corporation.

At the same time, the web has increased *technological* barriers to entry. Effective competition requires a well-designed site, regularly updated. It requires a presence on social networks and effective use of social media. It may require back-end databases with real-time inventory information. Today, the web is the primary vehicle for creating technological entry barriers.

Threat of Substitute Products or Services

A substitute product is *a product or service that meets the same need as an existing product or service*, often in an innovative way. MP3 downloads are a substitute for music CDs (Figure 2.7). Earlier, music CDs were a substitute for vinyl records and audio tapes. Airlines replaced long-distance bus and train travel. Computers with word processing software replaced typewriters.

Substitute products may not take over an entire market. Some people still buy vinyl, take trips by rail, use typewriters, and cross oceans by ship. Still, these substitutes took enough of the market to impact businesses that depended on the earlier product.

FIGURE 2.7 MP3 player superseding CDs.

A substitute product may apply to some, but not all, customers. The option not to buy bananas was used as an example of customer power above. The availability of substitute products affects this decision. If the bananas were to be sliced into sour cream for dessert, nectarines may be a good substitute. If they were to add potassium to someone's diet, tonic water might do the job. If they were needed as a prop in a stage play, there may be no acceptable substitute.

The response of a business to substitute products is a major factor in its competitive position. Music sellers who see their job as getting tunes into ears can prosper with MP3 downloads or anything else. Those who see their business as putting plastic disks into boxes may not.

A substitute product is not the same as a new source. A new pizza shop is not a substitute product for existing pizza shops. It offers the same basic product: pizza. A substitute product might be a new food that is designed to appeal to pizza lovers, just as margarine was designed to appeal to people who like butter. A new pizza shop is a new entrant in the existing market for pizza.

The Impact of the Internet and the Web on the Threat of Substitute Products and Services

Technology enables digital substitutes for information-based products. MP3 downloads replacing music CDs, mentioned above, is one example. People get news online instead of from the daily paper, download software instead of buying a CD-ROM, check maps.google.com instead of a road map, and visit dating sites instead of singles bars. The common feature is that the original product had physical form for lack of an alternative, not because it was inherently necessary.

In addition, technology can replace information-based services. People who need technical support are often asked to consult a database of common problems before calling a specialist—of whom there are fewer than before, but whose jobs are more interesting because they no longer need to answer simple questions. Online reservation systems have replaced many airline and hotel employees. Again, the remaining ones tend to have more interesting, and more challenging, work.

> ***Where you fit in:*** If you understand the five fundamental competitive forces and how information systems can influence them, you're in an excellent position to suggest ways to use information systems strategically. Organizations value people who can do this. Their contributions are visible to top management. Their career prospects are bright.

2.4 THE VALUE CHAIN

Just as competitive forces have existed for ages, but weren't analyzed until recent decades, so has the value chain. The *value chain* is the series of steps a company goes through to transform its raw materials into something of greater value. It was first formalized by the same Michael Porter who first analyzed the basic competitive forces.

Figure 2.8 shows the value chain for a typical manufacturing company. It is divided into *primary activities*, the foundation on which every organization rests, and *support activities*, which span the primary activities.

Primary activities differ from one industry to another. Their structure and effectiveness can create a difference between one organization and another in the same industry. Support activities are usually similar in concept, especially if two organizations are in the same industry, but can also be carried out better by one firm or another.

Value chains exist in every organization. A hospital transforms its inputs (ill and injured people, medicines, facilities, and the efforts of medical professionals) into healthy people. Since people or their insurers are willing to pay for this transformation, the output must be worth more than the total cost of the inputs. The right end of the value chain diagram, labeled "margin," reflects this increase in value. The primary activities in the value chain of a hospital are admitting, diagnosis, treatment, patient care, therapy, and discharge.

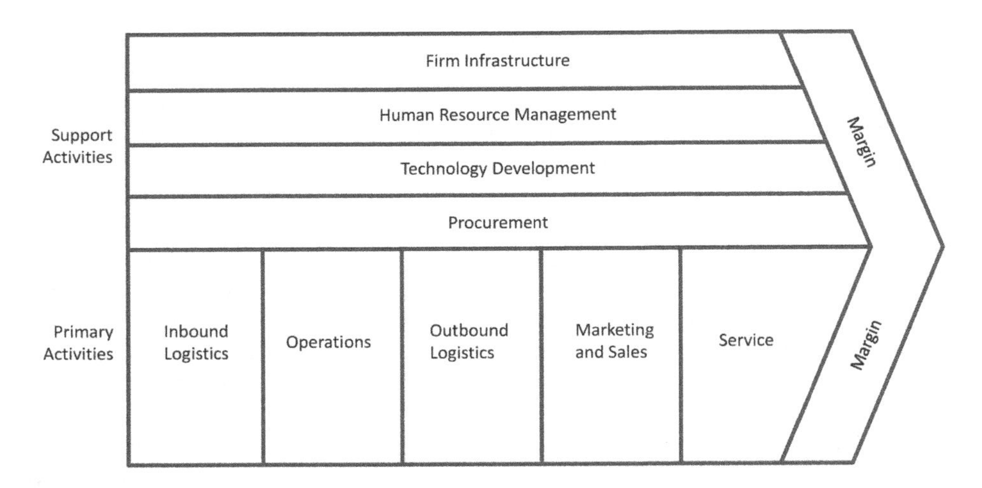

FIGURE 2.8 Porter's value chain diagram

A hospital can do things like these to apply information systems to its value chain:

- Admitting: A database of patient information, started for patients when they are first admitted, can be used throughout a patient's stay and as a starting point on later visits.
- Diagnosis: Doctors can research unusual symptoms for rare conditions online. (Today's diagnostic equipment uses computer technology internally, but that's a separate topic.)
- Treatment: A database of drug interactions can prevent prescribing a medication that will interact badly with others that the patient takes.
- Patient care and therapy: A shared database with patient information can improve communication among professionals in different specialties or on different shifts.
- Discharge: Billing can be handled electronically; information systems can provide information to visiting nurses or others who will care for the patient at home.

Most hospitals do most or all of these today. The first to do them obtained a competitive advantage. Today they are a competitive necessity, which benefits both hospitals and patients.

When you try to understand the value chain of any organization, such as one you might work for in the future, start with inputs and outputs. What are the elements that the organization uses in its work? What is its end product? Then, define what activities transform one to the other, and work from there to the other activities of the organization.

- In a school inputs are people who need education, facilities, and educators' efforts; outputs are educated people; and the basic operations are teaching and learning.
- In shipping the inputs are goods at point A and a means of transportation, the outputs are goods at point B, and the basic operation is moving them from A to B.

Where you fit in: The stages of the value chain show where an organization can apply information systems. Think about each of them in turn, asking questions such as:

1. How can the *cost* of this stage be reduced without reducing its value?
2. How can the *value* of this stage be increased without increasing its cost, or increasing its cost by less than its increase in value?

3. How can *communication* between stages be improved to increase their joint value or reduce their joint cost?

Then ask this central question: How can information systems help make this happen?

KEY POINT RECAP

There are several basic strategies: *low cost, differentiation, and niche, with variations on each. Every organization has one, though its employees might not be able to articulate it.*

As a businessperson, you must be able to tie information systems to business objectives. You can't do this unless you first understand those objectives thoroughly.

Five forces affect every competitive organization: *the threats of new entrants and of substitute products, the power of its customers and suppliers, and its rivalry with existing competitors.*

The organizations you will work for as a businessperson will all be affected by these forces. You must understand them in order to use them and counter their use by your competitors.

An organization's value chain is how it makes its output worth more than the sum of its inputs.
The organization you work for will use the value chain to make money (if you work for a profit-making company) or to contribute to society in another way. Understanding how it does this will increase your value to your employer.

Information systems can be applied most constructively where they improve the value chain, affect one or more of the competitive forces, and support organizational strategy.
As a businessperson, you will be in a position to allocate funds to information systems projects. Allocating them on this basis will usually yield the best results.

KEY TERMS

Bargaining power of customers (competitive force): The ability of an organization's customers to influence its success by their actions.

Bargaining power of suppliers (competitive force): The ability of an organization's suppliers to influence its success by their actions.

Competitive forces: The five forces that influence a company's competitive position.

Competitive strategy: The overall approach an organization takes to succeeding in its industry in the face of other organizations that also want to succeed and the fact that they cannot all succeed as well as they might wish.

Differentiation strategy: A decision on the part of a business that it will try to succeed by offering a product or service that differs from others in ways that apply to many potential buyers.

Low-cost strategy: A decision on the part of a business that it will try to succeed by offering a product or service that is tailored to the unique needs of a specific market segment.

Niche strategy: A decision on the part of a business that it will try to succeed by offering its customers the lowest possible prices.

Primary activities (in the value chain): Activities that contribute directly to making, selling, and supporting the organization's product or service.

Rivalry among competitors (competitive force): The ability of an organization's existing competitors to influence its success by their actions.

Secondary activities (in the value chain): Activities that contribute indirectly to making, selling, and supporting the organization's product or service.

Switching costs: The cost to a customer of changing from one supplier to another.

Threat of new entrants (competitive force): The degree to which an organization's success can be reduced by new organizations going into the same type of business.

Threat of substitute products or services (competitive force): The degree to which an organization's success can be reduced by different products or services that meet the same need.

Value chain: The sequence of activities through which an organization produces a product or service that is worth more than the total cost of producing it.

REVIEW QUESTIONS

1. What is the ultimate purpose of using information systems in organizations?
2. Identify the three ways in which information systems can make a big difference to organizations.
3. What are the three basic types of business strategies?
4. What are the five competitive forces that Michael Porter identified?
5. Describe, briefly, the competitive force of *rivalry.*
6. Describe, briefly, the competitive force of *customer power.*
7. Describe, briefly, the competitive force of *supplier power.*
8. Describe, briefly, the competitive force of *threat of new entrants.*
9. Describe, briefly, the competitive force of *threat of substitute products.* Explain how it differs from the threat of new entrants.
10. What does *value chain* mean?
11. In the context of the value chain, what is the difference between *primary activities* and *support activities*?
12. How are the value chain and the five competitive forces related to each other?

DISCUSSION QUESTIONS

1. Explain why the three ways that information systems can make a major difference to an organization's success, listed in Section 2.1, are more important for that purpose than knowing how to use all the features of Microsoft Office or a similar office suite.
2. Visit the links in the box at the end of Section 2.1 and answer the questions posed in its last paragraph. Give reasons for your answers.
3. Give examples of restaurants in your area that employ the three basic competitive strategies (one restaurant for each strategy). Explain why you think each fits its category.
4. When Japanese auto manufacturers began selling luxury cars in the U.S., they started new divisions with separate identities and dealerships: Acura (Honda), Lexus (Toyota), and Infiniti (Nissan). More recently, Hyundai did the same with Genesis. They did this even though it cost more than selling identical models through their existing dealerships. Using competitive strategy concepts, explain why they thought this was worth the cost.
5. An apartment cleaning service employs about ten people and cleans about 200 apartments; some weekly, some less often. Its rivalry with competitors involves how it compares with other apartment cleaning services in the same city. Describe, in a few sentences each, what the other four competitive forces mean to it.
6. Justin Thyme wants to start importing Chinese cars to the U.S., where they are not well known. Explain how his venture would be affected by the five competitive forces. (Some might not matter.)
7. Identify a well-known brand of consumer products, such as Fisher-Price toys. Is its position weak or strong with respect to each of the five competitive forces? Justify your answers. If you said it is strong with respect to a force, give an example of another firm that is weaker or of a hypothetical firm that would be weaker. If you said it is weak with respect to a force, say what could make it stronger with respect to that force.

8. Visit tgifridays.com and read about their Give Me More Stripes loyalty program. Which competitive force does it leverage? Do you think it is effective in motivating people to eat there more often than they would otherwise? Why or why not? What changes to it would you recommend, and why? If you answered "none," explain why you think it can't be improved.
9. Draw the value chain for your college or university. Start by defining its raw materials, or inputs, and its outputs. There is no single right answer to this question, so your answers will differ from your classmates'.
10. Draw the value chain for an airline. Start by defining its raw materials, or inputs, and its outputs. There is no single right answer to this question, so your answers will differ from your classmates'.

KHOURY CANDY DISTRIBUTORS: STRATEGY, VALUE CHAIN, COMPETITIVE ADVANTAGE

When Jake and Isabella signed in at the main KCD entrance the week after they had met Jason Khoury, they were given visitors' badges on lanyards. Their first names were in large letters that could be read at a distance, with their family names and the word "visitor" in smaller print below. They noticed that other people who came through the reception area wore similar badges, but without "visitor."

The receptionist called Jason, who came out to greet them a few seconds later. As they walked to his office, he said that Sandra Steere, KCD's vice president of business development, would join them. "She's got a better handle on some of what you want to talk about than I do," he explained.

Sandra was waiting in Jason's office when the three came in. After Jason introduced her to the students, the four sat down around a conference table, in the center of which was a large bowl of assorted candies. Jason opened with "Michael Porter had just published his competitive strategy concepts when I took my undergrad strategy course. Our professor mentioned them in passing. By the time I went to Standish for my MBA, they were on everyone's mind."

"And do they apply to KCD?" asked Isabella.

"Indeed they do," Jason replied. "Let's let Sandra tell us how."

"The thing about Porter's forces," Sandra began, then interrupted herself with "You *have* looked at them in school, right?" When the two nodded, she continued "is that they help us focus on the things that can make a difference instead of spreading our efforts around at random.

"For example, let's look at his 'bargaining power of suppliers' force. Normally, in our business, suppliers have the upper hand because candies have brand name recognition. If we want to sell Kit Kats to our customers, we have to get them from Hershey. Nobody else makes them, nobody else can. It's not like you're distributing auto parts: if you don't like one supplier's hose clamps you can carry someone else's. What we do to counter that is to try to get exclusive agreements with our customers. With those, we can say to Hershey 'if you want to sell Kit Kats to these stores, you have to sell them through us.' That levels the playing field a bit. Our customers win too, because when we have more power in dealing with Hershey, or anyone else, we can get better prices and pass on a lot of the savings.

"That ties into our value chain, which you may have also studied," she continued. "We add value by providing candy in quantities that retailers need, when and where they need it, without having to deal with several different manufacturers that all want to sell in larger quantities.

"Other than that, the concept is simple. A distributor buys large quantities of stuff from manufacturers, then sends out smaller amounts of several kinds of stuff to retailers. Replace 'stuff' with 'candy,' and you have us. Put in 'auto parts' and you get a different distributor. Think of us as a grid with products across the top and customers down the side. When we order we fill a column. When a customer buys from us they fill a row. The cell where they intersect is how many of that item we sold to that customer. You can think of each order as strung out behind it. Lots of data to keep track of, but computers make it easy."

"It might make sense to look at your existing systems next," Jake suggested. The others agreed, with Sandra suggesting that they start with KCD's hardware. "In that case," Jason concluded, "I'll set up a meeting with our CIO, Chris Evans, for next week. This time works for both of you, and I think it works for him. If that sounds good, I'll send you a confirmation email with the details as soon as I make sure."

Questions

1. Based on what you know so far, sketch your understanding of KCD's value chain. Start by defining its raw materials, or inputs, and its outputs. There is no single right answer to this question, so your answers will differ from your classmates'.
2. What do you think KCD's basic strategic approach is? Justify your answer.
3. Sandra Steere gave an example of how KCD deals with one competitive force in this episode. Give examples (one example per force) of how the other four affect it and what it should do about them.

CASE 1: USING SOCIAL MEDIA TO COMPETE

Nearly 40% of all people over the age of 13 (2.3 billion of 5.8 billion at the end of 2018) are Facebook members. Businesspeople can't help being aware of it—or of LinkedIn, Instagram, etc., though their membership numbers are smaller. We can't move without bumping into Twitter. We encounter those and other social networks at every turn.

No wonder businesspeople have been trying to figure out how to use these networks effectively. "Engaging with customers" is a fine slogan, but effectiveness requires more than that. Some companies have used social networks to advantage, though. Some examples:

Computer gamers are an obvious target demographic for social media interaction. When game company Electronic Arts launched its FIFA 13 soccer video game, they created a Facebook app called "Goals of the Week" through which users could upload their best goals. Over 13,000 goals were submitted. "Goals of the Week" compilation videos were viewed 4.28 million times in three months.

Breakfast biscuits are not, it would seem, an exciting product. Weetabix used the Vine six-second video sharing network to separate its On the Go biscuits from the crowd. It shot four series of clips on viewers' morning routines. The next Vine in each series was shot based on responses from over 260,000 people who watched the videos on Tweeter. The final video was retweeted almost 1,000 times, reaching an audience of over 600,000 Twitter users.

India is the world's second-fastest growing car market. Volkswagen, a new entrant in that market, needed to raise brand awareness. They decided to use LinkedIn Recommendation Ads, in which a user of a product recommends it and that recommendation is then displayed to other LinkedIn members in the user's network. Lutz Kothe, head of marketing for VW India, explains: "In a world where people spend an increasing amount of time at work, thinking about work, and interacting with their work colleagues, we believe it's important to foster discussion about Volkswagen products in a professional context. Our partnership with LinkedIn lets our customers learn about Volkswagen products and provides insights." VW India started its campaign with a goal of 500 recommendations, but got 2,700 in four weeks—and 2,300 new followers.

What's Dollar Shave Club? It takes five seconds to explain: "For a dollar a month, we'll send you high-quality razors right to your door." But how will people find out about it? Their answer is YouTube. DSC made a humorous, catchy video that had been viewed 26.4 million times by September 2019. More importantly for a business, 12,000 people signed up for their service within 48 hours of the video's launch.

HelloFresh, the recipe kit delivery service, focuses their efforts on Twitter. Senior Social Media Manager Clementine Berlioz says, "We post throughout the day to catch our audience at the right times. Twitter offers a unique sense of immediacy by allowing HelloFresh to build direct connections with our customers. We love rewarding customers for cooking our recipes in real time. We often use Twitter to experiment with new formats. When something works, we adapt it to other channels."

Berlioz extends her firm's reach by partnering: "We are looking for partners that will share our values and appeal to our audience. In February 2017, we worked with Animal Planet to promote The Puppy Bowl [a television program that mimics U.S. football using puppies] because we know how much our customers love their pets. We created brand synergies by offering our audience the recipe for a perfect Sunday: watching The Puppy Bowl and enjoying delicious food. The response was overwhelmingly positive; we reached more than 60 K people in one day."

A single social network does not stand alone. Social media strategy often involves more than one network, plus email. The Daily Skimm is an email newsletter with bite-sized news recaps that arm readers with information they need to hop into any conversation about current events. They incorporate email links that make it easy to share specific stories on social media and use social media for their fans, called "Skimm'bassadors," to spread the word. This enabled them to grow from their start in 2012 to over 8 million subscribers in 2018.

Five steps to developing a social media strategy are:

1 Define your goals (get sales leads? drive web site traffic? increase awareness?).
2 Define your target audience.
3 Choose your social media platform(s).
4 Carry out your program.
5 Audit its performance.

The process never ends, since audits will suggest changes or improvements.

Questions

1. Explain how each of the above use of social media affects, or why it does not affect, each of the five competitive forces. (Your instructor may assign specific examples to write about.)
2. Weetabix used six-second videos. Dollar Shave Club used a 90-second video. VW India didn't use videos at all. Do you feel that videos engage customers on social media? Why or why not? If it's not just yes or no, when and when not?
3. Are some of the competitive strategies described in Section 2.4 better adapted to social media than others? Which, why, and how?

CASE 2: PHARMACY PROCESSES AND FAXING

Faxing is *so* 20th century.* Fax machines gather dust in office corners and on store shelves—if they haven't been taken off those shelves to make room for products that people might actually buy. More business cards have Skype user names than fax numbers.

Yet in one corner of the world, fax refuses to die: health care. Privacy requirements, which affect information sent over the Internet, do not apply to fax messages. No third parties such as Internet service providers can examine a fax. A connection is set up before fax transmission begins, and is left alone until the transmission is complete, so encrypting messages is not necessary. Well over half of physician-to-physician information sharing is done by fax. Fax also accounts for a large fraction of communication among physicians, insurers, pharmacies, and regulatory agencies.

* Fax is actually even older than that. The first fax machine was patented in 1843. Commercial fax service between Paris and Lyon was offered in 1865, 11 years before the telephone was invented.

Fax has operational advantages. Fax can communicate complex information in human-readable form. It handles text, diagrams, and handwritten notes. It can be added to any office at low cost. It is harder to alter a fax message without detection than a digital message. Every fax machine in the world can communicate with every other one by using about 12 digits. And the cost of fax is essentially zero.

Fax has drawbacks, though. A message to an office may not reach the intended person. Faxes can be buried under other paper, delivered to the wrong person, or just lost. There is no convenient way to confirm that a person received a fax unless he or she chooses to reply. Fax equipment is subject to problems such as paper jams. And fax is not easily integrated with other applications.

Now, companies are merging fax and digital messaging technology in an effort to get the best of both worlds. Faxes can be created on paper or in digital form, and converted to an image of the page. Most text faxes can be converted to digital data via optical character recognition.

One such company is OpenText Corp.* of Waterloo, Ontario, Canada. Its RightFax product can use the content of a fax to access databases, can route them intelligently based on business rules, and can confirm that a person has seen a fax. By displaying faxes on computer screens, the cost of supplies is eliminated and the likelihood of mechanical failure greatly reduced. If the software can't figure out what to do with a fax, its original form is available for people to see.

One U.S. pharmacy chain used computer-integrated fax messaging to replace manual prescription entry into its central pharmacy system. Previously, a pharmacist would approve a prescription, add data to it, scan it, and send it to the chain's home office for manual entry. The home office separated faxes into new prescriptions, renewals, and others (such as insurance confirmations) before entering data. With a document capture and management system, approved prescriptions have the data added automatically—which is possible because the fax is in a computer, not on a sheet of paper—and are then added, again automatically, to the central pharmacy database.

With this system, lag in filling prescriptions went from hours to minutes. Customers benefit, because pharmacists spend less time scanning and routing fax forms. And the pharmacists can focus on what they prefer to do: serving customers and working with medications.

Another example of the benefits of computer-managed fax messaging: the central pharmacy of Denver Health Medical Center fills 700 prescriptions per day. They found that 30% of their requests were duplicates. The computer was able to flag potential duplicates by comparing the sending fax number and medical record number of an incoming prescription with those of prescriptions that arrived earlier. People who are alerted to possible duplication between two messages can compare those two easily, but people who wonder if an incoming message repeats one of the hundreds that came in earlier can't go through all of them to find out.

Questions

1. Draw the value chain for a pharmacy. Does approving a prescription and entering it into a database add value to the pharmacy's product, or not?
2. Discuss the competitive force of substitute products in the context of digital messaging competing with fax. How can fax, the existing service, retain its market? (It may not be able to keep all of it, but it might be able to keep more of it for longer.) How can digital messaging succeed in replacing it?
3. "Marketing myopia" is a term coined by Theodore Levitt to describe how providers of a product or service may see themselves as providing that specific product or service, not as meeting a customer need that can be met in other (perhaps better) ways. For example, a music publisher suffering from marketing myopia could see itself as selling CDs, not as providing music, and miss the shift to digital downloads. What is the customer need that fax meets? How can suppliers of fax equipment avoid marketing myopia?

* The use of OpenText as an example is not an endorsement. Other firms provide comparable products.

BIBLIOGRAPHY

Boeing Company, "Boeing, American Airlines sign Major order for 47 787 Dreamliners," April 6, 2018, www.boeing.com/commercial/customers/american-airlines/787-dreamliner-order.page, accessed September 25, 2019.

Horton, G., "Get rid of fax machines: Increasing the speed of health information exchange," OpenText Information Exchange, 2014, www.slideshare.net/faxsolutions/get-rid-of-fax-machines-increasing-the-speed-of-health-information-exchange, accessed September 16, 2019.

Larsen, J., "Viral video case study of Dollar Shave Club: What we can learn," *iMedia Connection*, August 27, 2012, blogs.imediaconnection.com/blog/2012/08/27/viral-video-case-study-of-dollar-shave-club-what-we-can-learn, accessed September 16, 2019.

LinkedIn Europe, "Volkswagen India case study," February 13, 2013, www.slideshare.net/linkedineurope/linkedin-volkswagen-case-study, accessed September 16, 2019.

MacLeod, I., "Weetabix finds out morning routine of public through series of Vines," *The Drum*, May 20, 2013, www.thedrum.com/news/2013/05/20/weetabix-finds-out-morning-routine-public-through-series-vines, accessed September 16, 2019.

Malone, N., "The Skimm Brains: 7 million people wake up to their newsletter, and their voice, every morning," *The Cut*, October 28, 2018, www.thecut.com/2018/10/the-skimm-carly-zakin-danielle-weisberg.html, accessed September 17, 2019.

Moth, D., "Seven useful social media case studies from 2013," *Econsultancy*, July 11, 2013, econsultancy.com/us/blog/63043-seven-useful-social-media-case-studies-from-2013.

OpenText Corp. web site, www.opentext.com, accessed September 16, 2019.

Perry, A., "Automating secure information exchange to optimize processes across the enterprise," October 10, 2016, https://blogs.opentext.com/automating-secure-information-exchange-to-optimize-processes-across-the-enterprise/, accessed December 6, 2019.

Porter, M., "How competitive forces shape strategy," *Harvard Business Review*, vol. 57, no. 2 (March–April 1979), p. 137; updated in "The five competitive forces that shape strategy," *Harvard Business Review*, vol. 86, no. 1 (January 2008), p. 78.

Sharma, G., "4 steps for creating a solid social media strategy," *Social Media Today*, July 15, 2019, www.socialmediatoday.com/news/4-steps-for-creating-a-solid-social-media-strategy-infographic/558752, accessed September 17, 2019.

Statista, "Number of monthly active Facebook users worldwide as of 3rd quarter 2018 (in millions)," 2018, www.statista.com/statistics/264810/number-of-monthly-active-facebook-users-worldwide, accessed November 7, 2018.

Williams, A., "3 ways TheSkimm perfectly marries email newsletters and social media," *Sociality Squared*, January 19, 2017, socialitysquared.com/content-marketing/3-ways-theskimm-perfectly-marries-email-newsletters-social-media, accessed September 17, 2019.

Window, M., "HelloFresh's sr. social media manager shares tips on developing your social strategy," *Twitter Business*, August 29, 2017, https://business.twitter.com/en/blog/HelloFresh-tips-on-social-strategy.html, accessed September 17, 2019.

YouTube, "DollarShaveClub.com—our blades are f***ing great," www.youtube.com/watch?v=ZUG9qYTJMsI, accessed November 7, 2018.

3 Information Systems Hardware

CHAPTER OUTLINE

WHY THIS CHAPTER MATTERS

You know about computer hardware. You've used it for years. Nobody has to tell you what a keyboard or a printer is, or that the cost of ink adds up if you print a lot.

What you know about hardware probably comes from using it. You've used personal computers: smartphones, tablets, laptops, desktops. Businesses also use other types of computers. Your ability to make business decisions that involve hardware depends on knowing about types of hardware that you probably haven't seen. Business computers use the same basic principles as the ones you know, but their pieces are organized in different ways for reasons that may not be obvious. When you make decisions that involve tens of thousands, or millions, of dollars, these differences matter.

Also, when we learn something by using it, we seldom take time to learn its principles. This leads to gaps in what we know. Gaps can cause problems when you make business decisions.

Finally, your fluency with computer technology concepts and with the vocabulary that technical professionals use will help you talk with them. If you use terms correctly, they'll respect your knowledge and listen to what you say. If you make what they think are elementary mistakes in terminology, they'll think you don't know much—even if you do. It's human nature. They'll feel that way even if what you're discussing doesn't depend on those concepts.

CHAPTER TAKE-AWAYS

As you read this chapter, focus on these key concepts to use on the job:

1 Hardware exists only to run software. Software (Chapter 4) should drive hardware choices.
2 Hardware evolves quickly. You must distinguish between basic principles and how products you're familiar with happen to work.
3 Effective management of hardware requires an enterprise hardware architecture. Individual hardware purchases should conform to it.
4 Selecting the right hardware requires understanding what that hardware will be used for.

3.1 COMPUTER STRUCTURE

As you read in Chapter 1, *computer hardware* is one component of an information system. This term refers to the physical parts of a computer: those you interact with, such as mouse, keyboard and display, and its electronic and mechanical components.

As you also read there, a system is a collection of components that interact with a purpose. Computer hardware is a system. We can break it down as in Figure 3.1.

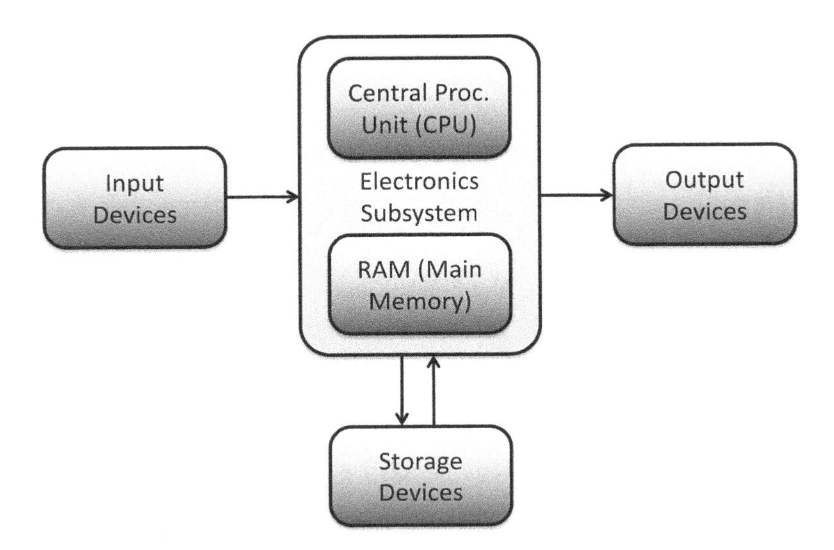

FIGURE 3.1 System-level diagram of computer hardware.

The elements of Figure 3.1 are systems, too. A mechanical keyboard consists of an enclosure, a few dozen plastic keys, springs and electrical contacts under each, and a way to tell a computer when a switch closes. Many of those can be broken down further.

Data enters computers via input devices. Keyboards are one. However, computers get more data from pointing devices such as a trackpad or mouse. We don't think of sending data when we move a mouse, but when a mouse moves, it sends data. That data doesn't end up in documents, but it helps us create them. Network connections (see Chapter 6) can also be input devices.

Output devices ship data to whomever, or whatever, needs it. Screens, printers, and speakers are familiar output devices, but many computers send output to other devices. The bill dispenser in an ATM is an output device. So are the gates that rise to let you leave a parking lot after you pay and the fuel injector in your car's engine. In today's cars, your foot doesn't control fuel flow. It sends input to a computer. That computer calculates when to send fuel to the cylinders, and how much.

Output can be also sent over a network. When you ask about a book on amazon.com or an album on the iTunes Store, that's how you get your reply. Many information systems send output to other computer systems, such as tax data to a government computer.

> ***Where you fit in:*** The characteristics of a system's hardware can affect how that system works. If a proposal can't be read on a tablet because the file is too big, or if a web page won't fit on a phone screen with text large enough to read, something has to change. Businesspeople must understand an entire system to make good decisions about its components.

Long-term storage is also part of most information systems. Your university's registration system stores your grades so it can produce your transcript, next week or in 2039.

Input, output, and storage components are called *peripheral devices* collectively because they surround a computer—that is, they are on its periphery. They are called that even when they're physically in a computer's main enclosure. The central electronics of the computer, in the middle of Figure 3.1, coordinates their activity. It contains circuits to manipulate the data it's working with and short-term storage for the data it's using at a given moment.

3.2 SWITCHES: THE BASIC HARDWARE BUILDING BLOCKS

The building blocks of all computers are switches that are on or off.* On a magnetic disk, they are spots that are magnetized in one direction or another. On an optical disk, they are areas where a non-reflecting pit has, or has not, been burned. In an electronic component, they are circuit elements at one voltage or another. In a fiber optic cable, they are lights that are on or off. The concept is the same: *a physical element is in one of two possible states.* A computer can set an element to one state or the other, and can later determine which state it was previously set to.

The data represented by a single switch is called a *bit*, short for *binary digit*. The two states of a bit are indicated by the digits 0 and 1. The representation of those states in a physical device—do 3.3 volts mean 0 or 1?—only matters to computer engineers.

Computers represent complex data by using more than 1 bit. Consider the letter H. We start with 1 bit: if it's 0, the letter is in the range A–P; if it's 1, in the range Q–Z. Since H is in A–P, its first bit is 0. A second bit narrows the letter down to A–H or I–P. H is in the range A–H, so the second bit is also 0. Adding each bit cuts the number of possible letters in half. With 5 bits, as in Figure 3.2, we can zero in on any letter of the alphabet. The figure shows the combinations that represent E–H, where 0 means "go left" and 1 means "go right."

Five bits can represent any of 32 choices. We can use 26 for letters and have 6 left over. In general, n bits have 2^n combinations:

- 4 bits have 16 combinations. That's enough for decimal digits plus a few other characters. Some decimal calculators use 4 bits to represent digits internally.
- 5 bits, 32 combinations, let us represent letters of the alphabet (as we just saw) but not much more. That's not enough for a useful computer. For one thing, we need digits, too.
- 6 bits allow 64 combinations. These can represent 26 letters, ten digits, plus common punctuation marks and symbols. Early computers used 6 bits to represent each character, but they didn't need lower-case letters. We wouldn't accept that limitation today.
- 7 bits allow 128 combinations: enough for upper-case and lower-case letters, digits, and common symbols. However, engineers find 7-bit storage units awkward. Powers of 2, such as 8, work better. Seven-bit characters were tried but never caught on.

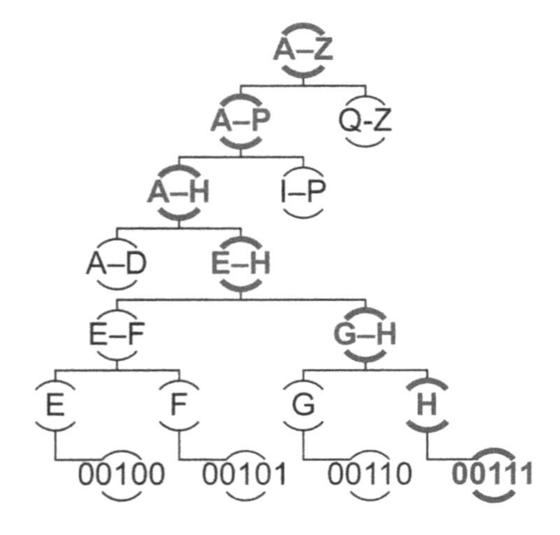

FIGURE 3.2 Five bits representing the letters E–H.

* *Quantum computers* are an exception, but they're not suitable for business information systems (yet). We won't discuss them here. You can learn more about them in Cusumano (2019) or other sources.

- 8 bits allow 256 combinations. These can handle all the characters in Latin-based languages. Characters are almost always stored in 8 bits today. An 8-bit storage unit is called a *byte*.
- Asian languages such as Chinese and Japanese have thousands of characters. Computers use *double-byte characters* for these languages. Two bytes, 16 bits, allow 65,536 combinations: enough for all written human languages combined.

The 96 positions of the *ASCII* (American Standard Code for Information Interchange) character set are shown in Figure 3.3. These 96, plus 32 more used to control devices, were the original 7-bit character set. Another 128 were added when computer designers realized that 7-bit storage units wouldn't catch on. Non-Latin alphabets use the second group of 128 characters. Those 128 places also provide for accented letters and additional symbols such as "curly" quotation marks.

Decimal value	*Character*	*Decimal value*	*Character*	*Decimal value*	*Character*
32	Space	64	@	96	`
33	!	65	A	97	a
34	"	66	B	98	b
35	#	67	C	99	c
36	$	68	D	100	d
37	%	69	E	101	e
38	&	70	F	102	f
39	'	71	G	103	g
40	(	72	H	104	h
41	)	73	I	105	i
42	*	74	J	106	j
43	+	75	K	107	k
44	,	76	L	108	l
45	-	77	M	109	m
46	.	78	N	110	n
47	/	79	O	111	o
48	0	80	P	112	p
49	1	81	Q	113	q
50	2	82	R	114	r
51	3	83	S	115	s
52	4	84	T	116	t
53	5	85	U	117	u
54	6	86	V	118	v
55	7	87	W	119	w
56	8	88	X	120	x
57	9	89	Y	121	y
58	:	90	Z	122	z
59	;	91	[	123	{
60	<	92	\	124	|
61	=	93	]	125	}
62	>	94	^	126	~
63	?	95	_	127	Delete

FIGURE 3.3 ASCII character set.

The 8-bit byte is today's basic unit of computer data. Its 256 combinations can represent a character, a number (from 0 to 255 or from −128 to 127), or part of a longer number. That number can, in turn, mean many things: the month in a date, the intensity of a color component at a point on a screen, part of a sound wave. Its meaning is determined by the program that will process it.

Multiple bytes can represent data that won't fit in a single byte. Postal Service codes for U.S. states use two bytes. Color information for a point uses three: a number from 0 to 255 for the intensity of each color component. Short word processing documents use 20,000 or so. Typical digital photographs use a few million. A full-length HD movie occupies a few billion.

Because many types of data occupy large numbers of bytes, metric system prefixes are used to describe large amounts of storage. The prefixes used to describe data sizes are given in Table 3.1.

Table 3.2 gives examples of how much space some types of data might occupy. Today's personal computers typically store a few TB of data. A large company might need petabytes of storage.

Near the upper end of databases, two authors working independently from public information in 2017 estimated that YouTube was growing by 76 PB or 93 PB per year: good agreement for estimates like this. Its total video storage probably exceeds an exabyte today.

The largest commercial database in the world is probably used by Google's search engine. Its size is a closely held secret, but it is surely larger than YouTube's. Some government security agencies may have larger databases, but those are even more closely guarded secrets.

TABLE 3.1
Metric System Prefixes for Multiples of 1,000

Prefix	Unit	Meaning
Kilo	Kilobyte (KB)	Thousand: 1,000 bytes
Mega	Megabyte (MB)	Million: 1,000,000 bytes
Giga	Gigabyte (GB)	Billion: 1,000,000,000 bytes
Tera	Terabyte (TB)	Trillion: 1,000,000,000,000 bytes
Peta	Petabyte (PB)	Quadrillion: 1,000,000,000,000,000 bytes
Exa	Exabyte (EB)	Quintillion: 1,000,000,000,000,000,000 bytes

TABLE 3.2
Data Storage Requirements

Data Item	Approximate Size
Typical word	About ten bytes
Typical printed page	2–3 KB
Low-resolution photo (on web page)	100 KB
High-resolution photo	2–5 MB
Three-minute song in high-quality sound	20–30 MB
A yard (about a meter) of shelved books	100 MB
Thirty-minute TV show in high definition	500 MB
All of Beethoven's music in high-quality sound	20 GB
The print collections of the U.S. Library of Congress	10 TB
All the words ever spoken by human beings	5 EB

A NOTE ON LANGUAGE

Table 3.1 gives storage units as powers of 10, using metric system prefixes. Computer designers find it easier to use powers of 2. 2^{10} equals 1,024. If you buy a 1 KB memory unit, it will hold 1,024 bytes, not 1,000. If it is necessary to tell the two apart, 1,024 is a *binary kilobyte*; 1,000, a *decimal kilobyte*. To a hardware designer, 1 MB is 2^{20} (1,048,576), 1 GB is 2^{30} (1,073,741,824). The difference between binary and decimal capacities seldom matters in practice, but you'll meet people who insist that one is correct, the other wrong. Arguing with them is never productive.

Where you fit in: It's important to know the terminology that people use when they discuss information system storage requirements and capacities. You must be comfortable with these concepts and terms to participate in those discussions intelligently. Besides, computing specialists often base their first opinion of a person on how well the person uses computing terms. If you're comfortable with them, they'll think you understand computers. If you use them awkwardly, they'll think you don't, even if you really understand a great deal.

3.3 COMPUTER SYSTEM COMPONENTS

Central Processor

The *central processor* or *central processing unit* (CPU) is the "brain" of a computer. It controls everything a computer does. When a calculation yields the answer "36," the central processor first calculates that number. It then figures out what dot pattern would convey it to a person, and where those dots should appear on the screen. Finally, it sets switches that control color at those dots to display one color, and those that control color around them to display a contrasting color.

This activity is directed by instructions located in the computer's main memory. Since everything in a computer is stored as bits, each instruction is also a series of bits. Different sequences of bits tell the computer to do different things. The sequence 00000001 might tell the processor "add two numbers," while the sequence 11101001 might mean "subtract."

The decision to use 00000001 for *add* is arbitrary. Designers could use any other combination of 8 bits, or a different number of bits. The way this and thousands of other choices are made define a computer's *instruction set.* A program in the form of bits will run only on a computer with the right instruction set. If you take a program for a computer that uses 00000001 to mean *add*, and load it into a computer that uses 101000 to mean *add*, it won't do anything useful.

Most personal computers today use the *Intel instruction set*, first designed by Intel in the 1970s and extended since then. Most CPUs with this instruction set are made by Intel or Advanced Micro Devices (AMD). Most smartphones and tablets use the ARM instruction set, with CPUs from several manufacturers. A program written for your PC won't run on your phone, and vice versa. Large computers and specialized processors often use other instruction sets.

The importance of this to businesspeople is understanding that software is tied to an instruction set, so the software you need to run will constrain the type of computer that you can use. (When an application runs on more than one instruction set, it is really two apps designed to be as much like each other as possible. The match isn't always perfect.)

Putting instructions together to carry out a complex job is not easy. Each instruction performs a tiny part of a task. Thousands, even millions, of them must be combined to do something useful such as processing words, making a game character jump, or sending an email message.

Hardware Evolution

Electronics components get smaller every year, making devices smaller, faster, and more capable. Intel co-founder Gordon Moore noticed a trend in the 1960s: the number of electronic elements on a chip tends to double every 18–24 months as engineers learn to shrink them. This became known as *Moore's Law.* It's more an observation, not a law, but it's too late to change its name.

As Figure 3.4 shows, technology has more or less followed Moore's Law for about half a century. Experts have predicted an end to this growth for decades, but every time one way to put more components on a chip reaches its limit, another is found. The end of Moore's Law is now foreseen for about 2022, but history suggests that it would be dangerous to bet on that. It has also been suggested that, before then, the historical three-way combination of "cheaper, smaller, faster" will become "choose any two." The reason is that making chips smaller and faster requires greater investments in design and production, offsetting savings from smaller components.

Why are smaller transistors good? Smaller transistors can be closer together. That reduces the time for electronic signals to travel from one transistor to another, allowing a device to be faster. Smaller transistors also consume less power and produce less heat. With more transistors on a chip, that chip can perform more activities in parallel, thus completing an overall job faster. Putting more transistors on one chip also allows one chip to replace several earlier chips, making devices even smaller and less expensive.

There are some difficulties with this concept. Smaller components require thinner connecting wires. Thinner wires are slower and heat up more. To maintain speed, components must be able to respond to weaker and weaker electrical signals. The tradeoffs are complex. Searching for the next advance keeps many intelligent people busy.

The latest step in putting more components on a chip is vertical stacking of circuit layers. This makes circuit design more complex and adds to the difficulty of heat removal, but the ability to stack

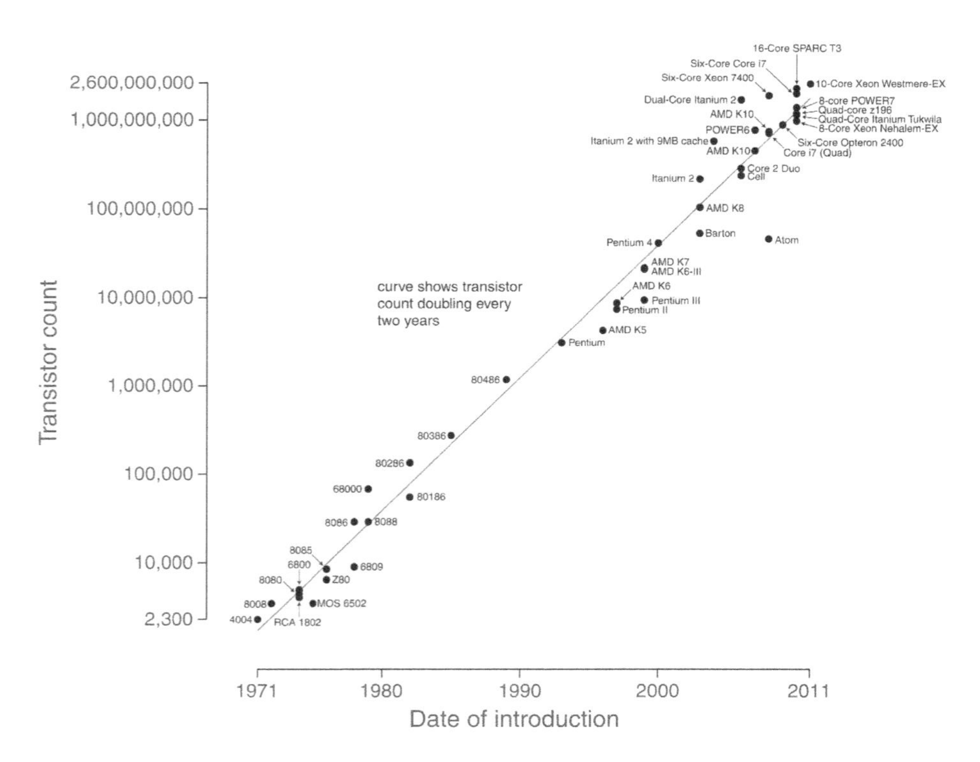

FIGURE 3.4 Trend in transistors per chip over the years (From: https://en.wikipedia.org/wiki/Transistor_count).

FIGURE 3.5 Livescribe Echo (Source: Livescribe).

many layers—each thinner than a human hair—on top of each other will be especially important as wearable devices are expected to do more and more. The benefits of having circuit components closer together suggest that layer stacking will become common in a few years.

Because computing circuits are becoming smaller and cheaper for a level of capability, they are showing up in more and more places as replacements for earlier devices. Today's aircraft have *glass cockpits*: computer screens instead of separate instruments. (The displays are often designed to look like traditional instruments.) Computer-driven instrument panels are found in automobiles and motorcycles. Even an ordinary-looking pen (Figure 3.5) can have a computer to record everything you write to read on a screen later. No more lost class notes!

Processor Clock Speed

For a processor to operate properly, signals must proceed from one component to the next in an orderly fashion. To ensure this, processors run under control of a central clock. Chip designers want this clock to run as fast as possible, so the chip will work faster—but if it ticks too fast, data may arrive at a component before that component is ready for it, and the chip may overheat.

The rate at which a processor's clock runs is one factor in its performance. This rate is measured in *hertz* (Hz). A speed of 1 Hz means the clock ticks once per second. The clock speed of early processor chips was measured in thousands of hertz: kilohertz, or kHz. That of today's processors is measured in gigahertz (GHz, billions of clock ticks per second). In 2019, most personal computer clocks run at 3–4 GHz. Most smartphones have clock speeds around 2–3 GHz.

Processors with faster clocks generate more heat. That heat must be removed, lest the system overheat and fail. Some processors can run faster than their rated speed for a short time, or while other parts of the system are idle so they aren't generating heat.

All else being equal, a processor with a clock speed of 4.7 GHz, such as the one in Figure 3.6,* ought to be faster than one with a 4 GHz clock. However, all else is seldom equal:

* Not a real shelf tag, but based on a real high-end desktop computer of late 2019.

Computer Company, Inc. Model 7000	
Processor	Intel® Core™ i9, 8 Cores
Processor Speed	4.7GHz
Graphics Processor	NVIDIA® GeForce RTX™ 2070
RAM	32GB DDR4 RAM at 2666MHz
Disk	2TB 7200 RPM SATA
Optical Drive	Tray Loading DVD-RW Read-Write
Keyboard	U.S. English

FIGURE 3.6 Sample computer shelf tag in a store.

- Faster components elsewhere in a system can compensate for a slow processor. A cost-effective system should be balanced. Money spent on a faster processor is wasted if a system's limiting factor is elsewhere. It might be better spent speeding up something else.
- A processor with a slower clock speed but more cores (next section) may outperform one with a higher clock speed but fewer cores. The workload determines if this is true or not.
- Different processors do different amounts of work in a clock tick. Comparing processors by clock speeds is like comparing the speeds of horses and mice by steps per minute, ignoring stride length. If Intel and AMD both sell Intel instruction set processors at 4.7 GHz, one may get work done faster than the other. We can't tell without extensive testing.

Multiple Cores, Multiple Threads

Moore's Law tells us that transistors in a central processor will continue to shrink. Some benefits of smaller transistors were discussed earlier. One was the ability to do more at a time by carrying out processing tasks in parallel.

For decades, this produced steady performance gains. In the 21st century, the benefits of more transistors per chip reached the point of diminishing returns. A new approach was needed.

One approach is to put multiple processing elements on a single chip. Each such element is a complete processor, but the term "processor" was already in use to refer to the chip itself. These processing elements were named *cores*. A processor with two cores can run two programs at the same time. A user's work can, potentially, be done twice as quickly—if the other components of the computer can get data on and off the CPU chip fast enough. Today's smartphone and laptop processors typically have 2–8 cores; larger computers have 16 or more. (The shelf tag in Figure 3.6 tells us that this computer's processor has eight.) Large computers can use more than one processor chip, further improving performance.

A core can keep even busier if it has several programs to work on. When one program uses one part of the core, a different program may need another part. The core can then keep most of its elements busy most of the time. This is *multi-threading*. Each program in a core is a *thread*.

The value of multi-core processors and multi-threading depends on what a computer is doing. Most computers can utilize two cores easily: one may run an application for the user, such as word processing, while the other does things like checking email. Beyond two, it's problematic. Unless a computer serves many users at the same time, each additional core adds less and less performance. By developing multi-core, multi-threaded processors, hardware designers solved their problem (what do we do with all those transistors?) but created a software problem (how can we use those cores and threads?). Solving *that* problem will take time.

As a step in that direction, applications that perform large computations with independent parts can be written to take advantage of multiple cores and multi-threading. Figure 3.7 shows how a user controls what Microsoft Excel does in this regard. By using the Manual setting, Excel can be told to create more threads than the computer has cores.

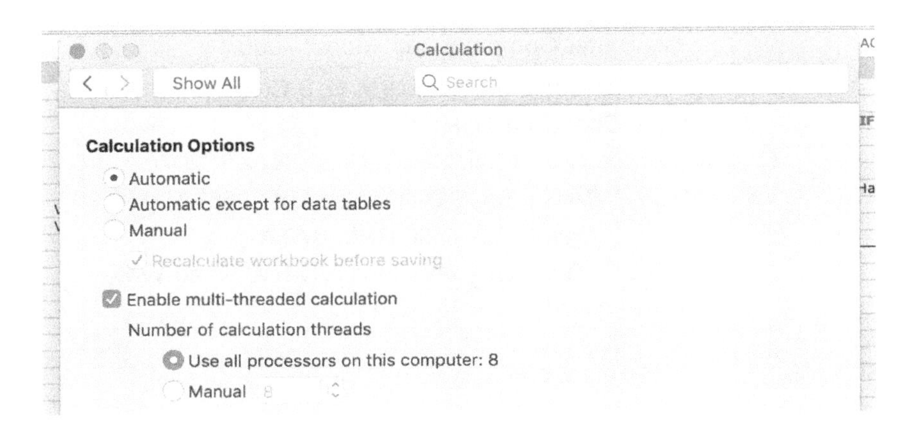

FIGURE 3.7 Screen shot of Excel preference pane.

MULTIPLE CORES MATCH IRIS PATTERNS

The largest biometric identification program in the world is the Indian government's Aadhaar project to collect fingerprints and iris patterns of its 1.3 billion citizens as means of identification. In mid-2018, over a billion people were in the program. Comparing new registrants' iris patterns with those already in the database means 500 trillion comparisons per day. One processor core can do about a million comparisons per second. At that rate, it would take about nine minutes to see if a person is in the database. By dividing the billion people into groups and assigning each group to a core, the task can be done in a few seconds.

The faster a chip's clock ticks, the more heat it generates. If some cores are idle, others can run faster before the chip reaches its heat limit. Some processor chips can speed up one core when others are idle. This "turbo" feature doesn't offset the benefits of multiple cores if software can put those cores to use, but when it can't, it lets a single thread run faster than it otherwise would.

Specialized Processors

Central processors such as the Intel Core series are jacks of all trades: they do most tasks well, but are not optimized to perform any one task outstandingly. That's fine for the main processor of a computer. One never knows what it may be asked to do. It may edit photos one day, compute molecular interactions the next, and forecast sales the day after that—while checking email, displaying web sites, and playing background music.

When the task is known in advance, one can design a specialized processor for it. This is only worthwhile when the specialized processor will be used in large numbers. The reason: economies of scale. When a product is sold in large volume design cost is spread over more units, reducing it on a per-unit basis. Also, production costs decrease as production volume increases: a firm can justify specialized production equipment, and there is a *learning curve* effect. Thus, a general-purpose processor manufactured in large numbers can be more cost-effective, especially when design cost is prorated, than a specialized processor with smaller production volume.

A few tasks are so common that specialized processors are routinely used for them. Displaying graphics is one such task. All but the least expensive systems augment their central processors with *graphics processing units* (GPUs) to allow the CPU to focus on other tasks. The computer of Figure 3.6 uses a graphics processor from AMD.

GPUs excel at performing the same operation on many data elements, one after the other. This enables them to carry out functions such as shading an area of the screen, moving a window or an icon from one part of the screen to another, or decoding compressed video efficiently. This ability

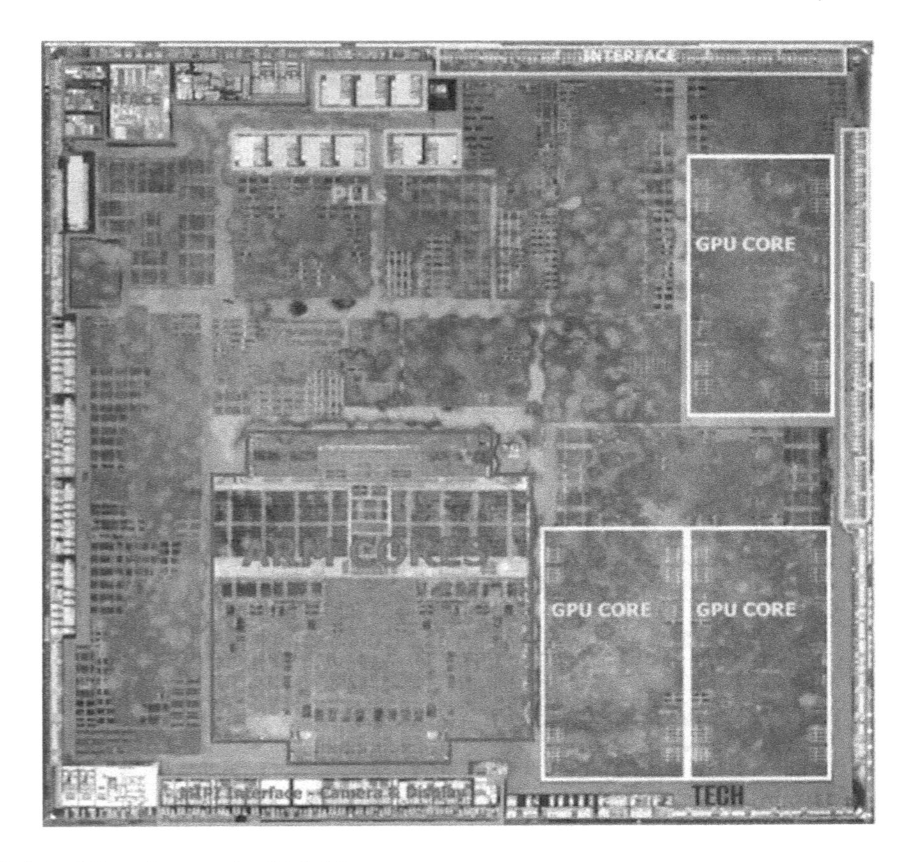

FIGURE 3.8 Microphotograph of A6 chip.

is not just useful for graphics. Scientific computers have been built using large numbers of GPUs, with a few CPUs used to control them. This approach works because many scientific calculations involve repeating a calculation for many elements of a matrix or points in a grid. The Summit computer at Oak Ridge National Laboratory in Tennessee, the fastest computer in the world in late 2018, uses about 26,000 Tesla V100 graphics processors from chip maker Nvidia, controlled by a smaller number of Power9 CPUs made by IBM.

GPU and CPU functions can be integrated on a single chip. These *integrated graphics* do not offer the same graphics performance as separate (or *discrete*) graphics processors but are a good tradeoff in overall system design when the highest graphics performance isn't required. Figure 3.8 shows the graphics processors integrated into Apple's A6 chip as used in some iPhones.

> ***Where you fit in:*** Whenever you buy a computer, for your personal use or for a business, it's important to be able to see past the buzzwords. Should your computer have more cores or a faster clock cycle? Is a discrete GPU worth its added cost? You must understand what these terms mean and how the computer will be used to make an informed decision.

Primary Storage

Computers need to store programs (instruction sequences to accomplish specific tasks) and data for those programs to work with. These programs range from games and video players through inventory management and production planning. Each program has its own data, which it may share with related programs. All programs and data share the same storage.

Storage is of two major types: *primary storage*, also called *main memory, random-access memory*, or *RAM*; and *secondary storage*. The difference corresponds to the difference between a desk and a file cabinet. You can't use a document in a file cabinet. It must be on your desk. However, desk space is limited; a file cabinet is much larger. You can get as much file space as you like, but the useful size of a desk is limited by the length of your arms and the distance at which you can read fine print. If your desk is overloaded, your work will suffer even if your file cabinet is half empty (Figure 3.9).

Primary storage must be immediately accessible, since the CPU uses it at every step of its work. It is built from fast components, and is placed near the CPU to minimize delays in sending signals from one to the other. Figure 3.10 shows them next to each other on a computer's circuit board.

FIGURE 3.9 Office worker with full desk (public domain from wpclipart.com).

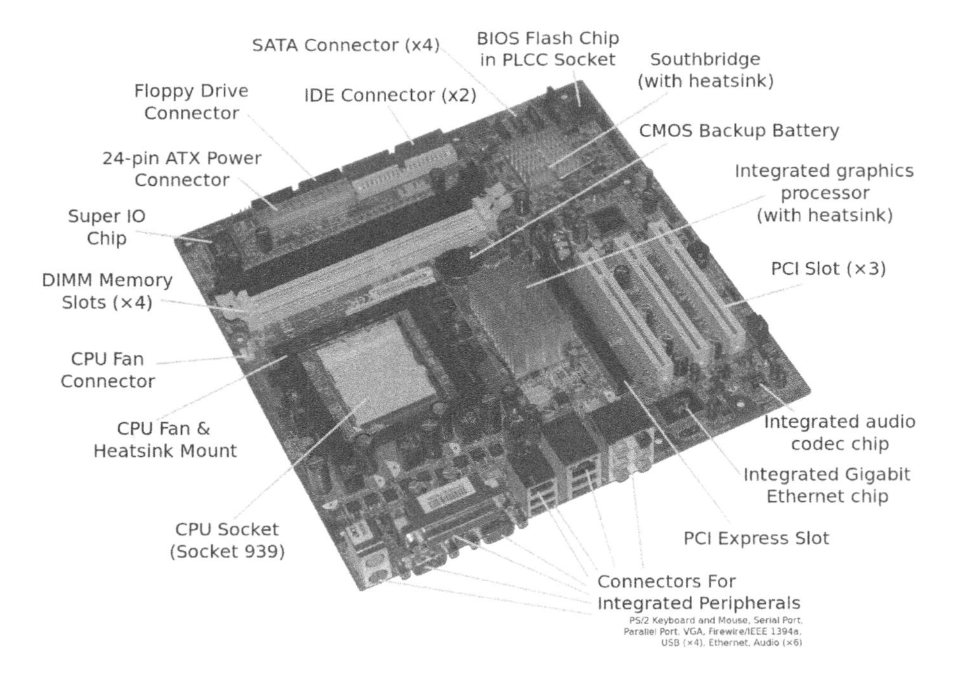

FIGURE 3.10 Motherboard (Source: Wikimedia Commons).

The capacity of a computer's primary storage is measured in the units of Table 3.1. Today's personal computers typically require at least 2 GB RAM. They may be able to run in less, but that is like asking an accountant to work on an airline tray table. Unfortunately, computers often come with minimal RAM to keep prices low. Adding RAM to a bare-bones computer is a good way to improve its performance. The ability to process data quickly is wasted if a CPU must wait for it.

A computer's maximum RAM is limited by its design. This involves business tradeoffs on the part of its manufacturer, based on its general knowledge of the uses to which computers of that type are typically put. The amount of memory you install should be based on your knowledge of the specific uses to which your computer will be put.

There are many types of RAM cards, with names like SIMM (Single In-line Memory Module) and DIMM (Dual In-line Memory Module), and different speed ratings. The computer whose shelf tag is shown in Figure 3.6 uses a type called DDR3 and is equipped with RAM chips that operate at 1.6 GHz. Each computer is designed to use a specific type of card.

When you buy a computer, try to get one that can accommodate at least twice as much RAM as you expect to need. That, in turn, will probably be more than the computer comes with. Limited expansion capacity may make it necessary to replace a computer sooner than you otherwise would. A more expandable computer may cost more, but that cost can pay off over time.

***Where you fit in*:** These decisions involve business tradeoffs. Make them with a business mindset.

Small computers, such as smartphones and tablets, have a fixed amount of RAM built in. It is always sufficient for standard personal applications. If you expect to run demanding business applications on a phone or tablet, make sure you get one with as much RAM as possible.

The trend is for computers to require, and have, more RAM. This trend is driven by the demands of increasingly complex software and users' desire to have more programs running at a time, and is supported by steadily decreasing RAM prices. 4GB RAM, marginally adequate for personal computers in 2020, would have been ample five years ago. It may be impractically small in 2025.

You may hear that RAM is *volatile*: when you remove its electrical power, its contents are lost. Most RAM technologies today are volatile, but that's not a basic principle. Using volatile RAM is a technology choice that computer designers make. There were times when non-volatile RAM was cost-effective and common. With new technologies, it may be common again in the future.

If an office worker's desk is too small, the worker can pile papers on top of each other, shuffling them to reach what is needed at any time. Computers do the same, moving data temporarily to secondary storage (next section) to free up RAM for other uses. This is called *paging*. By paging, a computer with 4 GB RAM can behave as if it had 8 GB or even more. However, there are limits to how much *virtual memory* can be added this way. If a computer with 2 GB RAM tries to act as if it had 16 GB, the delays of moving data back and forth will be unacceptable. It will perform slowly, like an office worker who spends so much time rearranging papers that he can't do much useful work. This slowdown is called *thrashing*. Thrashing is a sign that you need more RAM.

Secondary Storage

Secondary storage holds data that may be needed minutes or months later. The ideal secondary storage device would combine high capacity, fast access, low cost, and long-term reliability. That ideal doesn't exist. Different technologies offer different tradeoffs. The choice depends on:

- The amount of data that must be stored.
- Whether the data must be modified frequently, rarely, or not at all.
- How much data must be retrieved in a single access.
- The importance of speed in reading and speed in writing the data. (These may differ.)

- How long the data must be stored for.
- Your budget.

Rotating Magnetic Disks

Most 2020 secondary storage uses rotating *magnetic disks*. They are the least expensive type of storage that is fast enough to serve as a computer's only secondary storage. Other technologies are too slow for that purpose or more expensive. Today's magnetic disks have capacities up to 20 TB (from Western Digital). Drives in the 12–16 TB range are available from several vendors. Most personal computers come with smaller drives (typically 2–4 TB) to reduce the initial price, or with solid-state drives (see below) for performance, but it is usually possible to order them with larger ones or to replace an inadequate drive with a larger/faster one later.

Magnetic disks use platters of aluminum or glass coated with a magnetizable alloy. An electromagnet called a *read/write head* magnetizes a spot on the surface in one direction or the other to record a bit as 0 or 1. To write data, the device goes through these steps:

1. It selects, electronically, the surface on which the data will be stored. (High-capacity drives have several platters, one above the other, and therefore multiple recording surfaces.)
2. It moves the head assembly over the surface to the correct circular *track*. The head is at the upper-right end of the arm in Figure 3.11. It is moved by a positioning motor at the other end of the arm. The arm rotates about a pivot to move the heads in or out over the surfaces.
3. It waits until the desired section of the track, or *sector*, comes under the head.
4. It sends electromagnetic pulses to set the magnetization of the bits in that sector.

The first three steps are the same for reading data. Step 4 is different: the head detects the magnetic field that is generated as it passes over the surface and interprets it as 0 or 1.

FIGURE 3.11 Magnetic disk (public domain illustration from clipartlord.com).

FIGURE 3.12 USB port expander.

Magnetic disk drives are installed inside the main computer structure (*internal drives*) or attached to it (*external drives*). Their enclosure protects them from airborne contaminants. High-capacity drives require close tolerances. Even a small dust particle between the head and the recording surface can destroy them.

Some external drives need to be connected to a power supply as well as to a computer. External drives that are powered through their computer connections are called *portable drives.*

Removable magnetic disks are seldom seen today. Older computers used removable *diskette* storage. Diskettes, or floppy disks, use flexible plastic platters instead of rigid metal or glass platters but operate on the same principles as hard disks. Diskettes use larger magnetizable spots than rigid (hard) disks, so they need not be made to the same tolerances and can be exposed to airborne dust with little risk, but larger spots mean lower storage capacity.

The speed at which an external disk drive can transfer data to or from a computer depends on the type of connection. (The computer specified in Figure 3.6 uses a connection called SATA to connect its internal disk to the processor.) A computer's specifications say what types of interfaces it has connections, or *ports*, for and how many of each. Later, if you buy a peripheral device such as an external disk drive, you must get one that can connect to one of those ports. One can add ports to some desktop computers by inserting an *expansion card* into its electronics enclosure. Laptop and smaller computers rarely have this capability, but *port expanders* can give them more ports (Figure 3.12).

The interface between a computer and its disk drives (or between a computer and nearly anything else) resembles, conceptually, the interface between a tractor and a semi-trailer (Figure 3.13). As long as both are designed for a standard "fifth wheel," an operator need not be concerned that they won't fit. New ones may have features such as the ability to detect if a trailer is attached properly, but those features don't get in the way when they're not used. The owner of a 40-year-old trailer can buy a brand-new Kenworth T680 tractor and haul away. The main difference is that computer interfaces haven't been static for decades. New ones come out every few years. Someone who gets a new computer, and wants to keep using an old drive, may need to invest $20 or so in an adapter.

The cost of disk storage has gone down rapidly over the years. In 1990, the cost of 1 MB magnetic disk storage was about $10. In 2000, it was about 1¢. In 2020, it is under 0.01¢. Most of this decrease is due to increased capacity per drive. The cost of a single drive has dropped, but slowly.

Optical Disks

Optical disks resemble audio CDs or video DVDs. To record data, pits are burned into a reflective surface. The pits reflect less light than an unpitted surface. To read data, a low-powered laser is aimed at the surface. A photocell interprets the reflection as a 1 or a 0.

FIGURE 3.13 Tractor with fifth wheel seen from rear (Source: Wikimedia Commons).

Optical disks are *removable storage devices*. A drive may be built in, but the storage medium is not. One can store an unlimited amount of data on a shelf of CDs or DVDs, but what's available at any moment is about what a magnetic disk can hold. The time to access data on an optical disk is measured in tenths of a second, minutes if a disk must be inserted, rather than thousandths.

There are several types of optical media. Some require special equipment for writing. Personal software is sold on these when it's not downloaded. *Writeable optical disks* can be written, or *burned*, with almost any computer. Such disks are useful for backing up and sharing data. *Rewriteable optical disks* can be written, erased, and written again. Rewriting is slow, as it usually requires an extra step to erase previous data before new data can be written.

The speed of an optical storage device is often given as "8×," "16×," or some other multiple. The basis for this comparison is the rotational speed of a device playing audio or video.

Solid-State Storage

Magnetic and optical disks are slow compared to electronic components. Glass, metal, and plastic can't move as fast as electrons. Storage technologies with no moving parts are much faster than magnetic or optical disks.

Such *solid-state drives* are used in performance-critical applications. They use non-volatile technologies to retain data even in the absence of electrical power. They are more expensive than magnetic disks for a given capacity, but, thanks to Moore's Law, their cost is dropping rapidly.

Laptop computers often use solid-state storage instead of magnetic disks to save weight. So do smartphones and tablets, whose limited storage capacity—typically less than that of a laptop—reduces the cost penalty of an expensive storage technology. Lack of moving parts also renders them less subject to physical damage than a magnetic disk drive would be.

At the enterprise level, solid-state storage can be used for frequently accessed files. Usually, most of any computer's accesses are to a small set of files. Using solid-state storage for those, but magnetic drives for less frequently accessed files, provides most of the performance advantages of solid-state storage at a slight cost premium over a totally magnetic storage system.

Some disk drive vendors sell *hybrid drives* that take advantage of this concept. Drive electronics monitor what data is used most, and put that data in solid-state storage for rapid access. This is based on usage statistics, not on knowledge of file content, so it can't be planned as carefully as assigning files to different device types, but it is effective nonetheless.

Because solid-state storage devices are light and don't need sealed enclosures, they make excellent removable storage. Your flash drive is an example. So are your phone's SIM card and your camera's SD card. Their capacities increase rapidly each year, as their cost drops.

The highest-capacity solid-state drive available today is the Nimbus Data ExaDrive DC100. Its 100 TB capacity exceeds that of the largest available rotating magnetic disk drives by a substantial margin. However, this capacity comes at a price: well into five figures. The same capacity, which would have to be spread over five or six drives, is available in magnetic disks for about $2,000. If one needs maximum capacity and performance, though, there is no current alternative. According to Nimbus, companies such as eBay have large amounts of their storage.

Magnetic Tape Storage

Magnetic tape, an older technology, is still used at the enterprise level though it is rare on personal computers. A modern tape cartridge that conforms to the 2017 LTO-8 standard, such as the one shown in Figure 3.14, can store 12 TB of data. However, in 2017 IBM and Sony jointly announced a breakthrough that will permit storage of 330 TB in a single palm-sized cartridge. In addition, tape drives fail several orders of magnitude less frequently than magnetic hard drives, and have about 1/100 the rate of unrecoverable errors created when writing data.

The big drawback of tape is *sequential access*: to reach any data item, the drive must go over the tape from its present position to that item. That takes much longer than moving a read-write head across a disk, a *direct access* device.

Tape is used today for applications that read or write large amounts of data at a time: data backup, archival storage, transferring large files from one computer to another, or loading applications into a computer. For these applications, the combination of low cost, high capacity, and high transfer speed make it attractive, and the delay in initiating a data transfer is less important.

Enterprise Storage Subsystems

Compared to an ideal storage device, disk drives have several drawbacks. They are slower than one might wish, they fail from time to time, their interfaces are designed for attaching to a single computer, and they don't provide the mix of speed and cost that an organization with a lot of data requires. To overcome such problems, devices are combined into *storage subsystems*. You probably don't need these, but they are essential in large installations. There are three main types:

RAID

RAID stands for *Redundant Array of Independent* (originally *Inexpensive*) *Disks*. RAID systems use two or more disk drives plus electronics to mimic the behavior of a single drive:

- The capacity of a RAID subsystem can exceed the capacity of one drive, while it looks like a single drive to a computer. This can be important when very large files must be stored.

FIGURE 3.14 Tape cartridge.

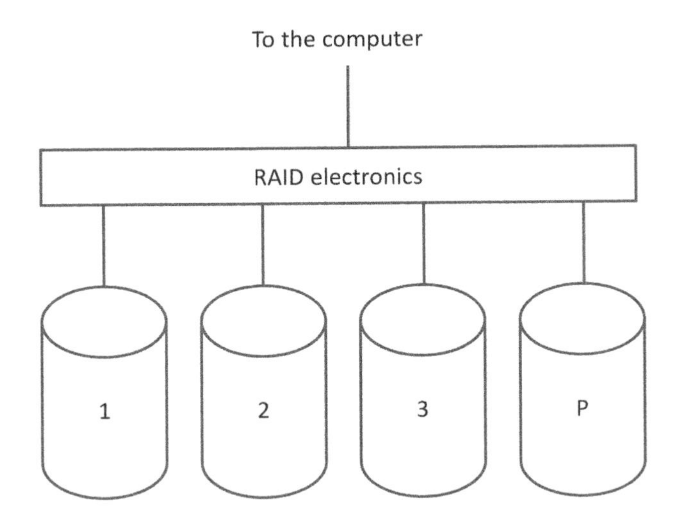

FIGURE 3.15 Diagram of RAID redundancy.

- By moving data to or from several disks in parallel, RAID can transfer data faster than a single disk drive.
- By using more space than the data require, RAID can protect against data loss if a drive fails.

How does this last feature work? In the simplest case, a RAID system can use two drives, storing two copies of the data. If one fails, the system uses the other until the failed unit is replaced.

This would, however, not be economical. A corporation with storage costs in six or seven figures would rather not double that cost. To an enterprise, twice as many drives mean twice as much floor space, twice as much power, and twice as much heat to remove by using twice as much air conditioning (and twice as much power for that). Fortunately, there is a better way.

Suppose we use four drives where three could hold the data, as in Figure 3.15. Drives 1, 2, and 3 each store a third of the data. The drive labeled P stores the sum of the data on the other three, treating data as numbers. (That's not exactly how it works, but it's the concept.) If drive P fails, it is replaced, and its data re-created by summing the data on the other three drives. If one of the other three fails, its data can be re-created by subtracting the data on the two remaining data drives from the sum on the P drive. RAID electronics do this on the fly. It's slower than accessing data directly, but not much, and far better than losing data. The added storage needed to use this method is 33.3%. It can be reduced even further by using more drives.

Different types of RAID offer different features. Using an extra drive as in Figure 3.15 is called *Level 5 RAID*. It protects against data loss at reasonable cost. It provides good performance and acts like a single drive with three times the capacity of each of the four drives it contains.

NAS

NAS, *Network-Attached Storage*, solves the problem of disk drives being designed to attach to a single computer. Disk drives in a NAS attach to a specialized computer inside the NAS. That computer also connects to one or more networks. The computers that use data from the NAS also connect to those networks. Those computers send requests to the NAS over the network rather than to their own drives. The NAS responds to the request like a locally attached storage device, reading or writing data as needed (Figure 3.16).

A NAS subsystem can use RAID instead of individual drives, providing the advantages of both.

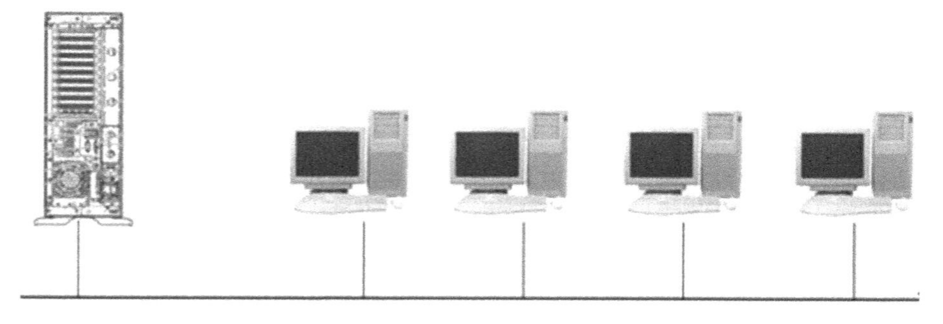

FIGURE 3.16 Network with NAS storage system.

SAN

A *Storage-Area Network* (SAN) is a self-contained system that includes storage plus one or more specialized computers to manage that storage. Those computers access the storage devices over a high-speed network inside the SAN.

A SAN makes storage available to computers that are connected to it. Those computers treat that storage as if it were their own. The SAN handles much of the work of managing it: monitoring their use of storage, optimizing the amount of storage assigned to each computer, migrating data to higher-speed or lower-cost devices based on usage, and managing backup and recovery.

SANs connect to computers much as disk drives do. In keeping with the high capacities of SANs and the correspondingly high demands of the computers that use them, their connections may use fiber optic links rather than copper wires. Fiber optics can cover distances of several miles or km, so a SAN can support a large data center or several data centers.

SANs can have very high capacity. A small SAN such as that in Figure 3.17 can accommodate 120 disk drives. If they store 8TB each, the SAN can store about a petabyte. Larger SANs can have

FIGURE 3.17 EMC Clariion CX500 SAN.

thousands of drives with a total capacity of several petabytes. While most SANs are smaller than that, SANs tend to be used with large-scale information systems.

Where you fit in: It takes time to install any storage subsystem and move an organization to it. A simple RAID system may not take long, but a complex SAN could take months. It's important to stay on top of changing storage needs so that system enhancements such as these can be planned.

Cloud Storage

You've probably heard about *cloud storage*. You may use a cloud storage service such as Dropbox, iCloud, Google Drive, or Microsoft's SkyDrive. You know that, for some purposes, cloud storage works the same way as your computer's secondary storage. But what *is* it, really?

Cloud storage isn't a hardware technology. Data "in the cloud" is stored on a device, usually a magnetic disk, somewhere. The term *cloud storage* refers to how we access it.

"The cloud" is an informal term for the Internet. The Internet is often pictured as a cloud that connects whatever is around it. Anything described by the word *cloud* is accessed over the Internet. When you use cloud storage you send your data somewhere via the Internet, often to another company. That company stores it for you on its storage devices. When you need it, they send it back—again via the Internet.

Compared to local storage, cloud storage has these advantages:

- The vendor enjoys economies of scale in buying many large storage devices. That gives it a lower cost per unit of storage than a person or a small company would have.
- The vendor has a mix of devices, and can migrate files to low-cost devices when appropriate. A small installation will have fewer device types and therefore less flexibility.
- The vendor provides professional administration, regular backups, redundant power supplies, and industrial-strength security. An individual or a small business probably can't afford these. If it wants them, it can't spread their cost over as much storage.
- Because cloud storage is remote from the user, disasters such as earthquakes are unlikely to affect both. Cloud storage providers keep copies of data in several locations to minimize the impact of a problem at one site.
- Because it is not attached to a single computer, cloud storage can be accessed from anywhere. It can be shared easily among multiple computers belonging to one or several people.

Cloud storage also has some disadvantages:

- There are potential loss of control and security problems in handing data to a third party.

  Case in point: Capital One, the tenth-largest bank in the U.S., revealed in August 2019 that criminals had gained access to the personal data of over 100 million customers that was stored in the cloud. FBI agents later arrested an ex-employee of a cloud hosting company that Capital One used. Reportedly, a firewall (see Chapter 6) misconfiguration gave her access that, even as an employee, she should not have had. Was this a cyberattack in the usual sense? Perhaps not, especially since that employee made no attempt to cover her tracks or to sell the data. Still, since the breach would not have happened had Capital One stored all its data on its own devices, it points out that cloud storage is not without risk.
- The provider has overhead and (it hopes) profit, offsetting some of the inherent economies.
- Access may be slow, since communication lines are often slower than storage connections.

Large companies can use *private clouds*. These offer the benefits of cloud storage without control or profit margin concerns. Private clouds aren't cost-effective for small or medium organizations, though the practical lower limit for a private cloud is dropping. A company with a private cloud can use a public cloud to handle less critical data or overflow needs, thus creating a *hybrid cloud*. This is called *cloud bursting*.

> ***Where you fit in:*** Is cloud storage for you? That is a business decision. Fortunately, it's not all-or-nothing. Most cloud storage users also have local storage. They keep frequently used files locally, using the cloud for sharing, backup, and archival storage of infrequently accessed data. Properly used, cloud storage is a valuable part of today's computing picture.

Input Devices

We're used to keyboards and pointing devices of several types: mouse, trackpad, touchscreen. As you've seen on phones and tablets, a touchscreen can replace a physical keyboard, though many people like the feel of keys that move when touched. Figure 3.18 shows an *ergonomic keyboard*. It permits the hands to take a more natural position than keys arranged in straight lines.

Many devices exist for entering images into computers. Still and video cameras, and scanners, are some of them. In a computer, images become grids of thousands or millions of dots, each having a color. Artists use *pen tablets* (also known as *graphics tablets*) where they draw with a pressure-sensitive pen, with their strokes transferred to drawings in the computer—often as mathematically described curves that can then be adjusted. Users can also draw on touch-sensitive screens or trackpads with a finger or a stylus (Figure 3.19).

Motion sensors send input to many personal applications. Game controllers use them. Astronomy applications on tablets rely on them to show the right portion of the sky. Health monitoring phone apps count how many steps you take. In business, a truck computer can monitor how smoothly a driver operates it and how fast it is driven.

Specialized systems have input devices designed for specific tasks. Store checkout computers have bar code readers to identify what was purchased, an example of *source data automation*. Bar codes can track merchandise in a warehouse or an automobile on an assembly line. As the cost of electronics continues to drop, bar codes are being replaced with *radio frequency identification* (RFID) tags that convey more information than a bar code and don't require positioning a product for scanning. RFID identifies cars on a highway, opens garage doors to authorized drivers, transfers office telephone calls to the room a person is in, times runners in road races, and enables utility workers to read meters without leaving their cars. Other types of short-range communication unlock your car and let you use your phone to pay for coffee or cruise tickets.

FIGURE 3.18 Ergonomic keyboard (Source: Wikimedia Commons).

FIGURE 3.19 Pen tablet in use.

Finally, both wired and wireless telecommunications links can provide input to computers. GPS input comes in wireless form—as does the RFID data that, as just mentioned, is replacing bar codes. You'll read more about communication links in Chapter 6.

Where you fit in: Today the cost of enabling an electronic device to communicate with a computer is small. Just about anything can become an input device. Home thermostats, athletic shoes, and more are all used as input devices. Be open to new devices that might provide input at work.

Output Devices

As with input devices, you're familiar with the most common output devices found on personal computers. Displays, from the size of a wristwatch to projectors that fill a movie theater screen, and printers are the major ones.

Displays (Monitors)

Most displays today use *liquid-crystal diode* (LCD) technology, where a backlight shines through a panel that transmits different amounts of light. These displays are flat, thin, and, thanks to high-volume manufacturing, reasonably priced. The screen of your laptop is probably an LCD display. Smaller devices such as smartphones may use *light-emitting diode* screens instead. These use less current, but age more quickly, and are more expensive than LCDs.

The main (but not the only) parameters involved in choosing a display are:

- **Size:** Usually expressed as a diagonal measurement in inches. For some purposes, *aspect ratio* (relationship of height to width) may be important.
- **Resolution:** The number of distinct dots, or *pixels* (short for "picture elements"), that a device can display. The most popular resolution has been 1,366×768 pixels for several years, in part because higher resolutions are not standardized. Software can display information at lower resolution. That enlarges items on the screen, but may make them look fuzzy.
- **Dot pitch:** How many dots per inch a device can display. A dot pitch under about 70 is unacceptable for close work. Dot pitches over about 300 appear continuous to the human eye. Even metric countries often use the measure "dots per inch."

FIGURE 3.20 Man wearing Oculus Rift headset.

- **Color gamut:** The range of colors that a device can show. For most business uses, including viewing photographs, video, and videoconferencing, any display gamut is fine. Applications such as photo editing may require special displays.
- **Cost:** Improvements in these and other parameters are not free.

Where you fit in: As a businessperson, when you buy things for your company, you must walk a line between scrimping to the point where it impacts employees' work and overspending for no benefit. Buying computer displays is an example. The technologies you'll look at in ten years will differ from those we use today, but the tradeoffs in that decision will be similar.

Beyond that, your business career will be full of tradeoffs among desirable factors that cannot be achieved simultaneously. The factors will vary, but the concept of tradeoffs will not change. You must look at all the factors together, whatever they are, to select the best overall combination.

There is no reason a display must look like a computer display. The Oculus Rift headset creates an impression of "virtual reality" by showing different pictures to its wearer's two eyes (Figure 3.20).

Some displays are intended for viewing by a group. These use a variety of different technologies, including projectors that can be mounted on the ceiling of a room or placed on a table. These technologies will all evolve during your career.

Some applications use special displays. A point-of-sale station can have a small customer-facing display and a larger one facing the clerk. A surgical computer can display the location of a tumor on the patient's body on special glasses. Such glasses can provide any type of information that a computer display could show, such as the history of a building the wearer is looking at or the name of an approaching person. They may be of value in overcoming one limitation of phones in general-purpose computing: their small screens. Google Glass is an early example of an eyeglass-mounted output device. The wearer in Figure 3.21 is operating Glass via the touchpad on its side.

Printers

Most printers use one of two technologies: toner or ink.

In a toner-based printer, a light passes over a drum and is turned on to magnetize the drum where toner is desired. (The alternate term *laser printer* comes from using a laser as this light source.) Toner is attracted to magnetized areas, deposited on paper, and fused there by heat. This is done

FIGURE 3.21 Man wearing Google Glass (Source: Wikimedia Commons).

for each color in the image, making alignment critical in color printing. Figure 3.22 shows a high-volume black and white laser printer suitable for heavy use. A printer such as this can print over 100,000 pages per month versus a few thousand for a personal printer.

In an ink-based printer, print heads spray ink onto the page. Multiple print heads, held in a single carrier to maintain alignment, provide color. The most common design uses current pulses to heat liquid ink and propel it onto the paper.

Specialized printers may use other technologies. Store receipts may be on paper that darkens when electric current heats it. Such *thermal printers* don't use ink or toner, so there is nothing to replace except a roll of paper—which cashiers can easily replace with little training.

Considerations in choosing a printer include:

- **Resolution:** Printer resolution is measured in dots per inch. Some printers claim resolution of several thousand dpi. However, ink-based printers cannot print *distinct* dots at their claimed resolution. Ink spreads. The amount depends on the ink and paper, but it always does. An ink printer that claims resolution of, say, 5,000 dpi can probably print a few hundred distinct dpi.

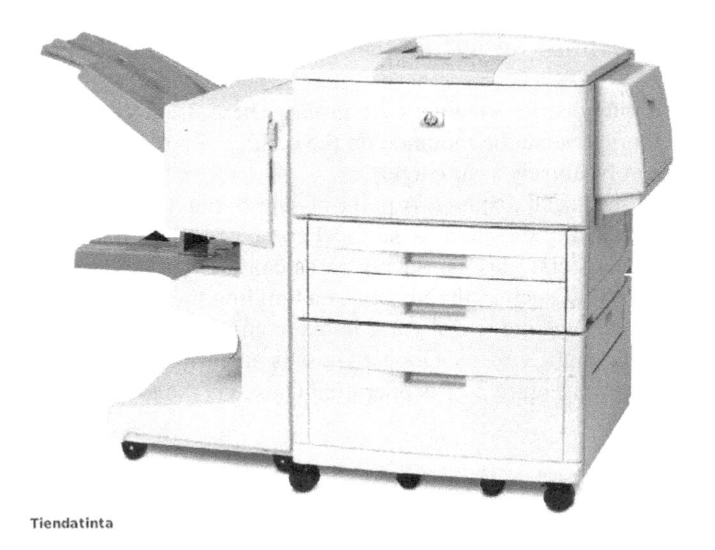

FIGURE 3.22 Office printer (Source: Tiendatinta.com).

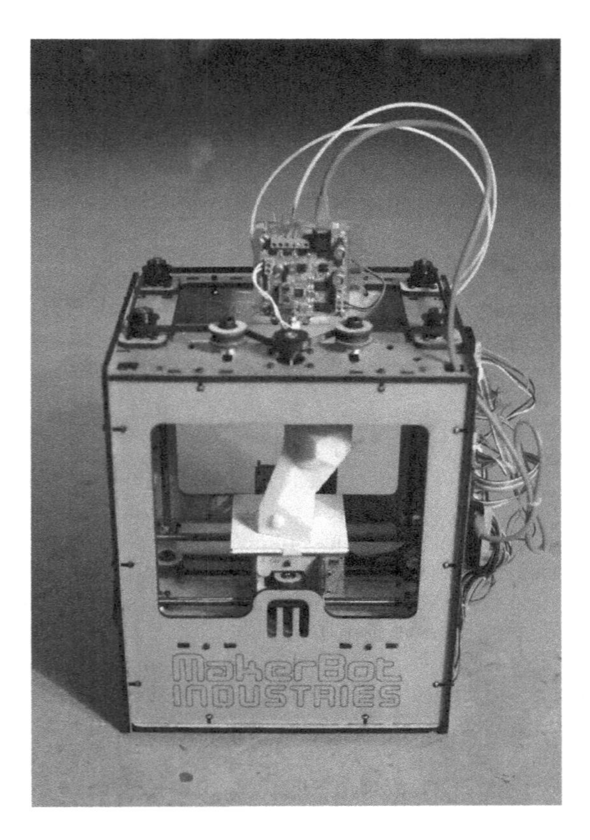

FIGURE 3.23 3D printer (Source: Wikimedia Commons).

- **Speed:** Printer speed is usually measured in pages per minute or seconds per page. The delay in printing the first page may be longer than this figure suggests. Higher speeds cost money.
- **Features:** At the personal level, added features on a printer include the ability to scan images and make copies. Businesses that do much of these use dedicated devices. Printer features to consider in business are collating and stapling, double-sided printing, user identification for cost accounting, and others that individuals may not need.
- **Cost:** The cost of supplies over a printer's lifetime can exceed initial cost. Toner usually costs less than ink, and toner-based printers can print more pages before supplies need replacing.

Three-dimensional printing, creating physical objects as output, is practical for small businesses in both cost and ease of use. One approach is to deposit thin layers of powder or liquid plastic on top of each other, via a nozzle that moves in two dimensions over the printer bed and rises for the vertical dimension. The plastic hardens to create the finished object. Inexpensive 3D printers can build objects that will not be subject to much stress, or prototypes of objects to be produced in final form later. More expensive 3D printers (Figure 3.23) can replace machining of many manufactured items.

Other

While the above are the most common output devices, there are others. Speakers provide sound output for visually impaired users, for telephony and voice conferencing, and for videos or music.

Haptic or *tactile* output applies motion to something a user can feel. A series of raised pins can provide Braille output to visually impaired users. Aircraft have "stick shakers" to warn pilots of impending stalls. Some cars vibrate the steering wheel or seat if the car departs from its lane without a signal, or if driver inactivity suggests that he or she might be nodding off.

Automobiles speedometers and gauges are, today, computer output devices. When a mechanic plugs a cable into a car's diagnostic port, the OBD-II (On Board Diagnostics II) scanner at the other end of that cable becomes a computer output device as well. Almost anything driven by electricity—motors in a prosthetic ankle, paint spraying equipment in a factory, a telescope in an astronomical observatory, a locomotive on a model railroad layout—can be an output device.

Where you fit in: The variety of output devices is limited only by the creativity of a person with an idea. That can be you.

3.4 COMPUTER CATEGORIES

Computers, like insects, are grouped into categories. As individuals we use the smaller kinds: smartphones, tablets, laptops, desktops. These are used in business as well. However, enterprise-level tasks require larger computers. You are less likely to have seen those.

Larger computers are categorized as servers, mainframes, and supercomputers.

Servers are computers that serve something—data, web pages, or the use of an application—to other computers. Those computers are called *clients.*

In theory, any computer can be a server. By analogy, any car can be a police car if it's painted blue, given a siren and lights, and registered to a police department. Still, if you know in advance that a car will be used as a police car, you'll equip it with that in mind. It's the same with servers.

Computers meant to be used as servers have features that personal computers don't need. They can handle multiple processors, since having many users creates tasks that can be done in parallel. Auxiliary processors handle communications, freeing the CPU(s) for other work. Multiple data paths between secondary storage and RAM speed access to data. More RAM allows more concurrent tasks. Multiple power supplies let them continue to operate if one fails.

Where you fit in: Features cost money. A higher RAM limit requires more slots, a beefed-up power supply to handle memory cards in those slots, and a larger enclosure to make room for them. Those cost money even if the higher limit isn't used—yet. Space and connections for multiple power supplies cost money even if only one is present. However, those features don't make the server look better in a comparison of speed and storage capacity. An uninformed person might ask, "Why does this server cost $2,000, if I can get a PC with the same specs for $600?" If you look carefully at a computer built as a server, the reasons for its cost will become clear.

Another difference is that servers, in organizations, are used in multiples. A large organization may have thousands. Research firm Gartner Group estimated that Google had about 2.5 million servers in 2016. Even an organization with hundreds can't just put them on desks. They must be arranged more efficiently, provided with power, and cooled. A room housing many servers is called a *data center* or, informally, a *server farm.*

Many server farms house servers in *racks.* A standard rack is 19″ (48.3 cm) wide. Rack-mounted components are multiples of 1.75″ (44.5 mm), a *rack unit* (RU), high. A standard rack is 73.5″ (1.87 m), 42RU, tall. A rack can hold at least 30 1RU servers plus power supplies and cables.

Blade servers increase the number of servers that can fit in a rack. A blade server is a modular server that can be inserted into a *blade enclosure.* The enclosure provides power, network, and storage connections to several blade servers at the same time. A rack can hold two to three times as many blade servers as it could hold individual 1RU-high servers. Figure 3.24 shows blade servers in a rack. Note the cooling vents in the floor.

Thousands of rack-mounted blade servers can fit in a large room. Power and cooling then become major concerns. Companies must consider climate and power availability in choosing the location of a large server farm. Google located a data center in The Dalles, Oregon, for its cool climate and nearby hydroelectric power. Europe's largest data center is in Kolos, Norway, for the same two reasons. Apple chose Maiden, N.C., because it had the space and climate for a 20-megawatt solar power installation.

The largest data center in the world is probably the Citadel by Switch. It is located in northern Nevada and is powered entirely by renewable energy: solar, wind, and geothermal. Its campus covers over 1,000 acres (400 ha). The first of six planned buildings, which opened in 2017, has an area of 1,200,000 square feet (110,000 square meters). The Citadel is a colocation data center: Switch clients put their computers in the Citadel rather than building their own facilities. Data centers operated by Facebook (Prineville, Ore.) and by the U.S. National Security Agency (Bluffdale, Utah) are not far behind it in size—and are not those organizations' only centers.

Some computer power use is avoidable. The *New York Times* found that as much as 90% of the power used by large data centers is wasted. One reason is that facilities are designed to handle peak loads, though peak loads exist only for short times. Since data centers worldwide use the power output of 30 nuclear power plants, this waste is significant on a global scale.

It may not take much effort to save power. Researchers from the University of Toronto found that running a data center only 1°C (1.8°F) above the usual 20–22°C (68–72°F) can reduce power use by 2–5%, by using less air conditioning, with no impact on equipment reliability or longevity. Worldwide, that is about the output of one nuclear power plant.

FIGURE 3.24 Blade server rack.

Where you fit in: Be sensitive to how much power your employer's computers use, including power to cool the rooms they are in. Saving energy is good for both business and the planet.

Mainframes are the largest commercial computers. (The name comes from the "main frames" that mid-20th-century telephone switching equipment was bolted to.) Early mainframe computers filled a room. Today, the difference between mainframes and servers is largely in their software. Mainframes run (mostly) applications that are descended from those old room-sized systems. They are used by companies that still have such older applications, called *legacy applications*. Servers run (mostly) software from the web era. Mainframe and server hardware look much the same today, though their processors may have different instruction sets.

The last step is to *supercomputers*. Supercomputers are the fastest computers, in the sense that race cars are the fastest cars. Race cars reach speeds of over 230 mph (370 km/h) but are not practical for normal use. Supercomputers use thousands of processors to carry out a computation. They are used in molecular modeling, developing pharmaceuticals, weather forecasting, oil and gas exploration, and analyzing data from sub-atomic particle experiments. They aren't equipped with communications and data transfer capabilities for airline reservations or running a network of ATMs. Figure 3.25 shows the Blue Gene/P supercomputer, made by IBM and installed at the Argonne National Laboratory west of Chicago, Ill. Note the floor vents for cooled air.

A new breed of supercomputer is being developed for data-intensive applications. The first of these to gain attention was IBM's Watson, which beat human champion Ken Jennings on the Jeopardy! quiz show in February 2011. A system such as Watson consists of specialized parallel hardware coupled with software to take advantage of what this hardware can do. IBM continued to enhance Watson after its Jeopardy! win for uses besides answering trivia questions. In 2020, an IBM division offers access to Watson via the web, as part of its IBM Cloud offering.

Supercomputers can't handle the types of data access that companies need in daily operations, but that is not their purpose. The ability of supercomputers to handle very large databases is growing. While you won't find a supercomputer in your office unless you go to work for a research lab, your company may use supercomputing capabilities from a company such as IBM.

FIGURE 3.25 Supercomputer (Source: Wikimedia Commons).

KEY POINT RECAP

Everything in a computer is based on switches that can be on or off.

The technology behind the switches varies from device to device and time to time, but the concept stays the same.

As a businessperson, your understanding of this concept will enable you to adapt to new technologies and to zero in on the basic issues in many technical decisions.

The potential capability of a computer system is determined by its hardware.

The limiting factors differ in different situations. The limiting factors of computers used in business are often not those we are familiar with from personal computing.

As a businessperson, you will be involved in decisions about business computers. Making those decisions on the basis of personal computing experience can lead to bad decisions.

Businesses use types of computers and subsystems that have no direct equivalents in the world of personal computing.

As a businessperson you will be asked to fund rack and blade servers, enterprise storage such as SAN and NAS, and more. You must understand these concepts to do this intelligently.

KEY TERMS

Aspect ratio: The ratio between the height and width of an object such as a display screen.

Bit (short for *binary digit*): The data represented by a single switch, often written as 0 or 1.

Blade enclosure: Electronic housing for blade servers with power supply, communications connections, and often shared secondary storage.

Blade server: Small server designed so that several can be housed in a single blade enclosure.

Burning: The process of recording data on an optical disk.

Byte: Storage unit consisting of 8 bits and allowing 256 combinations, often containing a single character.

Central Processing Unit (abbreviated CPU, also *central processor*): The part of a computer system that carries out computational processes.

Client: A computer that receives information or processing services from a server.

Clock speed (of a central processor): The rate at which it carries out internal computation steps.

Cloud bursting: Using a public cloud to handle overflow requirements from a private cloud.

Cloud storage: Accessing remote storage via the Internet.

Computer hardware: The physical components of a computer or information system.

Core: Processing element with all the functions of a central processing unit that shares a semiconductor chip with other processing elements.

Data center: Large room housing many servers with communication links, power, climate control, and other supporting services.

Direct access device: Secondary storage device on which two data items can be accessed without passing over all intervening data items between them. See also *sequential access device.*

Discrete graphics: Graphics processing chips that are separate from a computer's CPU chip. See also *integrated graphics.*

Diskette: Removable magnetic disk that uses a flexible recording surface.

Double-byte character: Character represented by two bytes, used for languages having more than a few dozen characters.

Ergonomic keyboard: Keyboard designed to minimize fatigue and repetitive strain injuries during long-term use.

Exabyte (abbreviated EB): Approximately 1 quintillion (1 billion billion) bytes.

Expansion card: Electronics component installed in a computer to increase its capabilities.

External drive: Secondary storage device housed outside the main electronics enclosure of a computer. See also *internal drive, portable drive.*

Gigabyte (abbreviated GB): Approximately 1 billion bytes.

Graphics processing unit (GPU): Processing unit whose instruction set is designed for the computations needed to display graphical elements on a screen.

Graphics tablet: See *pen tablet.*

Haptic output: Computer output that works with users' sense of touch.

Hertz: In a central processor, a unit of speed corresponding to one activity step per second. Often prefixed by *kilo*, *mega*, or *giga* (see Table 3.1).

Hybrid drive: Secondary storage system containing both magnetic and solid-state storage devices, with electronics to optimize the use of each element.

Input device: Any device, such as a keyboard, that sends data to a computer.

Instruction set: The way in which specific combinations of bits tell a CPU how to process data.

Integrated graphics: Graphics processing elements built into a CPU chip. See also *discrete graphics.*

Intel instruction set: The *instruction set* used by the CPUs of most personal computers, originally defined by Intel Corporation.

Internal drive: Secondary storage device housed within the main electronics enclosure of a computer. See also *external drive.*

Kilobyte (abbreviated KB): Approximately one thousand bytes.

Light-Emitting Diode (LED): A display technology using circuit components that emit light.

Liquid-Crystal Diode (LCD): A display technology using circuit components that pass, color, or obstruct light from another source.

Magnetic disk: Secondary storage device using a rotating magnetizable surface that passes under an electromagnet to detect or change the orientation of small areas.

Magnetic tape: Secondary storage device that records data on a long, flexible plastic strip coated with a magnetizable substance. Magnetic tape is a *sequential access device.*

Mainframe (or mainframe computer): A large, fast computer, often running older software.

Main memory: See *primary storage.*

Megabyte (abbreviated MB): Approximately 1 million bytes.

Moore's Law: The observation that the maximum number of transistors on a semiconductor chip doubles approximately every 18–24 months.

Multi-threading: The ability of a core, or of a single-core, processor to carry out instructions for more than one program in parallel.

Network-Attached Storage (NAS): Storage subsystem attached to a network, accessible by all the computers on that network.

Optical disk: Secondary storage device based on burning pits in a reflective surface to make it locally non-reflective.

Output device: Any device that receives data from a computer in order to present it to people or operate equipment.

Pen tablet: Input device that detects the position of a stylus resembling a pen.

Petabyte (abbreviated PB): Approximately 1 quadrillion (1 million billion) bytes.

Pixel: A dot on a display device. Short for *picture element.*

Port: Computer interface to which a device can be attached via a suitable cable.

Port expander: Device that plugs into one computer port, and provides several ports for devices.

Portable drive: *External drive* that is powered entirely by its connection to a computer.

Primary storage: Storage for instructions and data that a *central processing unit* is using at the moment. Data and instructions must be in primary storage to be used in a computation.

Rack: 19″ (48.26 cm) wide electronics enclosure with standardized mounting points.

Rack Unit (RU): Standardized height unit for rack-mountable components, equal to 1.75″ (44.45 mm).

Radio Frequency ID (RFID) tag: A small, inexpensive tag that can be attached to a product and can communicate information when in the proximity of a suitable reader.

Random-access storage (RAM): See *primary storage.*
Redundant Array of Independent Disks (RAID): Storage subsystem in which several disk drives behave as a single drive with improved capacity, reliability, and/or performance.
Removable storage device: *Secondary storage* device whose recording element can be removed from its housing and stored separately, to be loaded when needed.
Rewriteable optical disk: *Optical disk* on which data can be re-recorded several times.
Secondary storage: Long-term storage for instructions and data that a *central processing unit* does not need at the moment, but may need in the future.
Sequential access device: Secondary storage device on which two data items can only be accessed by passing over all intervening data items between them. See also *direct access device.*
Server: A computer that provides a service, such as web pages, database access, or the use of an application, to other computers.
Server farm: Informal term for a large *data center.*
Solid-state drive: *Secondary storage* device that uses semiconductor technology, having no moving parts.
Source data automation: Entering data into a computer without human data entry, such as reading a product bar code or the magnetic stripe on a credit card.
Storage-Area Network (SAN): Storage subsystem containing multiple storage devices and a high-speed internal network for routing data among them and the computers attached to the SAN.
Supercomputer: Very fast computer optimized for numerical calculations.
Switch: A physical element that can be in one of two states.
Terabyte (abbreviated TB): Approximately 1 trillion (1 million million) bytes.
Tactile output: See *haptic output.*
Thermal printer: A printer that creates images on special paper by heating its surface.
Thread: Program that is run in parallel with other programs in a single core or processor.
Three-dimensional printing: Creating physical objects as computer output.
Volatility: Characteristic of a storage device that its contents are lost when electrical power is removed.
Writeable optical disk: Optical disk on which data can be recorded once by a computer equipped with a suitable drive, without special production equipment.

REVIEW QUESTIONS

1. What are the main types of components in a computer system?
2. Define the term *bit* in terms of the states of a physical device.
3. Why are 8 bits usually used to represent a character in computers?
4. List the first five prefixes for multiples of bytes, starting with 1,000.
5. Define the term *instruction set.*
6. What is "Moore's Law?"
7. Why can adding RAM to a computer make it perform faster?
8. Describe how a magnetic disk works.
9. Other than magnetic disks, what are three secondary storage media?
10. Define RAID, NAS, and SAN.
11. Why would anyone store data "in the cloud," rather than on their own disk drives?
12. What is the *aspect ratio* of a computer display?
13. List several types of output devices other than displays and printers.
14. What is a *server*? Does a server have to be a special computer?
15. Give three ways in which the physical arrangement of servers in a room with hundreds of them differs from the way a few dozen computers are arranged in your lab.
16. What are supercomputers? Give one typical application for them.

DISCUSSION QUESTIONS

1. Sir Noten Downing is making a rare visit to London. Instead of being driven in his Rolls, he is with his granddaughter Kate. Like any young Londoner, she is thoroughly familiar with the Underground (U.S. subway) and often prefers it to the slow, congested streets. The last time Sir Noten used the "tube" was in the 1970s. Things have changed! First, Kate buys his ticket from a machine along the wall, using her debit card. He inserts it into a turnstile slot, after Kate shows him how to position its magnetic stripe. The ticket pops out to be inserted again when they leave, to confirm that he went no further than his fare allows. Instead of buying a ticket for herself, Kate taps a blue Oyster Card on a pad by the turnstile. It unlocks to let her in. She'll tap her card again as they leave. Her fare will be deducted from the card balance.
   a. Identify all input and output devices in this scenario.
   b. The Underground system has about 280 stations. An average station has ten gates, three ticket machines, and a booth with two attendants. (Quiet stations in the outskirts have fewer. Busy downtown stations have more.) How many people or devices might use its ticketing system at the same time? What sort of computers does this suggest it might use?
   c. A smartphone with short-range wireless can replace Oyster Cards. Discuss that change. Give at least two advantages and at least two disadvantages.
   d. Adding funds to an Oyster Card at a ticket vending machine adds to congestion and can delay other travelers. Discuss three ways to reduce or avoid this problem. Give at least one advantage and one disadvantage for each of your ideas.
2. Visit a local store that sells computers. Examine four desktop systems it has for sale, looking at high-end models if more than four are on display. Draw up a comparison table with the information on the shelf tag of Figure 3.6. Discuss what you found.
3. The electronic capabilities of a smartphone track those of laptop computers with a lag of about six to seven years. A 2020 smartphone has about the same processing power, RAM, and so on as a laptop of 2013–2014. Such a laptop is still a useful computing device and can, with enough RAM, run today's software. What keeps people from using phones to replace computers? What can be done about that? (You may want to do some research to answer the last part.)
4. Research the Dvorak keyboard layout. Its proponents claim that it permits faster typing with less fatigue and fewer errors than the standard layout. Are you tempted to learn to use this keyboard and to switch your phone and computer to its layout? Explain why or why not. (You won't need a new keyboard. Keys on an existing one can be reassigned electronically.)
5. As of early 2019, you could buy a BenQ GL2760H 27″ computer monitor for under $150. You can also pay over $1,500 for one like the Hewlett-Packard HP DreamColor Z27x G2. Why is there such a wide price difference between these devices? If you were a manager and one of your subordinates asked for a monitor such as HP's, how would you reply? (You may want to check their specifications before answering this question.)
6. Discuss possible business applications of headsets such as the Oculus Rift (Figure 3.20). A web search for videos about it might be a source of inspiration.
7. Find out how many disk drives your school has in its main data center and their total storage capacity. (You may want to ask your school's information systems department once on behalf of the entire class.) How much storage is this per student? Evaluate this amount in terms of information it has to store, keeping in mind that some stored information isn't about students.
8. The U.S. Library of Congress has about 20,000 disk drives. The mean time between failures (MTBF) of a typical disk drive has been estimated at about 50 years. Based on

those figures, it has about 400 drive failures every year, an average of over one a day. How should it plan to handle device failures when they occur? How is that different from the way a small business with five to ten drives should plan to handle device failures?

9. Suggest a type of computer to use for each of the following. Be as specific as you can with the limited information provided here. Explain your reasons.
   a. An accountant in an individual practice.
   b. A car dealership with about ten salespeople and about ten service bays.
   c. The membership service center of a professional organization with 100,000 members.
   d. A jet engine research department investigating how hot air flows through a turbine and where blades get hottest (and might become weak).
   e. A video sharing service such as YouTube with about 5 million members, each of whom watches an average of one 15-minute video per day, and a library of 2 million videos with about 100,000 new videos added per month.
   f. A professional baseball team managing and tracking ticket sales. The team plays 81–93 games per season in its home stadium, with attendance averaging 40,000 fans.
10. For each scenario in the previous question, suggest a storage device, devices, or systems that would make sense. Explain your reasons.

KHOURY CANDY DISTRIBUTORS HARDWARE

Some of the terms in this episode will be clarified in the next few chapters. They're in the index if you want to look them up before you reach them that way.

Chris Evans was waiting for Jake and Isabella the following week when they arrived at the KCD lobby. After the students received their visitor badges, saved from the previous week because they had told the receptionist they were likely to return, the three made their way to the conference room in the information systems section of the building. As in Jason Khoury's office, there was a large candy bowl on the table.

"I can see myself looking like a bowling ball with feet if I worked here," Jake laughed.

"After a while you don't even see them," Chris replied. "When my daughter worked at Bob's Ice Cream over on North Main during high school, they let everyone eat all they wanted and bring a pint home every day. For about two weeks it was a big deal. After that, not so much.

"Anyhow," he continued, "Jason and Sandra told me what you're here for, so we might as well get right into it. We have a couple of diagrams that we use to orient new IS hires into our overall architecture, one for our headquarters systems and one for the systems at our distribution centers. Distribution centers also include Springfield, so we have both of them here in this building. Here's the first one, for HQ."

With that, Chris pressed a key on a remote and Figure 3.26 appeared on a screen at the end of the room.

"As we see it, an architecture has three aspects: data, processes, and communications. Data's on the left. Our SAN holds our main ERP database, a data warehouse we're experimenting with, and a few assorted smaller files. Most of the time the SAN handles its own backup, but we send critical data to the cloud so a disaster that wipes out the SAN won't disable us. As Internet links get faster, we may go that way more, or keep backups at other distribution centers. We have to look at costs for those options and a couple of others. Speaking of backups, do you know what they say about two kinds of people?"

"No, what's that?" asked Isabella.

"There are people who have lost data because of a hardware failure, and people who will. If you're not in the first group, you're in the second. You do keep backups of important files, don't you?" he asked them.

Isabella replied promptly that she uses an external drive that she attaches to her laptop whenever she's in her dorm room.

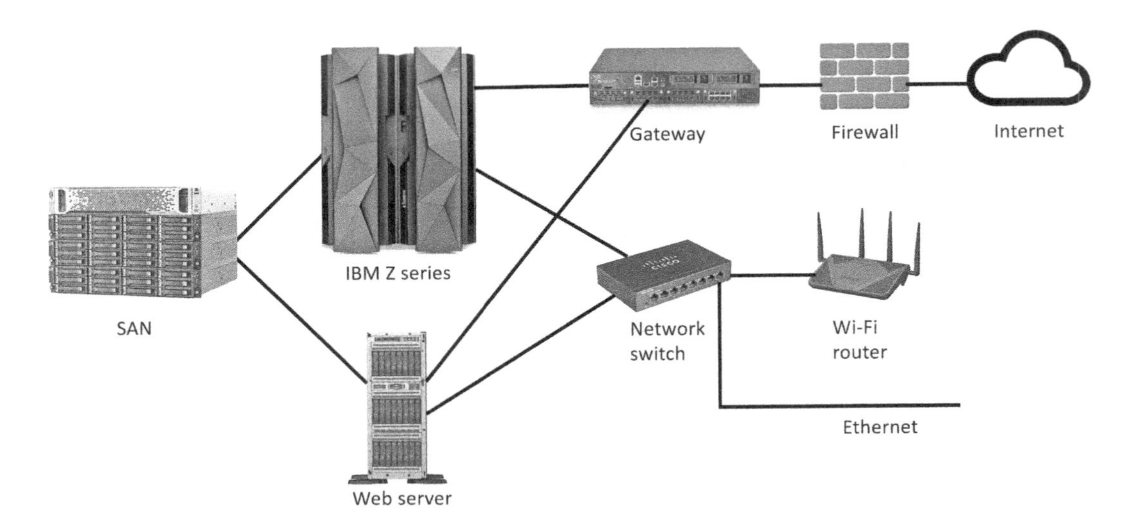

FIGURE 3.26 KCD headquarters system architecture.

Jake blushed and said "I will. Right away."

"This box in the middle is our IBM z13," Chris continued, indicating it with a laser pointer. "We got our first big computer in the late 1980s. It was an IBM 4300—Model 4381, I think, though I wasn't in IS at the time. Once we got into that family and its software, it made sense to stay with it. The Z Series is today's version. It's smaller than what we had back then, and gets way more done for less money, but you can trace a direct line from it back to the '80s and even earlier. Most companies stay with a family unless it stops meeting their needs. For example, I don't know if you two use Windows or Macintosh, but I'd bet a lot of money that whatever you have now is the same of those two as your last one."

"We won't take that bet, because we know you'd win!" Isabella said. "But, speaking of Windows and Mac, which do you use?"

"Mostly Windows," Chris replied. "Jason had MS-DOS on his PC/XT before Macs came out. Its software was way ahead of what Macs had for business when they were new, and it was a couple more years until Macs had hard disk drives. It wouldn't have been practical for us to use them until the late 1980s, and by then PCs were pretty well established here. That said, there isn't much you can't do on either today, so if someone prefers a Mac, they get one. So far that has created zero problems. Also, one of our distribution centers was an independent company that we bought. They were all Macintosh and still are. Their people help Mac users in the rest of the company. We even have a few folks who like to carry tablets around.

"Anyhow," Chris returned to the topic, "we don't use our z13 as a web server, though we could. We put our first site on a separate system and don't see any reason to move it. We run Apache web software under Linux, like most other firms. Our web server has a local disk for page templates. It fills them in with data, like inventory data, from the SAN.

"The communications side is on the right. Outside communication goes through this gateway. Web pages go out, data from distribution centers comes in, plus email, ordering, and the rest. The firewall watches over it to keep us out of trouble. Our mainframe can also be accessed online from where someone is: at a distribution center, on a business trip, or at home. You need a password for that, of course.

"Finally, at the bottom right are our local area networks inside HQ. People on the LAN can access an intranet. It acts like the web, but we don't share the information on its pages with the world. We can access the z13 through it, too.

"Which brings us," Chris continued, "to the distribution centers. This is the general idea," he said as he pressed a key on the remote. "The centers vary in details, but they all fit this framework. If we open a new distribution center, which we might if we expand further west, it will fit it too (Figure 3.27).

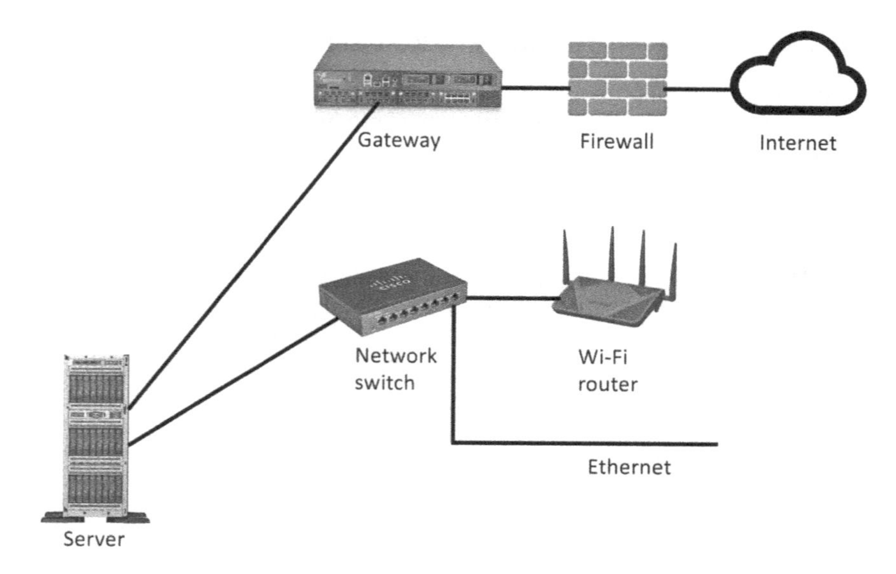

FIGURE 3.27 KCD distribution center system architecture.

"You can see that this is the headquarters diagram without the SAN and the mainframe. Also, the server has been relabeled because it's for data and applications, not web pages.

"Each of our distribution centers has a server with data for that center's inventory, employees, and so on. Those databases can act like one database, but local users just see their part and can access it quickly via their local area network. The gateway over here lets people in the center access the Internet, to connect to our other locations, to use the web, and to send emails. All these accesses go through this firewall"—he pointed to the upper right—"to keep bad stuff out of our computers, and keep people from updating their Facebook status or watching car crash videos at work. The gateway and firewall are the same in every center and at HQ. The servers are all from Hewlett-Packard, like the web server here at HQ, so we can buy in volume and have a single point of contact for support, but the details are different depending on what each center needs. The clients, as I said earlier, are just about anything.

"I suspect you're more interested in software than in boxes, though," said Chris as he finished describing the distribution centers. "I can set up a meeting at the same time next week for you on that. Make sense?"

The two students agreed enthusiastically and left, but only after taking a handful of candy from the bowl on the conference room table.

Questions

1. The headquarters architecture diagram (Figure 3.26) indicates an IBM Z server at the center, not the z13 model Chris said they have (though the picture happens to show a z13; it has to show something). Why is it not as specific as Chris was?
2. Both diagrams show a wired and a wireless local area network. (You'll read more about these in Chapter 6.) Both networks cover the office areas; the warehouse area in the distribution centers is covered only by the wireless network. The wired network is considerably faster than the wireless. Give two reasons why someone in an office might want to use the wireless network despite that.
3. Which of the two groups Chris mentioned midway through this episode are you in? What is your backup approach? Do you think it's sufficient? Do you plan to do anything about it? When? Why not sooner?

CASE 1: THE GREEN GRID MEASURES ENERGY EFFICIENCY

Information technology uses approximately 10% of the world's total electricity: roughly the same amount that lit the entire planet in 1985, the total power production of Germany and Japan, or 50% more power than aviation uses. Streaming a high-definition movie over a wireless network uses more energy than it would take to produce and ship a DVD of that same movie.

Given this level of consumption, many companies would like to improve the resource efficiency of their data centers. That calls for a way to measure how efficient those data centers are.

Enter TGG, The Green Grid: a global consortium dedicated to advancing data center energy efficiency. It hopes to influence data center design by developing and providing metrics, models, and educational resources that demonstrate best practices in sustainability.

TGG's measure of data center resource efficiency is Power Usage Effectiveness (PUE), expanded and enhanced in 2016. PUE is the ratio of total facility power to power used by computer equipment. The best possible PUE is 1 if all power goes to computers, not achievable in practice. An average PUE is about 2.5. (The U.S. EPA's Energy Star program reported 1.91 in a 2009 survey, whose participants were biased toward energy-conscious data centers.) The Uptime Institute says that most data centers can bring their PUE down to 1.6 without redesign by using efficient equipment and best practices. TGG suggests that companies should aim for 1.2 in new data center designs.

Power efficiency was front and center when Facebook designed its Singapore data center. This center, for which they broke ground in September 2018, is to be a leader in energy efficiency, with an expected annual PUE of 1.19: less than 19% of its energy input will be used for non-computing tasks. In comparison, Singapore has established 1.78 as a target average PUE. (PUEs are high in Singapore because of its climate.) Its roof will be covered by a solar energy array, further reducing the center's environmental impact.

Earlier, eBay used TGG measures and methods in its Project Mercury. This update to eBay's technology infrastructure consolidated 11 data centers into 3, deploying tens of thousands of servers in less than six months.

eBay saved energy by designing servers for different workloads. One was for compute-intensive work, a lot of computation on a small amount of data. The other was for analyzing "big data," with a higher ratio of data transfer to computation. eBay improved the efficiency of both over what they would have been with a single server design, yet by using only two designs still reaped most benefits of standardization.

Once eBay had finished, it measured PUEs averaging 1.35 across its site, with some data points under 1.1 in January when Arizona cooling requirements are at a minimum.

QUESTIONS

1. Aside from conserving our planet's resources and improving their public images, what specific business benefits might Facebook obtain from its new data center? Might eBay obtain from Project Mercury?
2. TGG's Data Center Maturity Model (DCMM) covers power, cooling, computing, storage, and networks. Each is at one of five effectiveness levels, providing an overall data center energy efficiency maturity level. As president of a company, write a memo to your CIO saying why you want your company to use the DCMM. (You can learn more about the DCMM at *www.thegreengrid.org/Global/Content/Tools/DataCenterMaturityModel*.)
3. Server power consumption at idle can affect vendor selection. Dean Nelson, VP of Global Foundation Services at eBay, says "if eBay had considered only the hardware procurement costs of the servers it used for Project Mercury, it would have chosen a different vendor." You must choose between two desktop computers. One uses 100 W when idle. The other, which costs $100 less, uses 150 W. They will be on 14 hours per day, in active use for four of those hours. They use the same amount of power in active use and are otherwise

comparable. At your electricity rate (use 20¢ per KWh if you don't know it) what is the cost difference for a year? If you plan to keep this computer for three years, which should you buy? Ignore the time value of money unless your instructor tells you to consider it.

CASE 2: MELTIN'POT DEPLOYS BLADE SERVERS

Meltin'Pot of Italy sells high-fashion denim clothes in Europe and Asia, primarily to consumers aged 18 to 25. It opened its first store in 2013, having previously sold only through independent stores, and has over 20 in Europe in 2019. To support stores and international expansion, the firm had to grow and modernize its technology infrastructure. It needed to track customer demand more closely in the fast-moving high-fashion market. It also needed more effective and cost-effective management of sales, orders and inventory, and improved customer service. This meant an IT infrastructure able to support a greater workload and more users than its previous one.

Meltin'Pot chose IBM BladeCenter HS22V blade servers after evaluating offerings from several vendors. Each HS22V can have two processors with six cores each, and up to 288GB of RAM, in a 1RU-high chassis. (Today's personal computers typically max out at 16GB.)

Storage on each server was two 1.8″ disk drives: smaller than most laptops, but the servers are meant to be used with shared storage devices. Local disks are needed only for startup, shutdown, and storing small amounts of temporary data. Here, the servers share an IBM disk array. This array, also housed in a standard rack, had a maximum capacity of over 0.5 PB per array. Some of its drives can be solid-state drives for improved performance, without a significant reduction in the maximum capacity of the array.

After selecting the hardware, Meltin'Pot decided to work with a local system development firm, H.S. Systems, to put together their system. H.S. Systems works with several hardware vendors, combining their products with custom software to develop a total system for each of its clients. On the hardware side, they generally use computers from Dell or IBM. They are an authorized business partner of both, with experience in projects such as Meltin'Pot's.

Reliability is important in systems that businesses depend on for daily operations. According to Roberto Felline, manager of Meltin'Pot's IT department, "The IBM infrastructure is extremely available and resilient, and we have been able to prove this through several quality tests. On one occasion, for instance, we tried to take one blade out of the BladeCenter server while the machine was switched on: we were absolutely amazed by the fact that operations continued smoothly. … If one blade fails, the workload is automatically shifted to the remaining blades, allowing the processes—and, therefore, our business and services—to continue working."

Compared to Meltin'Pot's previous system, its new servers provided a 300% increase in processing power while occupying 50% less floor space. This enabled Meltin'Pot to achieve the business objectives that it set out. It positioned them, as regards IT infrastructure, to continue their growth for several more years.

Questions

1. IBM is not the only supplier of blade servers. The comparison with Meltin'Pot's earlier system, four times the processing power in half the floor space, reflects primarily improved technology since that earlier system was designed. Most competitive servers would also have provided several times the processing power in a fraction of the floor space. That being the case, is it fair for IBM's case study to claim this improvement as a benefit of the new IBM system? Justify your answer.
2. Meltin'Pot first chose IBM blade servers. With IBM's assistance, it then chose H.S. Systems. Alternatively, it could have chosen a development partner on its own first, and then worked with that firm to select the hardware. Give two advantages of doing things in each order.

3. There are two aspects to selecting a system such as Meltin'Pot's: the specific items that are purchased, and a product family that the customer expects to stay with when that equipment is replaced. Growth potential is therefore important in this choice. Discuss two areas where a firm might pay more initially, to get a capability that it does not yet need, for this reason.

BIBLIOGRAPHY

Anthony, S., "IBM and Sony cram up to 330 terabytes into tiny tape cartridge," *Ars Technica*, August 2, 2017, arstechnica.com/information-technology/2017/08/ibm-and-sony-cram-up-to-330tb-into-tiny-tape-cartridge, accessed December 6, 2019.

Arch, "What's the largest hard drive you can buy? [July 2019]," *Tech Junkie*, July 19, 2019, www.techjunkie.com/largest-hard-drive-you-can-buy, accessed September 24, 2019.

Ayyar, K., "The world's largest biometric identification system survived a Supreme Court challenge in India," *Time*, September 26, 2018, time.com/5388257/india-aadhaar-biometric-identification, accessed March 1, 2019.

Bracken, C., "Capital One's AWS server data breach trips alarms over cloud security," *IT Toolbox*, August 7, 2019, it.toolbox.com/article/capital-ones-aws-server-data-breach-trips-alarms-over-cloud-security, accessed September 25, 2019.

Chugh, M., "What is the total capacity of YouTube storage?" *Quora*, March 24, 2017, www.quora.com/What-is-the-total-capacity-of-YouTube-storage, accessed December 23, 2018.

Clancy, H., "5 green data center tips from eBay's Project Mercury," *ZDnet*, May 11, 2012, www.zdnet.com/blog/green/5-green-data-center-tips-from-ebays-project-mercury/21188, accessed September 17, 2019.

Clay, L.T., "Rethinking data center design for Singapore," *Facebook Engineering*, January 14, 2018, engineering.fb.com/data-center-engineering/singapore-data-center, accessed September 17, 2019.

Collins, S., "The public eye?" University of Cambridge, February 6, 2014, www.cam.ac.uk/research/features/the-public-eye, accessed September 17, 2019.

Cusumano, M., "The business of quantum computing," *Communications of the ACM*, vol. 61, no. 10 (October 2019), p. 20.

Data Center Knowledge, "Google data center FAQ," March 16, 2017, www.datacenterknowledge.com/archives/2017/03/16/google-data-center-faq, accessed December 23, 2018.

El-Sayed, N., I. Stefanovici, G. Amvrosiadis, A. Hwang and B. Schroeder, "Temperature management in data centers: Why some (might) like it hot," *Proceedings of the 12th ACM SIGMETRICS/PERFORMANCE Joint International Conference on Measurement and Modeling of Computer Systems* (SIGMETRICS '12), London, U.K., June 2012.

Finley, K., "IBM bets $1B that its *Jeopardy* machine can rule the business world," *Wired*, January 9, 2014, www.wired.com/wiredenterprise/2014/01/watson-cloud, accessed September 17, 2019.

Glanz, J., "Google details, and defends, its use of electricity," *New York Times*, September 29, 2011, www.nytimes.com/2011/09/09/technology/google-details-and-defends-its-use-of-electricity.htm.

Glanz, J., "Power, pollution and the Internet," *New York Times*, September 22, 2012, www.nytimes.com/2012/09/23/technology/data-centers-waste-vast-amounts-of-energy-belying-industry-image.html, accessed September 17, 2019.

Greengard, S., "The future of data storage," *Communications of the ACM*, vol. 62, no. 4 (April 2019), p. 12.

Heath, N. and B. Vigliarolo, "Photos: The world's 25 fastest supercomputers," *TechRepublic*, November 12, 2018, www.techrepublic.com/pictures/top-10-supercomputers, accessed December 23, 2018.

Hruska, J., "Is Moore's law alive, dead, or pining for the fjords? Even experts disagree," *ExtremeTech*, July 11, 2019, www.extremetech.com/computing/294805-is-moores-law-alive-dead-or-pining-for-the-fjords-even-experts-disagree, accessed August 16, 2019.

H.S. Systems web site, www.hssystems.it (in Italian), accessed September 17, 2019.

IBM (on H.S. Systems site), "Meltin'Pot powers its business-growth strategy with IBM," December 14, 2012, https://www.hssystems.it/News/Articoli/tabid/346/page/15/Default.aspx, accessed September 17, 2019.

Krishna, S., "What is the total size (storage capacity) of YouTube, and at what rate is it increasing?" *Quora*, December 4, 2017, www.quora.com/What-is-the-total-size-storage-capacity-of-YouTube-and-at-what-rate-is-it-increasing-How-is-Google-keeping-up-with-the-increasing-demands-of-Youtube%E2%80%99s-capacity-given-that-thousands-of-videos-are-uploaded-every-day, accessed December 23, 2018.

Lawson, S., "Moore's law isn't making chips cheaper anymore," *Computerworld*, December 5, 2013, www.computerworld.com/article/2486035/moore-s-law-isn-t-making-chips-cheaper-anymore.html, accessed September 17, 2019.

Mellor, C., "MAMR Mia! Western digital's 18TB and 20TB microwave-energy hard drives out soon," *The Register*, September 4, 2019, www.theregister.co.uk/2019/09/04/wd_mamr_drives, accessed September 24, 2019.

Meltin'Pot web site, meltinpot.com, accessed September 17, 2019.

Menear, H., "Top 10 biggest data centres in the world," *Gigabit*, August 13, 2018, www.gigabitmagazine.com/top10/top-10-biggest-data-centres-world, accessed September 17, 2019.

Merritt, R., "Moore's law dead by 2022, expert says," *EE Times*, August 27, 2013, www.eetimes.com/document.asp?doc_id=1319330.

Monroe, D., "Electronics are leaving the plane," *Communications of the ACM*, vol. 61, no. 8 (August 2018), p. 17.

Myatt, B., "The Green Grid expands the 'PUE' metric," *Mission Critical*, August 9, 2016, www.missioncriticalmagazine.com/articles/88565-the-green-grid-expands-the-pue-metric, accessed September 17, 2019.

Nimbus Data web site, nimbusdata.com, accessed September 24, 2019.

Pratt, M., "Computerworld Honors 2013: Measuring the sustainability of data centers worldwide," *Computerworld*, June 3, 2013, www.computerworld.com/s/article/9239143/Computerworld_Honors_2013_Measuring_the_sustainability_of_data_centers_worldwide, accessed September 17, 2019.

Team Nuggets, "The 6 largest data centers in the world," June 22, 2017, www.cbtnuggets.com/blog/technology/data/the-6-largest-data-centers-in-the-world, accessed September 15, 2019.

The Green Grid web site, thegreengrid.org, accessed September 17, 2019.

Ultrium LTO web site, www.lto.org, accessed September 7, 2019.

Uptime Institute web site, uptimeinstitute.com, accessed September 17, 2019.

Walsh, B., "The surprisingly large energy footprint of the digital economy," *Time*, August 14, 2013, science.time.com/2013/08/14/power-drain-the-digital-cloud-is-using-more-energy-than-you-think, accessed September 17, 2019.

4 Information Systems Software

CHAPTER OUTLINE

WHY THIS CHAPTER MATTERS

Without software—instructions that tell hardware what it should do—a computer is an expensive paperweight.

Most information systems decisions you'll face as a manager will involve software. Some will be entirely about it: Should we get package A or package B? Should we hire programmers to write a program that will be just what we need when they finish next year, or compromise on a package that will cost less and is available now? Should we upgrade to the new version of a familiar app, or look for a better one? Such decisions require you to know more about software than you know from using it.

Other decisions may turn out to involve software. You may be considering computer A versus computer B, but the key issue may be what software each runs or how well it can run it. You may be concerned with data security, but protecting it is (in part) the job of software. Deciding where to host your web site may hinge on the software each hosting service uses.

This chapter will give you the background to deal with such issues. You'll also use this material in later chapters, when we discuss applications you'll see on the job and where they come from.

CHAPTER TAKE-AWAYS

As you read this chapter, focus on these key concepts to use on the job:

1. Hardware is useless without software, since software tells hardware what to do.
2. Software can be divided into categories: system/application, proprietary/open-source, and (for applications) horizontal/vertical. Understanding these categories gives you a head start on understanding software you'll encounter.
3. The most important piece of system software is the operating system. Its three functions are managing shared resources, defining the user interface, and providing shared services.
4. Software is complex, detail-ridden, and error-prone.

4.1 SOFTWARE CONCEPTS

Imagine your first day on the job after graduation. You're given a desk, a computer, paper, and pencils. That's it. Nobody tells you what your job is or what to do.

You'd probably take initiative, perhaps asking a manager what you should do. Maybe you'd hear "The orientation program starts after lunch. Until then, read up on the company on our web site."

Or, perhaps "Oh, I didn't know you were here yet. Analyze these sales figures to see if there's a problem in the Central region."

A computer is like a new graduate with no initiative. It can carry out instructions that can be combined in many ways to carry out useful processes. It has no idea which instructions to carry out.

The computer that hosts a shopping web site was not built knowing that it should use two letters in a box labeled STATE, in a U.S. shipping address, to access a table named TAX_RATES for the state sales tax rate. It has no idea that, if the state is MA, there's no tax on the first $175 of an item of clothing—or how to tell if item N-376A is clothing at all. Something has to tell that computer what to do, down to the tiniest detail. That's software.

Software is a collective term for the instructions that tell a computer what to do.

A NOTE ON LANGUAGE

"Software" is an *uncountable* noun. It has no plural. You can't have two softwares. You can have two programs, two applications, two software packages, or two CDs with software on them—but not two softwares. If you have one game and buy another, you have two games. You have more software, but you still don't have two softwares!

Where you fit in: As in other areas of technology, people will form opinions of you in part on how well you use terms. If you say *softwares*, they'll think you don't know much—and may try to "pull the wool over your eyes." If you use terms correctly, they won't draw that conclusion.

Writing Software

Sets of instructions that tell a computer what to do are called *programs* (*computer programs*, if context does not make that clear). A program is a series of instructions that, when executed by a computer, performs a task. Programs are written by people called *programmers*. Writing programs is *programming*.

In 2019 computers can't be programmed in English or any other human language. They can respond to questions posed in natural languages within limits, but they must have previously been given instructions about how to respond to those questions. Those instructions are a program, too.

A computer sees a program as 1s and 0s, because computers see everything as 1s and 0s. It is hard for people to write long sequences of 1s and 0s, so they developed *programming languages* with which to describe a computation to a computer. While not as easy as telling a computer what to do in English (or French, or Hindi), they're a big improvement over 1s and 0s. Programming languages combine words and symbols according to precise rules to specify a computation. Figure 4.1 shows a short program in one such language.

This program counts a data item called *i*, which it is told can only take integer values, from 1 through 2, 3, … up to 10. Each time *i* changes, the program displays the word "Line" followed by its then current value. It displays "Line 1," "Line 2," through "Line 10." This form of a program is called a *source program*.

A source program such as Figure 4.1 must be converted into 1s and 0s before a computer can run it. When you buy software, you get this *object program*. You usually don't get the source program, but few people care. You can't use it, and only a programmer can understand it.

Unfortunately, most programs aren't as simple as the one in Figure 4.1. Business programs can consist of several million lines of computer instructions. Instead of one data item called *i* they have

```
program LineNumbers(output);
begin
  for i := 1 to 10 do
    WriteLn ('Line ',i)
end.
```

FIGURE 4.1 Example Pascal source program fragment.

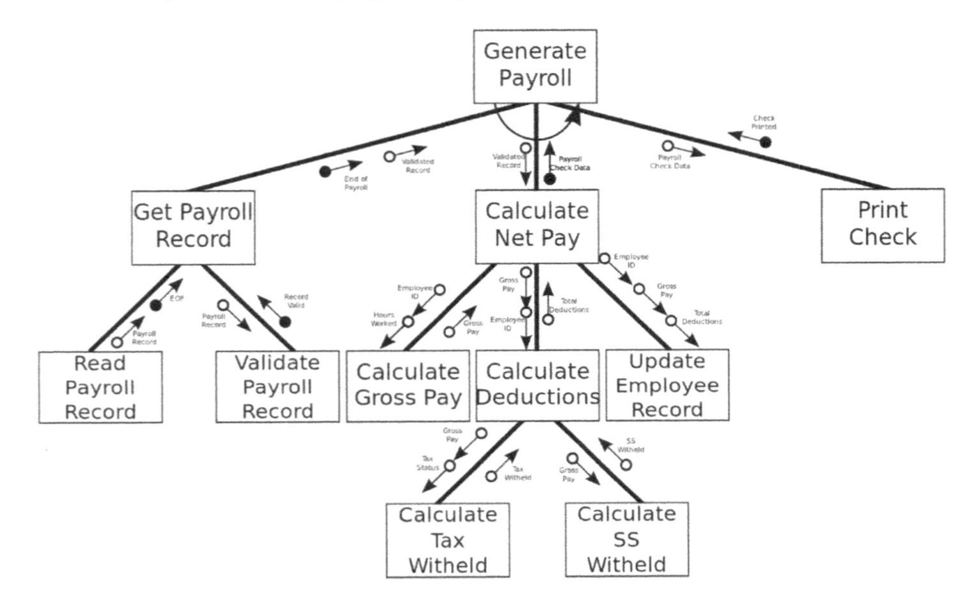

FIGURE 4.2 Structure of a payroll application (Source: Wikipedia Commons).

thousands, with arcane names that are gibberish to non-programmers, and which interact in unexpected ways. Their commands, rather than being ordinary words used in a not-so-ordinary context, are incomprehensible without a good deal of study.

These millions of lines of computer code aren't written as a unit. Just as a book is divided into chapters, paragraphs, and sentences, programs are divided into modules. Each module carries out a small, ideally self-contained, part of the overall task. Dividing applications into modules that can be written separately is part of planning its development, as you'll read in Chapter 11. Figure 4.2 shows a payroll application divided into modules, showing how data flow from one module to another. Modules are typically a few hundred lines long. If a module can't be written in that many lines, it is usually subdivided further.

A computer's ability to carry out millions of operations in a second is offset by needing millions of steps to do things we find simple. Consider word wrap. We all know that when a word doesn't fit on a line, our word processor starts a new line. We don't think about it. It just works.

To a programmer who writes this part of a word processor, though, word wrap is not simple. It involves these steps (and this description is, believe it or not, drastically oversimplified):

- Record where the next character on the line will start. (Reset this to the beginning of the line when a new line is started.) This requires choosing a unit of measure.
- Calculate, from page size, orientation, margins, and paragraph indents, how long the line is.
- When a character is entered, calculate its width from its font, style (bold letters are wider than regular), size, and letterspacing (regular, expanded, condensed, etc.).

- Add this width to the starting position to find where the letter will end. If that's before the end of the line, update the starting position for the next character, and wait for it.
- If the character ends past the end of the line, work back to find a place where the line can break: a space, a hyphen, or one of a few other characters.
- Collect the characters from that point on and move them to a new line.
- If that new line is past the bottom of the page (based on the height of its tallest character, line spacing, page size, and top/bottom margins), start a new page.
- If the current paragraph has "keep lines together" checked, move it all onto the new page.
- If the current paragraph has "widow/orphan control" selected, so one line of the paragraph should never be on a page by itself, move two lines to the new page.
- If that leaves just one line on the previous page, move that line too.
- If the whole paragraph is now on the new page and the previous paragraph has "keep with next" selected, move the last two lines of that paragraph onto the new page. Work back until you reach a paragraph that does not have "keep with next" checked or the page is full.

There's more, but you get the idea. Programming is detail-ridden. It is easy to make mistakes, and can be hard to find them. It's no wonder that programs have errors (*bugs*). It may be more of a wonder that they don't have more.

LITTLE ERRORS HAVE HUGE CONSEQUENCES

In February 2014 Apple discovered a flaw in programs that confirm the identity of a trusted site. It released a patch for iOS and Mac OS soon after. The error was in this duplicate line:

```
...
if ((err = SSLHashSHA1.update(&hashCtx, &serverRandom)) != 0)
     goto fail;
if ((err = SSLHashSHA1.update(&hashCtx, &signedParams)) != 0)
     goto fail;
     goto fail;
if ((err = SSLHashSHA1.final(&hashCtx, &hashOut)) != 0)
     goto fail;
...
```

It's easy to spot the error in seven lines when someone circles it. It's not nearly as easy in a 2,000-line program, itself a tiny part of a much larger system, with no reason to suspect this section as the problem.

To make things worse, *software is not constrained by the physical world.* Architects and aircraft designers are constrained by what can be built. They can push limits, but physical constraints are always in their minds. Software has no physical constraints, so its plans may be more ambitious than anyone can build with the available time and resources. "One more little feature" turns out to affect many things, and the list of "more little features" never stops at one!

Where you fit in: You need to know what programmers do in order to make informed decisions about what they should do. Chapter 11 describes the process of writing software and puts it into its business perspective, focusing on how it will involve you as a businessperson.

Open-Source Software

Much of the software you use is written by companies whose business is selling software. For example, Microsoft makes money by selling Windows, Office, and other programs.

Other software is written by organizations for their own use. Amazon has a program that suggests purchases by looking at a customer's earlier purchases and recommending items that customers with similar tastes bought. Amazon won't sell this program because it gives them an advantage over other sellers. They make money by using it themselves.

Those are *proprietary software*: software that belongs to whoever paid to have it written. This organization or person expects to benefit by selling it or using it.

Alternatively, people can band together to produce software for anyone to use. If you need office software but don't want to buy Microsoft Office, go to *www.openoffice.org/download/index.html* to get OpenOffice. It's not a copy of Office, as copying Office would infringe on Microsoft's intellectual property rights, but it includes programs with comparable features (Figure 4.3).

OpenOffice is *open-source software*, developed by volunteers: once hobbyists, today mostly professional programmers. It is free to anyone. You can download it and use it. You can read its source programs and modify them as you wish. You can examine its programming methods and use them in a program of your own. You just can't pass it off as your own work or charge for it.

Technically, the term *open source* means that anyone can see the software's source code. It need not mean that the software is free. Companies can make source code public while retaining the right to charge people to use it. This has been done, but not often. As a practical matter, open-source software is almost always free software. The reverse is not true, though: free software is often not open source. Microsoft's Edge web browser is free, but it is not open source. Microsoft won't show you the Edge source code. It's their intellectual property.

Why do people spend time and effort producing software that won't earn them money? For the technical challenge, the sense of contributing, the feeling of participating in a community, to get software for their own use, and more. Open-source projects involve thousands of people from around the world. Major companies use open-source software in their business. You may have heard of Linux as a free alternative to Windows or Mac OS. The open-source Apache program is the most popular package for sending web pages over the Internet. It is used by large businesses that can afford to buy any web server software they want.

Using open-source software poses risks. Since (usually) nobody benefits economically when you use it, nobody feels compelled to deal with your problems unless they have the same ones. That doesn't mean you won't get help, but it means the people who help you won't have helping you in their job descriptions. Open-source software users confirm that things usually work out well, and software company bankruptcies prove that buying proprietary software doesn't ensure support either, but this is still a concern. Conversely, savings in license fees can cover a good deal of paid support.

Apache OpenOffice Product Description
Compatible with other major office suites, Apache OpenOffice is free to download, use, and distribute. Download it now, and get:
Writer a word processor you can use for anything from writing a quick letter to producing an entire book.
Calc a powerful spreadsheet with all the tools you need to calculate, analyze, and present your data in numerical reports or sizzling graphics.
Impress the fastest, most powerful way to create effective multimedia presentations.
Draw lets you produce everything from simple diagrams to dynamic 3D illustrations.
Base lets you manipulate databases seamlessly. Create and modify tables, forms, queries, and reports, all from within Apache OpenOffice.
Math lets you create mathematical equations with a graphic user interface or by directly typing your formulas into the equation editor.

FIGURE 4.3 OpenOffice description page.

Some companies have made businesses out of open-source software, such as Red Hat with Linux. They can't charge for the software—but they can charge for collecting all the bits an organization needs, packaging them together, configuring the package for easy installation, making it available as a CD-ROM or as a download, and providing documentation and support. People who want to use open-source packages may be willing to pay for these. This willingness gave Red Hat revenue of $2.9 billion in its 2018 fiscal year, up 21% over fiscal 2017, and keeps 12,000 employees busy. In January 2019, Red Hat shareholders approved a plan by which the firm will merge with IBM.

4.2 SYSTEM SOFTWARE

It has been said that "system software is software that doesn't do anything useful." That's not literally true. It would, though, be correct to say that "system software doesn't produce a directly useful end result." Not as good a sound bite, but closer to the truth.

System software has *indirect* value: it makes it easier to create and use software that provides business benefits. It supports tasks that application software carries out, or supports people who use it. You can visualize the place of system software, and of other types of software we'll discuss soon, as shown in Figure 4.4.

The most important type of system software is the *operating system*. It sits just above the hardware in Figure 4.4, below the rest of the software and the user.

Operating System

You've used at least one operating system (OS): Windows, Mac OS, or Linux. Your phone has Android or iOS. You use this OS to start programs, delete files, and set your screen background image. You know what an OS does for a user. Here, we'll look at what an OS does inside, and at differences between OSs you're used to and OSs for business.

An *operating system* is a control program that manages what other programs in the system do. It accepts requests from users to perform tasks such as starting an application or copying a file. It also accepts requests from programs to perform tasks on their behalf, such as saving data on a disk drive. Its work falls into three main areas:

Defining the User Interface

In Figure 4.4, layers of software separate the user from the hardware. The top line in that figure is the *user interface*: how we, a computer's users, see software and data. We use the user interface to start programs, open files, organize them in folders, minimize windows, and much more.

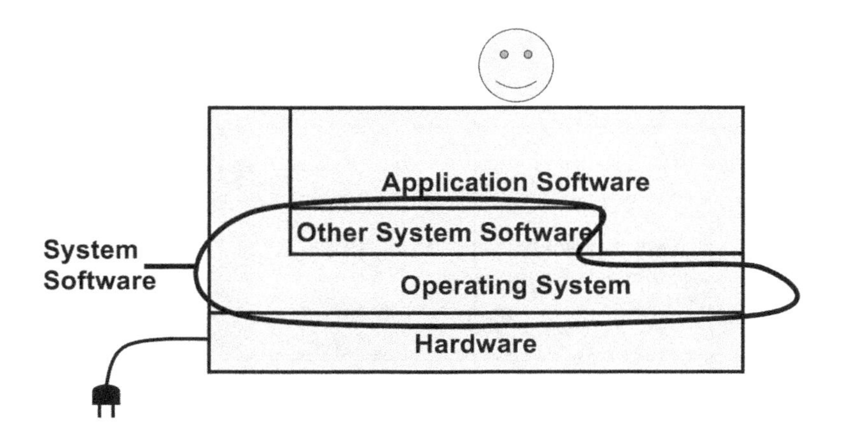

FIGURE 4.4 Layered diagram of computer system with software.

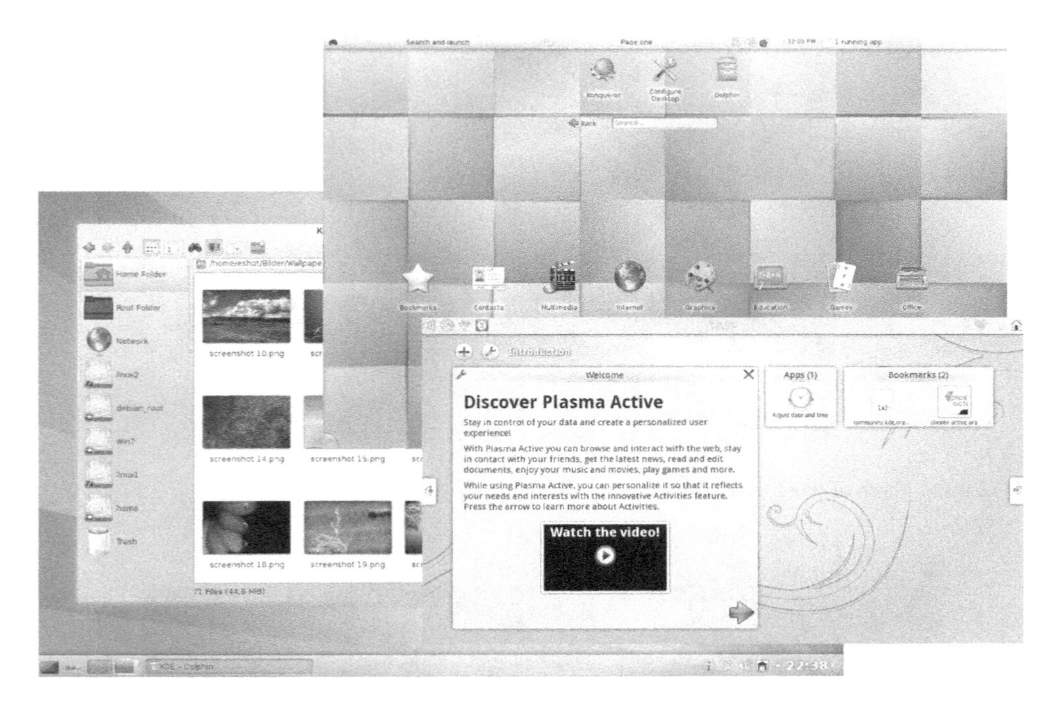

FIGURE 4.5 Screen shot of KDE graphical user interface (Source: Wikimedia Commons).

Most computers today use a *graphical user interface* (GUI). Programs run in windows, filling the screen or part of it. Small images called *icons* identify programs and files. Menus along the top of screens or windows, or that drop from a screen object, let users control what a program does.

While GUIs differ on the surface—to close a window, do we click its top right or left corner?—their underlying concepts are similar. They follow work by Doug Engelbart at Stanford Research Institute in the 1960s, refined at Xerox's Palo Alto Research Center, commercialized in Apple's Lisa computer in 1983. Their concepts haven't changed in decades. Figure 4.5 shows the KDE Plasma GUI, available for many computers.

In the 21st century *touch-based interfaces* emerged: on phones, then on tablets, and later, because people know and like them, on computers. Touch-based interfaces use gestures to control a computer. Today they don't offer all the control of GUIs, but early GUIs were limited too. Touch interfaces will undoubtedly become more capable with time (Figure 4.6).

A third type of interface is a *command-line interface,* where users enter commands via keyboards. Consider how we delete a file in a GUI. First, we find the file's icon or its name in a list. We select it by clicking on it or navigating to it with arrow keys. Then we drag its icon to a trash can or recycle bin, select "delete" from a drop-down menu, or press the Delete key while holding another key down. In a command-line interface, there is no need to find the file. We need only know its name and where it is. If you're working in a folder named *photos* and want to delete the photo *surfer.jpg*, type *rm surfer.jpg* and it's gone. (*rm* is short for *remove.*)

Figure 4.7 shows a command-line interface. The OS waits for the user to enter a command after its "C:\>" prompt, which also reminds the user what drive or folder it is using. The user command "dir" asks the system to list (display DIRectory of) whatever is in that drive or folder. The "w*" after the command means that the user wants to see only files and folders whose names begin with *w.* Those folders, WIN32APP, WINNT, and WUTemp, are listed below the summary information. At the next prompt, the user types "cd" for Change Directory, followed by the name of a drive or folder to switch to: the first one whose name starts with WIN3, WIN32APP. The last line shows that the OS is now working in that directory.

FIGURE 4.6 Photo of touch-based interface.

```
CMD.EXE
Microsoft Windows 2000 [Version 5.00.2195]
(C) Copyright 1985-2000 Microsoft Corp.

C:\>dir w*
 Volume in drive C has no label.
 Volume Serial Number is 8CB9-A0E0

 Directory of C:\

06/10/2003  10:28a      <DIR>          WIN32APP
10/04/2005  02:40p      <DIR>          WINNT
05/24/2005  12:09p      <DIR>          WUTemp
               0 File(s)              0 bytes
               3 Dir(s)  45,221,155,840 bytes free

C:\>cd win3*

C:\WIN32APP>
```

FIGURE 4.7 Screen shot of command-line interface.

Command-line interfaces provide the most control over a computer, because keyboards can provide more information than clicks, taps, or menu choices. They are faster for those who know them because typing is quicker than pointing. However, they are less intuitive: it is harder to remember something than to recognize it. Computing professionals use them. Others seldom do.

Where you fit in: As an informed information systems user, you should be aware of command-line interfaces. Personal computers have them as an option for experts. And should you ever see a computing professional control a server, typing a lot without touching a mouse, you'll know why!

Managing Shared Resources

The operating system's second job is coordinating the use of a computer's resources: RAM, disk space, screen space, communication lines, devices such as printers, and more.

Consider disk space. When you tell a word processor to save a term paper, it saves it in secondary storage. The term paper must go into space that isn't already in use. Identifying vacant space requires a central repository of storage use information. Without one, your term paper might overwrite your music, your photos, or the spreadsheet with your Finance homework.

This concept applies to all shared resources. Data from a communications link must be routed to an email program, browser, chat program, music download, or Skype call. If you start printing one document before another finishes, something must hold the second file until the first is done. If you have several windows open on your screen, something must know which is in front so as to know what to do when you click.

RAM is an important shared resource. Few computers have enough RAM for all the programs you want to run at the same time. Operating systems move less-recently-used data from RAM to disk to make room for programs or data that are needed right away. This creates the illusion that the computer has more RAM than it does. You read about this *virtual memory* in Chapter 3.

Coordinating the use of hardware resources is difficult enough in a personal computer, where you control the computer directly and tell it what to do. An enterprise computer has many users, they can be anywhere, and they control the computer indirectly. That's one reason enterprise operating systems are more complex, and more difficult to use, than those of personal computers.

> ***Where you fit in:*** Being aware of the types of resources that a computer system consists of, and the conceptual complexities of managing them for multiple users, will help you understand the issues your company faces with its large computer systems.

Providing Shared Services

Application programs perform some tasks in similar ways. Consider creating a screen window or responding to a mouse click. What the window displays or what the program does after that click will differ, but the underlying activities are always similar.

Operating systems provide standardized shared services for common tasks, with these benefits:

- Applications do them the same way. If we can print in one application, we can print in any.
- Applications are easier to create. Instead of drawing a window on the screen, minimizing it, resizing it, and more, a program just tells the OS "I need a window this big, with scroll bars. ..." The OS does the rest. Since the OS is aware of that window, it can manage screen space.
- Tasks are done better. People who program a shared service know that many applications will use their work. They do as complete a job as they can, and test their programs thoroughly.

Because every operating system's shared services are accessed differently, a program written for one OS can't run under another. This parallels how a program written for one type of processor can't be run on another. Both the processor and the OS must suit the program you want to run. This combination of processor and operating system is called a *platform*. The platform may be as broad as Microsoft Windows, or it may be just Windows 10.

Multi-user operating systems are more complex than single-user operating systems. The OS of your personal computer, tablet, or phone is designed for use by one person at a time. It can't handle multiple users, but businesses depend on multi-user systems. The most popular multi-user software is Unix, or the closely related Linux. Unix or Linux are used by about two-thirds of all web sites, with most of the rest using Windows. Linux also runs all of the fastest 500 supercomputers in the world, including IBM's Watson system that you read about at the end of the last chapter. The last non-Linux systems fell off the Top 500 list back in 2017.

Where you fit in: You don't have to be an expert on multiple-user operating systems or be able to use them, but you must know they are vital to the information systems that organizations rely on.

TRIVIA

The name *Linux* combines Linus, the first name of its originator Linus Torvalds, with Unix, a multiple-user OS on which its concepts are based. Torvalds felt that the best way to learn about operating systems was to write one. That formed the core of his 1997 computer science master's thesis. Torvalds offered Linux software freely to anyone. People used it, liked it, and contributed enhancements. He continues to coordinate the Linux development community.

Torvalds's given name is pronounced LYE-nus, with a long *i*. The OS is LINN-ux, with a short *i*.

Other System Software

No operating system meets the needs of all information system users all the time. As people see what each is missing, software firms develop packages to fill its gaps. When OS suppliers see customers use outside packages for a missing function, they build it into the next version of their OS. This reduces the market for separate packages. This impacts suppliers of such packages, but is generally considered to be part of software evolution.*

The most common type of system software that supplements an operating system is *database management*. Chapter 5 is about this.

Data deduplication software is common in large organizations, though you don't need it at the personal level. When an organization has thousands of computers, chances are many of them store copies of the same files. Suppose 50 managers download and save the CEO's 8MB strategic plan presentation. The nightly backup run saves 50 copies of *plan.pptx*, 400MB all told. If this company keeps a week of daily backups, a month of weekly backups, and a year of monthly backups, that's 9.2 GB for 1,150 copies of one file. Nine GB isn't a lot of storage for a large firm, but that's one file of thousands. The total cost of storing duplicate data can be staggering.

Data deduplication software replaces duplicate files with pointers to a master copy. It can pay for itself many times over in storage cost savings. Data deduplication can also be used in an email server to save one copy of a file that was attached to different messages to different people, in repositories to which several people may upload the same file, and more.

System software includes *performance monitors*. They measure utilization of hardware elements, response times, and identify system bottlenecks. A monitor might find, for example, that most disk accesses are to the same drive. Reassigning files to different drives can improve performance at no cost. Or, it might find that processor load has been trending up and will soon impact response times. This provides lead time to upgrade before users notice.

Your university probably uses *central system administration* software. Instead of installing a new version of its word processor on every campus computer, a technician installs one at a central site and clicks "Distribute." Such software can upgrade the OS, check for virus infections and clean them out if they happen, and assign computers to printers—anything one could do at a keyboard.

Other system software is used for security, detecting attempts to penetrate a system and recording information that may lead to identifying the attacker; uses statistical methods to make files

* Adding features to an operating system is not the same as using operating system market share to promote the use of other programs. That has been held to violate U.S. and European Union antitrust laws.

smaller, so they occupy less disk space and take less time to transmit; diagnoses problems and pinpoints reasons for device failures; and more.

Where you fit in: Using add-on software to augment a firm's existing operating system can be an alternative to upgrading to a new version. Technical professionals might prefer to upgrade, since moving to the latest version helps them keep their skills current, but business considerations may argue the other way. As a manager, you must make sure both options get fair consideration.

4.3 APPLICATION SOFTWARE

Application software produces results that are of direct value to the organization that uses it. This covers everything from tiny programs that monitor Twitter feeds for your favorite hashtags up to huge ones that coordinate the production schedules of General Motors.

The third part of this book, Chapters 7–9, is about application software that provides the benefits listed in Chapter 1: linking parts of the organization, connecting organizations to customers and suppliers, and making better decisions. Chapters 10 and 11 discuss how organizations get their applications. Here, we'll cover a few general topics.

Horizontal versus Vertical Applications

Economists talk about *vertical segments* of the economy: manufacturing, agriculture, health care, transportation, distribution (wholesale and retail), hospitality (hotels, restaurants, etc.) and more. A *vertical application* is an application that is useful only in one such segment (Figure 4.8).

For example, student registration is a vertical application for education. You probably use it at least twice a year, but airlines and dairy farms don't need it. A dairy farm could use software to track milk production by cow, but an airline wouldn't need that.

A *horizontal application* is used in a range of industries. Productivity applications, such as word processing and spreadsheets, are an example. Universities, airlines, and dairy farms all use word processors. They process different words, but they use the same software. Other horizontal applications handle accounting, payroll, customer relationship management (Chapter 8), human resources, and other tasks that cross industry lines. These are *support activities* in the value chain of Chapter 2.

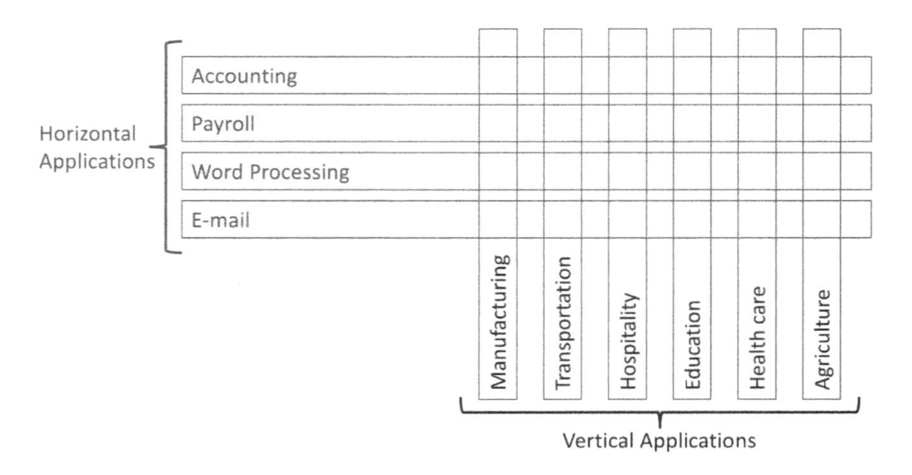

FIGURE 4.8 Diagram of vertical and horizontal applications.

Off-the-Shelf versus Custom Applications

Most personal computer users buy programs that someone else wrote and use them as is: *off-the-shelf software*. That term is still used, though today software is more likely to be downloaded than taken off a store shelf.

At the other end of the spectrum, companies can use *custom software* that was written for their use. This takes more time and money, but can provide an advantage over competitors. Custom software can be written by a company's own employees, or *outsourced* to a firm that is in the business of writing programs. The concept of custom software doesn't depend on who writes it.

Between these lie *customized packages*, used when needs are similar but details vary. For example, expense approval is nearly universal, but one firm may let managers approve expenses of all employees at lower levels, while another allows approval only by one's own supervisor. An expense-reporting application must be set up for one or the other before use. Customization, like programming, may be handled by a company's employees or outsourced. Some firms specialize in customizing complex applications, such as ERP you'll read about in Chapter 7.

Choosing between custom and off-the-shelf applications often leads to heated debates. Should a company get an application now, with good assurance that it works, but knowing that it's not exactly what it wants? Or should it go for what it really wants, knowing that this will involve added time, cost, and risk? The business tradeoffs among these types of software, and a way to approach the choice when you're involved in it on the job, are discussed in Chapter 10.

Open-Source Applications

Like open-source system software, open-source applications exist as well. They tend to be horizontal, as most vertical segments can't support an open-source community. Open-source productivity applications such as spreadsheet programs, word processors (such as OpenOffice, mentioned earlier), and graphics editors are popular, and are considered off-the-shelf packages. At the enterprise level, Apache web server software is used by over half of all web sites, including those of many large corporations. A web server application is a customized package.

The opposite of open source in this context is *proprietary*. A proprietary application belongs to someone who controls who can use it and how much they have to pay to do so. The cost can be zero, as it is today for the Chrome browser, but Google owns Chrome. Google could charge for Chrome if it wanted to.

> ***Where you fit in:*** These terms and concepts are often used in business discussions on software. Knowing what they mean will enable you to contribute to these discussions.

4.4 EMBEDDED SOFTWARE

Everything is getting smarter. Your microwave oven has settings to cook frozen food and reheat cold food; you just have to tell it how much the food weighs or how many servings it consists of. Your freezer will soon know what you take out and will tell the microwave wirelessly. The microwave will ask "Cook the frozen peas?" and will know what to do if you reply "Yes."

Embedded software makes products smarter. It is, informally, "software in anything that doesn't look and act like a computer." Embedded software will affect your career:

- If you work for a company that makes products, many types of products increasingly contain embedded software.
- If you work for a company that uses almost any type of equipment, much of that equipment increasingly contains embedded software.

- Things you use in your personal life contain embedded software: TVs, phones, cameras, cars, watches, athletic shoes. More will in the future.

Embedded computer applications are designed to do one thing well. Consumers don't want stoves to play TV shows, even if their screens and electronics could. Confusing controls are dangerous. Confusing automobiles cause accidents. Confusing medical devices harm patients.

Embedded computers use operating systems for reasons given earlier. Their user interface is specialized, but it's there. Touch screens like those of tablets control stoves and audio systems. Medical and manufacturing devices have screens and keyboards. Your present car may have a screen; your next one will. Managing shared resources and providing shared services are as vital to an embedded computer as they are to a laptop, though the resources and services are different.

Some embedded systems evolve into true computers. Mobile phones, which once just made calls, now do everything a computer can do. Android and iOS are full operating systems in every sense. This trend will continue where it does not conflict with a preference for single-function devices.

Embedded systems connected to networks are part of enterprise information systems. Store checkout stations connect to databases to get prices and update inventory. Medical equipment connects to hospital-wide information systems. Painting equipment in automobile plants accesses production plans to learn what model will arrive next, to spray only where that model needs paint.

Beyond enterprise networks, embedded systems can connect to the Internet. Your refrigerator doesn't have to be connected directly to your microwave in the frozen-peas scenario. If both are on the Internet, they can talk to each other. This leads to the *Internet of Things* (IoT). You'll read more about the IoT in Chapter 6.

Connectivity of embedded systems raises security concerns. Criminals attacked Target via point-of-sale systems and stole information on over 40 million credit cards. Those systems could access Target's main database and were not well protected against criminal activity. Researchers have been able to control a car's steering, engine, and brakes, with its driver unable to stop them.

Embedded systems can be more vulnerable than systems that look like computers. Their software is often difficult or impossible to update, so vulnerabilities may remain even when identified. Responsibility for fixing a security issue may lie with a chip vendor, software supplier, or device manufacturer, none of whom is motivated to fix older products. Some embedded software was developed before security became a concern. Embedded systems may not be under a company's IS department, so it may not know what's in them or how they work. And who installs security patches in a TV set?

Where you fit in: The importance of embedded systems will grow during your career. You must be aware of their potential to improve your employer's operations and of security concerns in their use. Target's data breach originated in a technology problem, but people throughout the company suffered (to say nothing of its reputation and its customers). It will be your job to make sure that embedded systems in your areas of responsibility are adequately protected.

4.5 TRENDS IN COMPUTING

Advances in computing technology, including hardware, operating systems, and the worldwide communications network known as the Internet, are leading to changes in how we use computers.

Virtualization

To understand virtualization, consider what an operating system does. It creates an environment in which application programs can run. The environment an OS creates doesn't work exactly like a computer. For example, telling the operating system to print a page isn't exactly like telling the printer to print a page. The application, instead of seeing the actual printer, communicates with a "virtual printer" that the OS defines. An application can tell this virtual printer "print three copies of this," and it will. A real printer would have to be sent each copy, or each page, separately.

A virtual printer can, if one wishes, be made to work like a real printer. A program would send this virtual printer the same commands it would send to a real one. The OS would then handle the actual printing. That virtual printer could be used by a program that was written to use a real one.

The same can be done with an entire computer. Programmers can create a software-defined environment that mimics the behavior of the real hardware down to the smallest detail.

Why would anyone do this? Because some programs were written to run on real hardware. Aside from specialized needs, such as testing a new computer before it has an operating system, we all use one such program: the operating system itself.

An operating system that mimics the behavior of a real computer is called a *hypervisor* or *virtual machine monitor* (VMM), because it creates imitations of real machines called *virtual machines*. A VMM (Figure 4.9) can run several operating systems (called *guest operating systems*) at the same time, just as an operating system can run several applications at the same time. This offers several benefits:

- Applications are isolated from each other. You've seen programs "crash." Your system stops working, with everything it's running. This isn't a disaster on a personal computer. We restart and hope we didn't lose much work. On a server, this is unacceptable. To prevent a crash from taking other applications down, organizations put applications on separate servers. Most don't need a whole server, so servers are underutilized. Research firm Gartner Inc. estimates that typical corporate server utilization rates are under 10%, primarily for this reason.

  With virtualization, each application runs in its own virtual machine. A crash affects only that VM. New applications are added by creating new virtual machines: faster, simpler, and less expensive than buying new computers. Using virtualization, Northrop Grumman eliminated 3,000 servers, about 80% of its total. That is a big part of how they consolidated 100 data centers into three, ultimately reducing CO_2 emissions by over 13,000 tons per year.
- A computer can run different operating systems at the same time. An organization that uses Windows 10 might have one old application that can't use it. With virtualization, a computer can run Windows XP and 10 at the same time. The same idea lets a Macintosh run Windows.

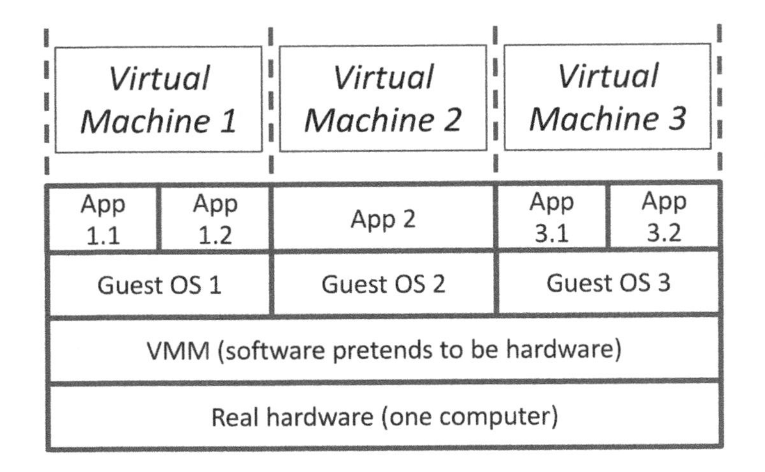

FIGURE 4.9 Layer diagram of VMM with three virtual machines.

- An organization can test a new version of an operating system while continuing to run the old one. Or, it can test new versions of applications in isolation from today's versions.
- The cost of enterprise software often depends on the hardware it runs on. A license for one computer with virtual servers may be less expensive than a license for that many real servers.

A virtual machine monitor resembles an operating system in that it controls the activities of the virtual machine or machines that it supervises. Regarding the three basic functions of an OS:

- The VMM allocates real resources to each virtual machine. From there on, each virtual machine's OS manages the resources that the VMM gave it.
- The user interface of the VMM is simple, since users interact with an OS that runs under the VMM. The VMM's user interface need only control the VMM itself.
- Since guest OSs believe they control a computer, a VMM doesn't provide them with anything that looks like shared services. However, guest OS hardware commands may activate shared services. When a guest OS changes the color of what it thinks is a screen pixel, the VMM intercepts the command. The VMM figures out where that virtual machine's window is on the real screen, or if it is minimized or hidden. If it is visible, the VMM changes the pixel that corresponds to what the guest OS wants to change on the screen it thinks it has.

VIRTUAL THIS, VIRTUAL THAT

You read earlier about the concept of virtual memory, making a given amount of RAM look larger. Virtual machines make one computer look like several. Both terms include the word *virtual*, both of them make one thing look like something else, but these are separate concepts.

Using virtual machines affects backup and recovery. It's simple if each VM handles its own. However, software that knows a single computer underlies several VMs can be more efficient. Companies that sell products that support one approach to virtual machine backup are quick to point out the advantages of their approach, but slow to concede that other approaches have advantages as well. It can be difficult to sort out the claims.

Where you fit in: This seemingly technical decision has a business aspect: pay once for unified backup and recovery tools, or pay over time by using less efficient methods? Decisions that seem technical often raise business questions. In making such a decision, don't accept what a vendor says without question, either in favor of that vendor's product or in favor of its approach. You must always ask: "What is this person *not* telling me? What would a competitor say?"

Client/Server Computing

Computing tasks involve the steps you read about in Chapter 2: accepting input, accessing data, processing data, and so on. Some of these activities focus on the user. Others take place behind the scenes. With modern communication links, those two parts of a task can use different computers. An organization can use one computer to handle the back end of its applications, with a database that all users share, while the user interface is on the user's desk (or lap, or sofa, or coffee shop table). This is called *client/server computing.*

A NOTE ON LANGUAGE

"Client/server" is always written with a slash, not with another separator such as a space or a dash. Don't write "client-server," "client server," or any other variation.

Where you fit in: As with other terms, information technology professionals form opinions of you based in part based on whether or not you use (in this case, write) terms like this correctly.

At least two computers are required for client/server computing:

- The *client*: It provides the user interface.
- The *server*: It performs computations and sends results to the client. If those results are a web page, the server is called a *web server*. If the server's main job is accessing shared data, it is called a *database server*.

Figure 4.10 shows client/server computing with all the clients connected directly to the server. The concept is general, though. It applies no matter how the computers communicate.

Some applications use both a web server and a database server. The database server obtains data from a database. The web server uses that data to create a web page, which it sends to a client. The client then displays the data. *Three-tier computing* (Figure 4.11) is common in large-scale systems.

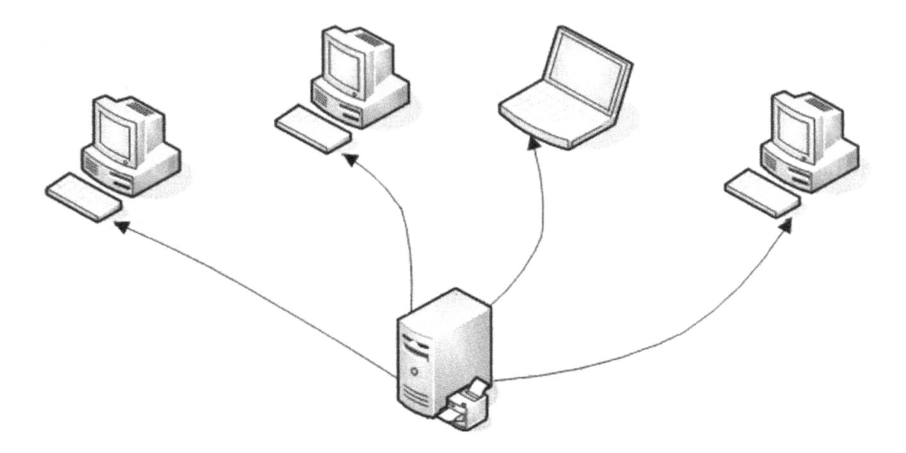

FIGURE 4.10 Client/server computing diagram.

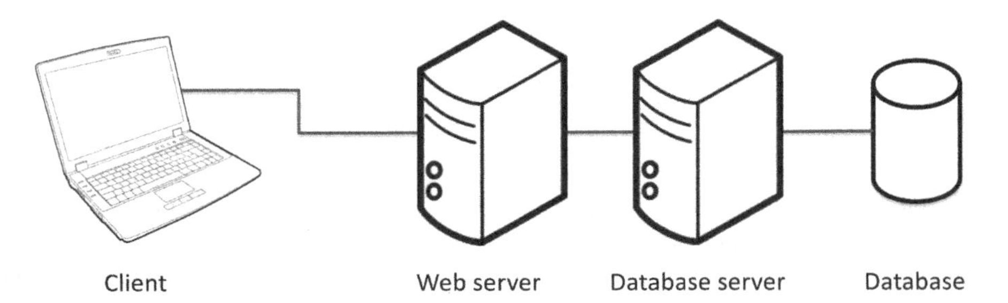

FIGURE 4.11 Three-tier computing diagram.

There can even be four tiers: the database at the back end, an application, a web server that puts application output onto a web page, and finally the client computer with its user.

Dividing an application into tiers requires planning. As clients grow more capable, the trend is to push more of the task onto them. This reduces the number of servers, which saves cost and makes it simpler to coordinate them. In a system designed a few years ago, a client might display a chart that was sent as a graphic. Today, the client may create that chart from the server's data.

The desire to use *thin clients* works against this. A thin client has just enough hardware to connect to a network and run a web browser. Everything else is on servers. Thin clients are less expensive than full computers. They are easier to administer in because users can't customize them, add applications, or ignore security. Users, however, may resent the restrictions that thin clients impose. Today's best-known thin client is Chromebook, which runs Google's Chrome OS. (When a distinction is necessary, a full-featured computer used as a client is called a *fat client*.)

Client/server computing is a software concept. Clients and servers use standard hardware, though high-end servers have, as you read in Chapter 3, added features. Personal operating systems keep *applications* out of each other's way, but server operating systems keep *users* out of each other's way. This complicates a server OS, but today's online world could not exist without them.

Cloud Computing

In the last chapter you read about cloud storage, accessing data over the Internet. Applications can also be accessed over the Internet. Results reach users as web pages or as data for a client to process. That is *cloud computing* (Figure 4.12).

Google Docs (part of Google Apps) and Microsoft Web Apps (part of Office 365) provide personal applications such as spreadsheets, word processing, and presentations as cloud services. Any platform that runs a web browser can use these applications. Because they store data in the cloud, any authorized user can access their documents. That facilitates team collaboration.

Cloud computing is a form of client/server computing where servers are located on the Internet rather than locally. Two potential benefits: cloud servers may be operated by a company which specializes in providing computing services, and that company often has enough extra capacity to

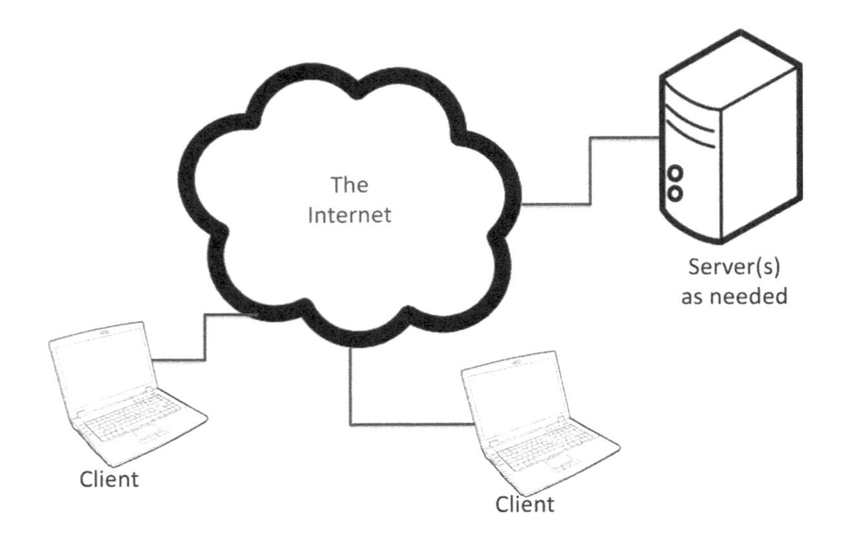

FIGURE 4.12 Cloud computing diagram.

provide more at peak loads because different customers have peak loads at different times. There are several variations on cloud computing, with more surely to come:

- *Software as a Service* (SaaS) provides access to specific applications.
- *Platform as a Service* (PaaS) provides access to a platform for customers' own applications.
- *Infrastructure as a Service* (IaaS) provides access to virtual machines, storage, and network capabilities for users to create their own platforms.

A cloud (the environment of cloud computing) may be *public* or *private*. A large organization may use cloud computing enough to justify its own private cloud, though it may pay someone to set it up and run it. Smaller firms can use a public cloud such as Amazon Elastic Compute Cloud (EC2) or IBM SmartCloud. An organization can supplement a private cloud with a public cloud, creating a *hybrid cloud*. SaaS application providers use private clouds.

Cloud computing affects operating systems. PaaS and IaaS extend the OS to provide access to cloud computing services. Their interfaces make it easier to develop applications for cloud computing, but tend to tie a user to one provider because they are not standardized. The choice of a cloud computing provider can be expensive and time-consuming to change.

One potential drawback of cloud computing is that response time can suffer if resources are many Internet links away from their users. To prevent this, companies may turn to *edge computing*, with resources located near the majority of their users. This is helpful in uses such as autonomous vehicles, which are viable only if response is rapid, and in the *Internet of Things* (see Chapter 6).

- Edge computing is unrelated to Microsoft's Edge browser.
- A related concept is *fog computing*, so called because it is like a cloud but closer to users (like atmospheric fog). Fog computing is more centralized and closer to cloud resources than edge computing, but is not quite as responsive.

Where you fit in: These trends will affect how your employer uses computers. You won't just have a computer on a desk connected to a database and the Internet. To participate in the business world of 2025 and beyond, to make business decisions involved in selecting which of these concepts to use, you have to know about them and what they mean.

Artificial Intelligence

Artificial intelligence (AI) has been described as "making computers do things that, if we saw a person do them, we'd say the person was using intelligence." AI is useful in many information systems areas today. It will surely become more so in the future. Most AI uses involve decisions, so we cover it in Chapter 9.

4.6 SOFTWARE LICENSING

The copy of FileMaker Pro at your campus store has a license that states what you may do with it:

- You may install FileMaker Pro on two computers, as long as you just use one at a time.
- If your computer supports multiple virtual machines, you may still run only one copy.
- You may copy FileMaker Pro once for backup. The copy must retain all copyright notices.
- You may not transfer your FileMaker Pro license to another person without the consent of FileMaker International.

FileMaker can limit your rights because you don't buy the software. You buy the box it's in and the plastic disk it's on (if you don't download it), you buy the right to use it—but FileMaker owns it. You may use it only as your license allows. Most software is licensed, not sold.

Enterprise systems have a wider range of licensing options, and enterprise customers generally have more leverage than individuals in persuading vendors to modify their standard terms. An example of pricing—simpler than many!—is in the box below. This topic is discussed further in Chapter 10.

Oracle database software is available in three "editions": Personal, Standard, and Enterprise. Each has a per-user price, a minimum number of users, and a per-processor-unit price:

Edition	Price/User	Min. Users Paid	Price/Proc. Unit	Max. Procs.
Personal	$460	1	n/a	2
Standard	$350	5	$17,500	4
Enterprise	$950	5	$47,500	No limit

Enterprise Edition processor units are calculated by counting the number of processor cores in all systems that the license covers, multiplying by a speed factor that depends on processor type, and rounding up to a whole number. For other editions, a processor unit is a processor socket, which generally holds one processor chip (which may have multiple cores).

Customers may pay per user, using the higher of the actual number of users or the minimum in the above table, or per processor. Customers pay per processor where users can't be identified, as in an Internet environment, or where there are so many individual users that a per-processor license is more economical. Customers who upgrade their hardware past the processor limits of their current edition must license a higher edition and pay the corresponding license fee.

Software vendors understandably want people who use their software to pay for it. BSA: The Software Alliance (www.bsa.org) exists to protect members' software copyrights. It finds unlicensed use of software and brings legal action against infringers. It has recovered a lot of money for members, in penalties for illegal use and in license payments going forward.

BSA estimated the commercial value of unlicensed software in use at $46.3 billion in 2017. (This values software at full retail price. Since many users of unlicensed software would do without it if they had to pay for it, and since discounts are widely available, industry revenue losses are less.) They also found that unlicensed software carries a higher risk of malware than licensed software.

Even free software has licenses. The Apache web server license reads in part, edited for brevity:

You may reproduce and distribute copies of the Work or Derivative Works thereof, with or without modification, in Source or Object form, provided that You meet the following conditions:

1. You give other recipients of the Work or Derivative Works a copy of this License;
2. You cause modified files to carry prominent notices stating that You changed the files;
3. You retain, in the Source form of any Derivative Works that You distribute, all copyright, patent, trademark, and attribution notices from the Source form of the Work, excluding those notices that do not pertain to any part of the Derivative Works;

4. If the Work includes a "NOTICE" text file ..., then any Derivative Works that You distribute must include a readable copy of the attribution notices contained within such NOTICE file, excluding those notices that do not pertain to any part of the Derivative Works, in at least one of the following places. ... The contents of the NOTICE file ... do not modify the License. You may add Your own attribution notices within Derivative Works that You distribute ... provided that such attribution notices cannot be construed as modifying the License.

These and similar terms in the licenses of most open-source packages prevent commercial entities from using free software in their products without informing their customers that they did so.

Where you fit in: As a manager, you will be involved in decisions and negotiations about software licenses for your firm. You have to know what your options are.

KEY POINT RECAP

Software is detail-ridden and complex to write. *Errors are hard to avoid, are often hard to find, and may be hard to correct.*

As a businessperson, you will be responsible for having software written for your company's needs. Supervising this activity effectively requires understanding what it consists of.

Software can be proprietary or open-source. *The choice between them is a business decision based largely on technical factors.*

As a businessperson, this is another decision you will be faced with. You must know enough to consider the pros and cons of both.

System software starts with the operating system and continues to other packages that support the organization's computing needs. *The main functions of the operating system are to define the user interface, to manage shared resources, and to provide shared services to other software.*

As a businessperson, you must understand operating system concepts well enough to tell the difference between the way something must work versus the way a particular operating system happens to work. Don't confuse principles with appearance.

Virtualization, client/server computing, cloud computing, and embedded computing *are among the trends that will affect information systems over the coming years.*

As a businessperson, you must keep on top of computing trends to avoid being blind-sided by a competitor who takes advantage of them first!

Application programs *get the job done. You should be familiar with horizontal vs. vertical applications and with custom versus off-the-shelf applications.*

As a businessperson, you will have to be familiar with the applications your company uses. As a manager, you will have to be familiar with other applications that it could use and make decisions about using them. Later chapters of this book will discuss many application categories in depth.

KEY TERMS

Application software: Collective term for computer programs that carry out tasks which computer users find directly useful.

Artificial Intelligence (AI): Informally, software that enables computers to do things that we consider to require intelligence when people do them.

Bug: An error in a computer program.
Client: A computer that receives information or processing services from a *server*. (From Chapter 3)
Client/server computing: A method of carrying out a computing process in which one computer, the *server*, provides resources such as data or web pages to another, the *client*.
Cloud computing: Accessing computing resources over the Internet.
Command-line interface: Controlling a computer by entering commands via a keyboard.
Computer program: A series of instructions that, when executed by a computer, perform a useful function.
Custom software: Software that is written to meet the needs of one organization. Compare *customized package*, *off-the-shelf software*.
Customized package: An application program that is publicly available but requires modification before use. Compare *custom software*, *off-the-shelf software*.
Database management system: System software that organizes and manages data.
Database server: *Server* that accesses a database on behalf of *clients*.
Data deduplication: Saving storage space by identifying and removing multiple copies of a file.
Edge computing: Cloud computing in which resources, accessed over the Internet, are located in close proximity to their clients.
Embedded software: Software built into a device to control it, and whose sole purpose is operating that device.
Fat client: *Client* computer with sufficient resources to function independently when not attached to a supporting network. Compare *thin client*.
Graphical user interface: Controlling a computer by using a pointing device to access on-screen menus and images.
Guest operating system: The operating system of a *virtual machine*.
Horizontal application: An application program that is broadly useful in many segments of the economy. Compare *vertical application*.
Hybrid cloud: *Cloud computing* environment which consists partly of a *private cloud*, partly of a *public cloud*.
Hypervisor: See *virtual machine monitor*.
Icon: A small on-screen symbol that represents a file, a program, or a command to a computer.
Infrastructure as a Service (IaaS): A form of *cloud computing* in which users access basic computing resources over the Internet and can create their own platform on it.
Object program: A computer program that has been translated into the form of 1s and 0s for execution by a computer. Compare *source program*.
Off-the-shelf software: Programs that are publicly available and require little or no modification for use. Compare *custom software*, *customized package*.
Open-source software: (a) Software that is produced communally and is available for anyone to use free of charge. Contrast with *proprietary software*. (b) Software whose *source program* is open to public view.
Operating system: A program that manages activities and resource usage in a computer.
Outsourcing: One company's engaging another company to carry out activities on its behalf that it would otherwise do itself.
Performance monitor: System software that measures the load on a computer with a view to optimizing its performance.
Platform as a Service (PaaS): A form of *cloud computing* in which users access a platform over the Internet and can run their own applications on it.
Private cloud: *Cloud computing* environment operated for the benefit of one organization and not generally available to others. Compare *public cloud*, *hybrid cloud*.
Program (noun): See *computer program*.
Program (verb): To write computer programs.
Programmer: A person whose job is writing computer programs.

Programming language: Set of key words and rules defining a way to write computer programs.

Proprietary software: Software that is legally owned by and controlled by a single company. Contrast with *open-source software*, sense (a).

Public cloud: *Cloud computing* environment which is available, often at a price, to anyone. Compare *private cloud, hybrid cloud.*

Server: A computer that provides a service, such as web pages or database access, to other computers (from Chapter 3).

Software: A collective term for computer programs. This noun has no plural form, ever.

Software as a Service (SaaS): A form of *cloud computing* in which users access an application over the Internet.

Software license: Agreement between a buyer and a seller of a computer program regarding the terms under which the buyer may use that program.

Source program: A computer program in the form that it was written by a programmer, prior to translation into 1s and 0s for execution by a computer. Compare *object program.*

System software: Collective term for computer programs that carry out tasks that are not directly useful, but that facilitate the development or use of other software.

Thin client: *Client* computer with minimal resources, designed to be used only when attached to a supporting network. Compare *fat client.*

Three-tier computing: An information processing environment that includes a database server, a web server, and a client.

Touch-based interface: Controlling a computer by pressing on its screen or another surface.

User interface: The way users interact with software.

Vertical application: An application program that is useful only in a specific segment of the economy. Compare *horizontal application.*

Virtual machine: A software-controlled environment in which programs run as if they were running directly on the hardware of a physical computer.

Virtual machine monitor: A control program that manages *virtual machines.*

Virtualization: The process of creating *virtual machines.*

Web server: *Server* whose main function is to provide web pages to *clients.*

REVIEW QUESTIONS

1. What is *software*?
2. Explain the difference between a *source program* and an *object program.*
3. What is the difference between *proprietary* and *open-source* software?
4. Describe the three fundamental functions of an operating system.
5. Give three examples of system software other than operating systems.
6. What is a *vertical* application? How does it differ from a *horizontal* application?
7. Define off-the-shelf, custom, and customized software.
8. Give two examples of embedded software that were *not* used in this chapter.
9. How does a virtual machine monitor differ from an operating system? Resemble one?
10. Give an example of client/server computing from your own experience.
11. What are three types of cloud computing?
12. Can you usually negotiate the license terms of personal computer software?

DISCUSSION QUESTIONS

1. Choose any two versions of Windows, Mac OS, or Linux that are intended for general use on laptop and desktop computers. (Don't choose a phone/tablet OS such as iOS or Android.) By searching the web for reviews or by any other method, identify three differences between those two versions of the same OS. Discuss who these

differences would matter to, how much you think they would matter, and how typical this person is of computer users overall. Also, describe how and where you found your information.

2. Resource allocation isn't just for operating systems. Consider the resources of your college or university. Identify one university resource that is allocated in each of the following ways:
   a. To one user for an extended period of time, like file space on a disk.
   b. To one user for a specific activity and then to another, like a printer.
   c. To many users who come and go up to its capacity, like RAM.
   d. To many users apparently at the same time by cycling through their needs rapidly, like central processor time.
3. User interfaces are generally designed for people with full vision. Not all people have that. Discuss a user interface you might design for people in these two categories:
   a. People who can recognize general shapes on a screen but cannot read text.
   b. People with no vision at all.
4. Read Jack Wallen's article "10 open source software projects that are leading innovation" at *www.techrepublic.com/blog/10things/10-open-source-projects-that-are-leading-innovation/3770*. Pick one of them for this question. (Your instructor may assign one.) Visit the web site given at the beginning of its description for more information. Then write a two-page report on that software. Cover at least the following points:
   - What does the software do? Explain its function in business terms, not technical terms.
   - What business need does it fill? What is the value to a business of filling that need? How can a business meet that need, other than by using software such as this?
   - Should businesspeople trust this software to be of professional quality? Why or why not?
   - What about this project attracts people to participate in it? Would you enjoy participating in it, if you had the background?
   - Find a commercial package that meets the same business need. Give one advantage of the commercial package over the open-source software, and one in the other direction.
   - State two criteria businesspeople should use in choosing between these packages. (Don't evaluate them. Just give your evaluation criteria and why they are important.)
5. Visit *www.mozilla.org/en-US/contribute* to see how you could get involved with the Firefox open-source project. Click on "Get Involved" and see how people can contribute to this project. Identify those for which you are qualified. For each of them:
   - Do you think you would enjoy working on this project? Why or why not?
   - Do you think you'd be good at it? Why or why not?
   - After you graduate and are hired into a professional position in your field, would it be to your employer's advantage to let you participate in such a project for about half a day per week? Why or why not?
   - If you didn't think it would be to your employer's advantage, when do you think it would be, in terms of types of employers and the types of jobs that new professionals do?
6. Your employer gives you a thin client. Write a memo to your boss explaining why you need an exception to the company's thin client policy and should get a full computer. Make any assumptions you wish about your work, but be sure to state them in your memo.
7. Your university needs a new student information system. Its present system runs on its own servers, in an on-campus data center that also houses its other multi-user computers. It is considering obtaining its new system via SaaS. Discuss advantages and disadvantages of SaaS in this specific context. Do not mention general advantages or disadvantages without relating them to this specific situation.
8. Your job is to select a program that will prepare U.S. income tax returns. Proceed as follows:

a. Develop a list of six to eight factors that are important in evaluating such a program.
b. Assign an importance weight to each of these factors, so that more important factors have proportionately higher weights. Weights should add up to 100.
c. Identify three programs as potential candidates.
d. Evaluate each of the programs on each factor. The best program on each factor should get 10 points. The others should be scored relative to that program. (In case of a tie, more than one program may get 10 points.)
e. Multiply each program's score on each factor by the importance weight of that factor to get the number of points that program earns for that factor. The importance weight of a factor is the same for all three programs, but their scores differ, so the number of points each program earns on each factor will also differ.
f. Add each program's points for all factors to get its total score.
g. Discuss: Do you think this is a reasonable selection process? Do you agree that the program with the highest total score is the one to buy? If you were choosing such a program for your own use, would you use this process or a similar one? Why or why not?

9. People in different countries need different software to prepare income tax returns. The U.S. and some other countries have several different packages for this purpose, but some have one or none. Discuss the factors that make a country attractive as a market for such packages. If you want to start a personal tax return software business, and you need to pick a country, what characteristics will you look for? The obvious markets for such packages already have one or more, and people who use a program often want to stay with it.
10. The white paper "Infrastructure Economics: What IT Leaders Can Learn from the CFO Organization (and Vice Versa)," *www.smp-corp.com/wp-content/uploads/2017/01/What-IT-leaders-can-learn-from-CFOs-and-vise-versa_whitepaper.pdf*, accepts the suitability of cloud computing for some of an organization's computing workload but argues for retaining much of it in house. It asserts that businesses that use cloud computing spend three times as much on managing cloud services as they do on the services themselves. This paper was sponsored by Hewlett-Packard Enterprise, which sells primarily products for on-premise computing. If your CEO read this paper and asked you what you think of this claim, how would you reply? (Besides reading the cited paper, do some additional research before you write your reply.)
11. Software company A is widely considered to have complex and inflexible contract policies. That doesn't stop people from using its software, since that software is very good, but it is frequently the subject of negative industry analyst reports and of customer complaints.
    a. You work in A's marketing department. What, if anything, would you do about this? If there isn't enough information here to answer that question, how would you decide?
    b. You work in a competitor's marketing department. Your company's contract terms are much simpler than A's. Industry analysts don't point them out as bad examples. Your customers don't complain about them. How can you benefit from this?

KHOURY CANDY DISTRIBUTORS SOFTWARE

The following week Jake and Isabella signed in at the receptionist for a meeting with Lakshmi Agarwal, KCD's manager of application software. Chris had set up the meeting, so they were expected. They went to the same conference room as the previous week. The candy bowl on the table still had some appeal, but its pull wasn't quite as strong.

After introductions and a few pleasantries, Lakshmi opened the discussion with "What do you two know about software?"

"Not much more than anyone knows, I suppose," Isabella replied for the two. "We've used it, of course, but we haven't studied it much. We'll study programming next term, but that doesn't help us now."

"You'd be surprised," Lakshmi replied. "You know there's a thing called programming, and that it's how software is made. That puts you way ahead of most people. The danger in taking one programming course is thinking that industrial-strength software is like the small programs you work on there, just more of the same. It's not. It's a whole different creature."

"How do you mean that?" Jake asked.

"Let me give you an example. Chris said he told you a bit about how we spend a lot of time and effort on backup and preparing to recover our database if we ever have to do that."

"Yes, he did," Jake confirmed. "He told me I should get my act together. You can tell him I signed up for cloud backup last Thursday. The Internet connection in my dorm is really fast, so uploading files doesn't take long. I got an app that uploads new files, or files that I change, automatically. I feel a lot better now."

"Good," said Lakshmi. "I'll tell him in our staff meeting this afternoon. Anyhow, the point is that we have to worry about things like reversing a transaction that turned out to be wrong. You don't have to do that on a PC—you just put in the right data in place of the wrong data and you're set. That sort of thing takes up a huge effort in an enterprise-wide system because actions leave footprints all over. For example, if we record a sale on March 1 instead of February 28, it might reduce the discount that customer gets for March. If we change the sale date, the software has to know it should figure a new discount, then go through all the March invoices and apply it. Things that never come up in personal computing, like checking for valid input data, turn out to be a lot more work than you might think!

"Anyhow," she continued, "this is the headquarters system architecture diagram that Chris showed you last week. As regards software, the z/OS operating system on the z13 does what your Windows or Mac OS does, but it does it for dozens or hundreds of people at the same time. That makes it enormously more complicated. It has to keep everyone out of everyone else's way while they all get their share of its attention, and make sure that two people don't inadvertently try to change the same data at the same time. And that's not even getting into some tools, like instrumentation to measure how each part of the system is used, that don't exist in nearly the same way on lower-end systems."

"Isn't software like that expensive?" Isabella asked.

"It sure is!" Lakshmi responded. "It's expensive for IBM to develop, so it's expensive for us to license. If they charged us the same as they charge Citibank, we couldn't afford that system. Fortunately, they don't. Also, they have what they call 'container pricing' that makes the cost of developing a new application on the z13 competitive with what it would cost, in terms of hardware and software, to develop on a smaller computer. Plus, we can take advantage of all the large system's features."

"Why would IBM cut its prices this way?" Jake wondered.

"Oh, I don't think they do it out of the goodness of their hearts," Lakshmi laughed. "IBM is a business, not a charity. They know that, if they don't do something like this, people will put applications on less expensive platforms, and the market for Z systems will dry up. They want to slow that down. That's smart business. It's working, at least with us. We'd rather keep our central applications in one place. With container pricing, it doesn't cost us more to do that. We don't pay IBM as much as we might, but we pay them a lot more than we would if we got rid of our z13!

"This web server over here," she continued, "is separate from the z13 because we wanted to use the Linux operating system and the Apache web server package in their native environment. The HP servers in the distribution centers run Linux, too, and Oracle database management software—the same database management package that we use on our z13. Oracle came out with Version 18c of their DBMS last year, so we'll probably move to that soon. By the way, here's today's bit of geek trivia for you: the version before 18c was 12c. They totally skipped 13 through 17. Don't ask me why. But would you like to come back next week to look at our databases in more detail?"

"That would be great," Jake and Isabella replied in unison.

"I figured you'd say that," said Lakshmi. "I warned our database administrator, Visal Phan, that we might need him next Tuesday and confirmed that this time works for him. Does it still work for you? Good!"

Questions

1. Consider the layer diagram of an information system (Figure 4.4). Where in this diagram do (a) z/OS, and (b) the "instrumentation to measure how each part of the system is used" that Lakshmi mentioned, fit?
2. Is (a) an application that forecasts candy demand by region horizontal or vertical?
3. Watson Walker, Inc. (*watsonwalker.com*) is a consulting firm that advises its clients on using IBM z systems and the z/OS operating system effectively. They offer a three-day workshop on containing z/OS software costs (*watsonwalker.com/education/containing-z-os*). Attending the workshop costs $2,100 for the first person from an organization, $1,050 per person after that. When do you think it would make sense for a company to send someone to such a workshop? Would it make sense for KCD? Make any assumptions you need, and be sure to state what they are.

CASE 1: SECURITY THROUGH OPEN SOURCE

"In 1815, the Duke of Wellington observed that the Battle of Waterloo had been won on the playing fields of Eton. Today, there's a war being waged every hour of every day on computer screens. And Nexor is on the front line."

—Nexor Ltd., Nottingham, U.K.

Nexor provides information technology and email services to defense, intelligence, and other government organizations in the United Kingdom and around the world, including organizations such as NATO. Nexor Sentinel is a secure email service that protects user organizations by validating inbound and outbound electronic messages to conform to a user's security policies. Its IT platform protects sensitive information on the web. In addition to high availability and security, both the email service and IT platform have to comply with international standards for computer security.

Nexor found it increasingly difficult to deliver the services that its customers expected in the second decade of the 21st century using an older operating system on older hardware, neither of which was still supported by its original supplier. Staying with that system was not an option.

Recognizing the limitations and ongoing costs associated with staying with that platform, Nexor decided to move to Red Hat Enterprise Linux. This is the open-source Linux operating system, packaged with additional support software and documentation by Red Hat Inc. of Raleigh, N.C. It uses the SELinux (Security Enhanced Linux) version, which is based on concepts developed by the U.S. National Security Agency (NSA).

"The main benefit for us [of the move to Linux] is that we have been able to move off a specialized, proprietary platform," says Colin Robbins, technical director at Nexor. However, Nexor could have done the same by going to a proprietary operating system such as Windows. Why choose Linux? Primarily because, after evaluating it, they found that Linux met the tight security requirements of their environment better than the other two alternatives it considered. Once they settled on Linux, they picked Red Hat Enterprise Linux because it had been formally evaluated at Level 4 (EAL4) of the international Common Criteria security specifications.

Nexor found several benefits to using Linux:

- **Efficiency.** Their email product runs faster. Internally, the time it took to build an application (to combine all its software components so it would operate as a unit) decreased from three days to a few hours.
- **Cost savings.** Training customers, for example, is much less expensive on a modern operating system with familiar concepts than it was on the older system.

- **Customer satisfaction.** One example is that security updates can now be passed to customers automatically, rather than waiting for them to be delivered. "The result is a simple, smooth and convenient model for product update delivery," says Robbins.

With its Linux platform, Nexor feels ready to continue its mission of providing information systems and information transfer to security-conscious customers worldwide.

Questions

1. Why shouldn't Nexor use an old operating system? Focus on reasons that apply to Nexor's situation. Some reasons that individuals upgrade, such as wanting to run current versions of off-the-shelf applications, do not apply to Nexor since it writes its own software.
2. You are a senior manager at a company such as Nexor and must make the final decision about the operating system that it will use going forward. Three of your subordinates have evaluated operating systems and recommend three different choices: Red Hat Linux, Linux with components compiled by Nexor directly from the source programs, and Windows 10. How would you approach the problem of resolving their opinions? (This is about decision making, not about the answer. Nexor's choice of Red Hat Linux is irrelevant.)
3. Most information systems don't raise the same security concerns that defense organizations such as Nexor have. Should that be the case? Or should organizations that have any type of information system heed the lessons of how carefully Nexor addresses security?

CASE 2: RESCUING FOOD

> "There is no reason for any American to go hungry with our country's ample food supply. We simply need better logistics."
>
> **—Kevin Mullins, co-founder, Food Rescue US**

The goal of Food Rescue US* is to end food insecurity in the United States. This isn't an easy task, but they have the assistance of hundreds of volunteers who sign up for "food runs." The volunteers drive to a supermarket, grocer, or restaurant that has surplus food and bring it to an agency that helps feed the hungry. This eliminates most costs associated with conventional food redistribution, a fleet of trucks with drivers to move the food, and warehouses to store it in.

That's a fine concept, but implementing it is not trivial. Food Rescue US is headquartered in Norwalk, Conn. In addition to Connecticut's Fairfield and New Haven Counties, the organization also operated in Albuquerque, N.Mex., and Columbus, Ohio, when the first edition of this book was written in 2014. Five years later it has added California, northwest Connecticut, the District of Columbia, Florida, Indiana, Kansas, North Carolina, two additional Ohio locations, Oregon, South Carolina, Utah, and Virginia. One office couldn't hope to coordinate all the food runs for those locations, let alone the added locations to which it will surely expand in the future.

Enter Go Rescue, a web application that Food Rescue US's partner WhenToManage developed for them. WhenToManage develops software for restaurants. It knows how many restaurants have food left over at the end of the day and would rather see it eaten than go into landfills or compost. Go Rescue has three modules:

- For restaurants, grocers, or other food providers: list surplus food they want to donate.
- For volunteers with vehicles: see the schedule of food rescues and select deliveries.
- For agencies such as community kitchens and food pantries: post needs and receive food.

* Referred to by its original name, Community Plates, in the first edition of this book.

"We built Go Rescue on WhenToManage's Peach platform, which allows us to quickly develop and update applications to meet the specific needs of the industries we serve," said Jeff Schacher, CEO and founder of WhenToManage. "This environment also enables us to easily continue modifying Go Rescue to further enhance the food rescue process, and over time, meet other nonprofits' needs."

Peach is a data management system that enables users to access data in customized ways from any device over the web. Thus, volunteers can tell Community Plates where they are—on a regular basis, in advance for one run, or at the last minute—using smartphones. Go Rescue matches them to a place with food and another that needs it.

Food Rescue US estimates that more than half of the 4 billion tons of food produced worldwide annually goes to waste one way or another. They rescued the equivalent of 21.3 million meals, saving 32 million pounds of food from landfills, by early 2018. As they expand, the totals will surely grow. The Go Rescue application will help that happen.

Questions

1. In terms of the software categories discussed in this chapter, is Go Rescue a horizontal or a vertical application? Off-the-shelf or custom? Open source or proprietary?
2. Two operating systems are involved in any use of Go Rescue: one on the user's smartphone (or other device) and one on the computer that runs Peach. Summarize what each of them does when running Go Rescue, in terms of the three main operating system functions.
3. Mullins was fortunate in finding a company like WhenToManage that was willing to put its resources behind his idea. Had he not known of WhenToManage, how do you think he might have proceeded? Suggest at least three possibilities, ranked in the order that you think he should consider them.

BIBLIOGRAPHY

BSA: The Software Alliance, "Software management: Security imperative, business opportunity: BSA global software survey," June 2018, www.bsa.org/~/media/Files/StudiesDownload/2018_BSA_GSS_Report_en.pdf, accessed December 30, 2018.

Community Plates web site, communityplates.org, accessed August 16, 2013, on September 17, 2019, it redirected to Food Rescue US (see below).

Datanyze, "About Apache HTTP server," www.datanyze.com/market-share/web-and-application-servers/apache-http-server-market-share, accessed January 4, 2019.

de Fremery, R., "Edge computing is changing the way IT pros think about infrastructure," *IT Toolbox*, March 26, 2019, it.toolbox.com/articles/edge-computing-is-changing-the-way-it-pros-think-about-infrastructure, accessed June 25, 2019.

Eanes, Z., "Red Hat investors approve $34B merger agreement with IBM," *The News and Observer* (Raleigh, N.C.), January 16, 2019, www.newsobserver.com/news/business/article224619300.html, accessed March 7, 2019

Food Rescue US web site, foodrescue.us, accessed January 2, 2019.

Golson, J., "Jeep hackers at it again, this time taking control of steering and braking systems," *The Verge*, August 2, 2016, www.theverge.com/2016/8/2/12353186/car-hack-jeep-cherokee-vulnerability-miller-valasek, accessed December 30, 2018.

IBM Corp., "Container pricing for IBM Z" white paper, October 2, 2018, www-03.ibm.com/support/techdocs/atsmastr.nsf/5cb5ed706d254a8186256c71006d2e0a/558202750925795285258160001b84ae/$FILE/ContainerPricing%20WhitePaper%20v2.2b.pdf, accessed December 26, 2018.

IBM Corp., "IBM z/OS Version 2 Release 3" data sheet, April 2018, public.dhe.ibm.com/common/ssi/ecm/zs/en/zsd00998usen/zsd00998-usen-32_ZSD00998USEN.pdf, accessed December 26, 2018.

Kassner, M., "Target data breach exposes serious threat of POS malware and botnets," *TechRepublic*, January 9, 2014, www.techrepublic.com/blog/it-security/target-data-breach-exposes-serious-threat-of-pos-malware-and-botnets, accessed September 17, 2019.

Lanneau St. Leger, P.-Y., "Google, Amazon and Facebook embrace open source software as future," *IT Toolbox*, November 19, 2018, it.toolbox.com/articles/google-amazon-and-facebook-embrace-open-source-software-as-future, accessed January 1, 2019.

Nexor web site, www.nexor.com, accessed January 2, 2019.
Opto 22, "Fog computing vs. edge computing," info.opto22.com/fog-vs-edge-computing, accessed June 25, 2019.
Oracle Corporation, "Database licensing," December 1, 2015, www.oracle.com/assets/databaselicensing-070584.pdf, accessed January 1, 2019.
Oracle Corporation, "Database licensing information user manual," November 2018, docs.oracle.com/en/database/oracle/oracle-database/18/dblic/database-licensing-information-user-manual.pdf, accessed January 1, 2019.
Oracle Corporation, "Oracle Technology global price list," August 12, 2019, https://www.oracle.com/assets/technology-price-list-070617.pdf, accessed August 17, 2019.
Peachworks (successor to WhenToManage) web site, peachworks.com, accessed March 11, 2019.
Pratt, M., "Computerworld Honors 2013: Mobile app helps get fresh food to people in need," *Computerworld*, June 3, 2013, www.computerworld.com/article/2497383/computerworld-honors-2013--mobile-app-helps-get-fresh-food-to-people-in-need.html, accessed September 17, 2019.
Pratt, M., "Northrop Grumman: Virtualizing 3,000 servers," *Computerworld*, October 24, 2011, www.computerworld.com/article/2550211/northrop-grumman--virtualizing-3-000-servers.html, accessed September 17, 2019.
Red Hat Enterprise Linux web site, www.redhat.com/en/technologies, accessed January 4, 2019.
Red Hat, Inc., "2018 Annual report," investors.redhat.com/~/media/Files/R/Red-Hat-IR/Annual%20Reports/RHT%20FY2018%20Annual%20Report.pdf, accessed January 2, 2019.
Schneier, B., "The internet of things is wildly insecure—And often unpatchable," *Wired*, January 6, 2014, www.wired.com/opinion/2014/01/theres-no-good-way-to-patch-the-internet-of-things-and-thats-a-huge-problem, accessed January 4, 2019.
Vaughan-Nichols, S., "Linux totally dominates supercomputers," *ZDNet*, November 14, 2017, www.zdnet.com/article/linux-totally-dominates-supercomputers, accessed January 4, 2019.
Vaughan-Nichols, S., "Supercomputers: All Linux, all the time," *ZDNet*, June 26, 2018, www.zdnet.com/article/supercomputers-all-linux-all-the-time, accessed January 4, 2019.
Venkatraman, A., "Case study: Nexor dumps ageing proprietary operating system for open source OS," *Computer Weekly*, July 31, 2013, www.computerweekly.com/news/2240202849/Nexor-dumps-ageing-proprietary-operating-system-for-open-source-OS, accessed September 17, 2019.
Worstall, T., "Business Software Alliance gets it wrong on software theft again," *Forbes*, June 3, 2012, www.forbes.com/sites/timworstall/2012/06/03/business-software-alliance-gets-it-wrong-on-software-theft-again, accessed September 17, 2019.

5 Data, Databases, and Database Management

CHAPTER OUTLINE

WHY THIS CHAPTER MATTERS

Computers do not make businesses successful. Intelligent use of information helps make businesses successful. Computers are a tool in using information intelligently. Databases are a key technology in making today's intelligent uses of information practical.

To get the information we need when we need it, that information must be properly organized. Proper organization makes it possible to retrieve information efficiently, without a lot of extra work; to control access to information, making it available to those who need it but not to those who shouldn't have it; and to use this information in new, creative ways.

As a businessperson you must be on the lookout for new ways to use information. Knowing how computers organize information will help you do this. Knowing this will also help you work with professional database designers to get the databases you need to do your work.

CHAPTER TAKE-AWAYS

As you read this chapter, focus on these key concepts to use on the job:

1. Databases help organizations become effective by allowing people and applications to share common organizational information.
2. Most databases today use the relational model. Relational databases are flexible. They can handle applications that were not foreseen when the database was first set up.
3. Applications that involve data analysis and decision making may use other database models.
4. Using databases requires database management software to store and retrieve data on behalf of users and applications.

5.1 THE DATABASE CONCEPT

A *database* is an organized collection of data about related real-world items or concepts.

At the lowest level data consists of bytes, because bytes are the basic storage unit of computers. Data made up of characters, such as names and addresses, fits naturally into bytes (Figure 5.1). Much business data consists of character strings. Other types of data are stored in bytes as well.

These bytes are grouped into *fields*, or *data elements*. Each field in a database contains a data item that has meaning to the organization. A student's family name may be a field in a university

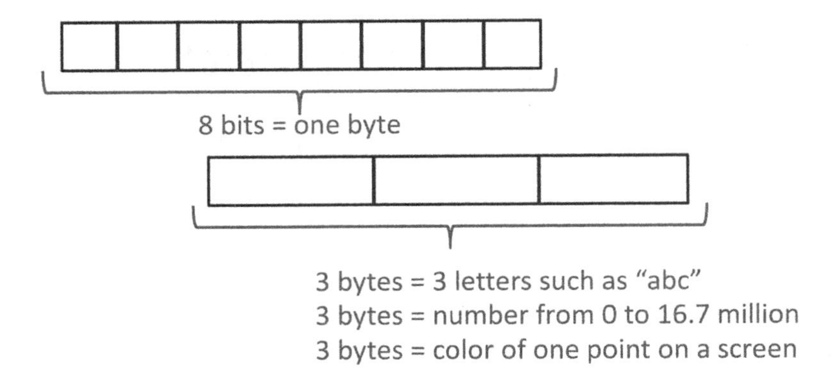

FIGURE 5.1 Bits forming bytes forming a character string and a pixel.

database. Other fields in that database may store the student's major, expected year of graduation, home town, and the other information items that universities keep about their students.

> ***Where you fit in:*** The best way to organize data into fields is not always obvious. Should a person's name be stored as a unit, or as family name and given name* separately? In a shoe database, should size *8C* be stored as one field, or as length (*8*) and width (*C*) separately? This decision should be made by people who know where the data will come from and how it will be used. When that's you, understanding database concepts will help you make good choices.

A data field might not consist of characters. A photograph in a driver's license record is a field. So is a recording of a person's voice or a scan of the vein pattern in a person's retina, both of which can be used for identification. Most software that manages databases can't decode these, but it can retrieve them on behalf of applications that can. Such database fields are often called *blobs*, for **B**inary **L**arge **Ob**jects.

Data fields are grouped into *records.* A record's fields describe a concept or object the database stores information on. That concept or object is an *entity.* For example, an airline database has records for the entity *flight.* Each record has fields for departure and arrival airports, scheduled and actual departure and arrival times, employee IDs of its crew, and more. Each record has this data for one flight, such as Delta flight 32 from Dallas/Fort Worth to Atlanta on October 17, 2019.

When the distinction is important, a specific flight is an *entity instance.* Flights overall are an *entity class.*

> ***Where you fit in:*** Database designers don't know what an organization needs. Is it important to save both scheduled and actual departure times? A database designer doesn't know, but someone at the airline does. If a pilot is replaced, should the database record both the original pilot and the replacement? Every database raises similar questions. Only people who know how the data will be used can answer them. That will be you.

All the records of a specific type form a *table.* Delta's flight table contains information about Flight 32 from Dallas/Fort Worth to Atlanta on October 17, Flight 58 from Boston to London on August 10, and many more. A database table can look like a familiar spreadsheet table.

* It's best to avoid "first name" and "last name," since name order varies from one place to another.

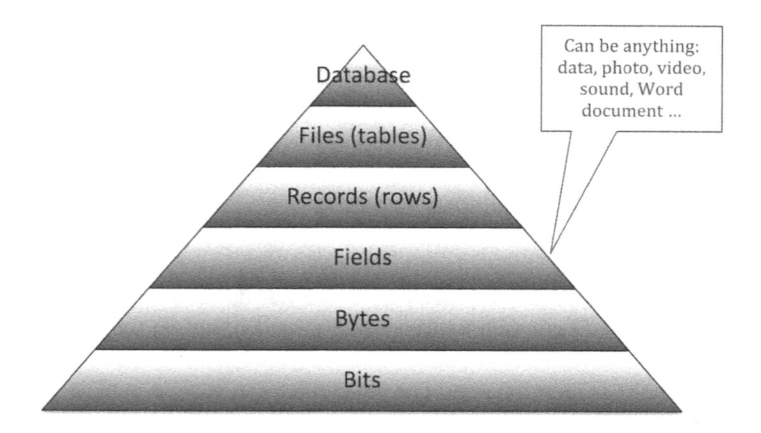

FIGURE 5.2 Database component hierarchy.

A table is sometimes called a *file*. However, the word *file* can also refer to something that the OS treats as a named unit. Your computer probably has files with homework, such as *MIS 301 Project 1.docx*; photographs, such as *Ronnie.jpg*; music, such as *Thriller.mp3*; video, and so on. These files are not part of a database. Using the term *table* is less confusing.

A *database* consists of multiple tables about an area, organized so their data can be used together. If your instructor calls in sick, a database can use a course table to find that instructor's classes on that day. It can then use an enrollment table to find the students in those classes, and finally use a student table to get their cell phone numbers and email addresses. An application can then send texts or emails that class is cancelled. A database can do this because it has tables for instructors, courses, course registrations, and students, and can connect data in one table to data in another.

These concepts are shown in Figure 5.2. There are many ways to realize them in software. The best way depends on how the data will be used. The two main uses are transaction processing (operational) and business intelligence (decision-making) uses. They use data differently:

- Transaction processing applications access individual records and must be able to update them. A system may process thousands of transactions a minute. You'll read more about transaction processing in Chapter 7.
- Decision-making applications access tens, thousands, or millions of records for analysis, but do not update them. Fewer people use these databases concurrently: often just one, at most a few dozen. You'll read more about using computers to help make decisions in Chapter 9.

Because the access patterns are different, different database structures are best for the two areas.

5.2 OPERATIONAL DATABASES

Operational databases are called that because they support business operations. In a bank, an operational database stores customer data. It is updated when a customer withdraws money at an ATM. In a hotel chain, an operational database stores room data. It is updated when a guest checks in. In a supermarket, an operational database stores inventory data. It is updated when a register scans a barcode. Operational databases are transaction processing databases.

Such databases serve multiple purposes. The bank database is also updated when a customer deposits a paycheck, moves to a new address, or changes to a different type of account. These actions affect only that customer. The same is true of hotel and supermarket databases: *business events affect only one record.*

Where you fit in: No matter what job you have after graduation, you will almost certainly use one or more operational databases. The better you understand how they work, the better you will be able to use them effectively and suggest new ways they can help your employer.

Relational Database Concepts

In 2020 nearly all operational databases use the *relational* approach. (A few older ones don't, but they are rare today and are generally replaced by relational databases when they no longer meet business needs.) The relational approach to organizing data is based on these concepts:

- Database tables look, as mentioned earlier, much like spreadsheet tables.
- Each record in the table, called a *row*, stores data about one real-world item. *Example: Each row of a student table stores information about one student.*
- Each field in the table, called a *column*, stores the same data for all the rows. Columns are also called *attributes* of the entity that the table represents. *Example: A column in a student table might store the student's major. "Major" is an attribute of the entity "student."*
- Every row in every table has a unique way to identify it, called a *primary key. Example: Student ID number.* There may be no duplicate keys. (Databases can enforce this rule.)
    - Student name is not a good primary key. Names are not necessarily unique.
    - Phone number is not a good primary key. A student might not have a phone, or might share one with another student. (Likely? No. Possible? Yes.)
- Keys connect records in different tables with each other. *Example:* An athletic team roster, instead of storing team members' majors, stores their student IDs. These IDs identify records in the student table. Those records have the students' majors.

Suppose students' majors were in both the student table and the volleyball team table. Now Laura switches from English to History. The registrar's office will update her registration record. The team data will be wrong, perhaps for a long time.

Storing the same information more than once, such as having Laura's major in two tables, is called *redundancy*. Redundancy leads to errors. Good database design minimizes redundancy.

A NOTE ON LANGUAGE

You may hear that these databases are called *relational* because they relate tables to each other. They do, but that's not where this name comes from. It comes from mathematics and means, in effect, that all the fields in a record relate to one real-world entity. Had Dr. Edgar Codd, who invented relational databases at IBM in the 1970s, foreseen how much confusion this term would cause, he might have picked another—but it's too late to change. We're stuck with that name.

Facebook, for example, is a giant database at heart with a "front end" program that takes data from that database and arranges it on web pages. Think about the information Facebook has. It can be broken down into several categories:

- Your profile, with information such as your real name, your date of birth, your username and password, your privacy settings, your gender and relationship status, and much more.
- Events in your timeline, with times and descriptions.
- Photos, which can be linked to events or to members.

- Friend connections, which link two members in a bi-directional fashion: if member A is a friend of member B, member B is a friend of member A. Both members must agree to this.
- Subscriptions (following), which also link two members but in one direction: member A follows member B. Member B does not have to agree to be followed.
- Likes, which one member sends to a status update, comment, photo, or other content of another member. The other member need not agree to be Liked.

Each of these, and many more, is a database table—or several. These tables are large, since Facebook has over two billion members. If members average ten friends each, which is lower than many estimates, there are 20 billion friending relationships in Facebook's database.

Where you fit in: Twenty billion records is large for a single table. Still, you can see that databases you'll find on the job are much bigger than those you may see in school.

Let's see how we might put together a database for a smaller social networking site.

An Example: MyVid Video Site

MyVid is a (fictional) video sharing service similar to YouTube. Its main table stores videos. Each row of this table has a unique Video ID. It also has the video's title, length, when it was uploaded, by whom, and how many times it's been viewed, for a total of seven columns. This table, along with the others described in the next few paragraphs, is shown in Figure 5.3. (This figure was created with Microsoft Access. The tables would look much the same in any other package, such as FileMaker Pro.)

The information about who uploaded a video is a Member ID. In the Video table, it's called a *foreign key*; it links a row of this table to a row of another table. A Member ID can show up only once in the Member table, since primary keys must be unique. It can appear more than once in the Video table, since a member may upload several videos. This is a *many-to-one relationship.*

The first video, "Birthday," was uploaded by member 1163. Who is that? The Member table tells us that it's Stuart Jones.

MyVid's Member table has information about each member: name, email address, date of joining, and more. Jones joined MyVid on July 16, 2019; he's a Bronze level member who pays monthly, and his email address is *sjones@isp.com*.

Since MyVid assigns member IDs, it can guarantee that every member has one and that there are no duplicates. MyVid can also guarantee that every uploaded video gets a unique Video ID.

The Member table gives each member's class of membership. Free basic membership entitles a member to upload one video a week, up to 5 minutes in length, and to comment on his or her own videos but not on other members'. Three premium membership levels, Bronze, Silver, and Gold, have higher limits and allow comments on any video. Details of each level are in the Level table.

Premium membership is billed monthly or annually (at ten times the monthly rate). A member's choice of billing term is in the Member table. A Bill table stores the date a billing period starts, whether it is for a month or a year (the member's choice may change after a bill is issued), the amount of the bill (rates may change after a bill is issued), and, for a paid bill, when and how it was paid. (Only the last four digits of credit card numbers are kept to reduce risk.) This table has Bill ID as its primary key. Member ID is a foreign key. The Bill and Member tables are in a many-to-one relationship: a member may have many bills, but a bill is for only one member.

Finally, the MyVid database has a table for comments on videos. Besides the text of the comment and a timestamp, this table has two foreign keys: Video ID of the video that the comment is on, and Member ID of the member who posted it. It has a Comment ID as its primary key. It needs this because none of the other fields or their combinations is necessarily unique.

Video

VideoID	VideoTitle	VideoLength	VideoUpload	VideoViews	VideoVideo	VideoMembe
47	Birthday	4:37	7/20/2019	16	[video file]	1163
66	Rose Parade	4:12	1/2/2020	46	[video file]	1163
75	Auto Show	8:23	1/18/2020	207	[video file]	1764
106	Disney World	15:06	1/5/2020	324	[video file]	2620
117	Trains	23:04	1/18/2020	125	[video file]	4705

Member

MemberID	MemberName	MemberEmail	MemberLevel	MemberBillT	MemberJoinDate
1163	Stuart Jones	sjones@isp.com	Basic	Month	7/16/2019
1764	Melissa Ohanian	mel@server.net	Bronze	Month	12/13/2019
2620	Carla Hunter	fashi@gmail.net	Gold	Year	12/28/2019
4705	Daniel Rosenberg	danrose@photos.com	Gold	Month	1/3/2020
7116	Ed Gazbono	nikonfan@hotmail.net	Silver	Year	1/30/2020

Bill

BillID	BillMember	BillStart	BillPeriod	BillWhenPaid	BillHowPaid	BillCCInfo
34	1764	12/13/2019	Month	12/13/2019	Visa	7163
46	1764	1/13/2020	Month	1/11/2020	Visa	7163
87	1764	2/13/2020	Month	2/10/2020	Check	
112	2620	12/28/2019	Year	12/28/2019	AmEx	4163

Comment

CommentID	CommentVideo	CommentMember	CommentReplyTo	CommentText	CommentPosted
2	47	4705		Next time you might want t	1/8/2020
5	66	1764		Very nice!	1/3/2020
24	66	1163	5	Thank you!	1/4/2020
67	66	7116		I like the colors of the flowe	2/12/2020
83	75	4705		Nice angle on the Corvette.	1/21/2020

Level

LevelID	LevelPerMo	LevelPerYr	LevelMaxLen	LevelMaxPer	LevelCanCom
Basic	$0.00	$0.00	5:00	1	☐
Bronze	$3.95	$39.50	10:00	3	☑
Gold	$12.95	$129.50	23:59	25	☑
Silver	$7.95	$79.50	15:00	5	☑

FIGURE 5.3 MyVid tables with sample data.

When a comment is a reply to another comment, the new comment has a third foreign key: the Comment ID of the earlier one. This enables MyVid to show conversations as threads to make them easier to follow.

Look at Figure 5.3 and trace all the foreign keys to their matching primary keys. Be sure you understand what these connections mean.

Now, someone at MyVid has an idea: "Let's let members rate videos." MyVid employees choose a 1–5 scale for ratings. For computing and updating averages, they decide to add five columns to the video table: one column per value, with the number of times the video was given that rating.

Then someone asks "Should we let members rate a video more than once?" After discussing how unlimited ratings could abuse the system, the group decides to impose a one-rating limit.

RatingID	RatingMem	RatingVid
1	1163	106
2	1163	117
3	1764	47
4	1764	117

FIGURE 5.4 MyVid video rating table with sample data.

What does this mean for the database? It must record each time a member rates a video. A rating is a new entity. We need a new table with a row for each time a member rates a video. This table has Member ID as a foreign key, Video ID as a foreign key, and a timestamp for the rating (not necessary, but easy to add). This table doesn't need a separate primary key, since the combination {Member ID, Video ID} is unique. If a new rating comes in that matches both, this member has already rated this video so the database management systems (DBMS) rejects the new rating. The rating table is in many-to-one relationships with the video and member tables.

Should the rating table also store the rating, or should MyVid only accumulate totals in the video table? That is a business decision. A database can do either. Database professionals do not have the business knowledge that this sort of decision calls for.

In the interest of protecting privacy, MyVid decides not to save individual ratings, but only the fact that a member rated a video. The new ratings table is shown in Figure 5.4. It shows that Stuart Jones rated Carla Hunter's Disney video and Daniel Rosenberg's Trains video. It doesn't show what he thought of either one. Be sure you understand how it shows this, and what the other two rows of the table mean.

Should members be allowed to rate their own videos? Again, this is a business decision. MyVid decides that they should not. That rule, too, can be enforced automatically.

> ***Where you fit in:*** Decisions such as whether or not to store individual ratings cannot be made on technical grounds. Businesspeople must make them. Businesspeople who don't participate in database design may not end up with the databases they need. They will have nobody to blame but themselves. Throughout your career, you will have opportunities to make sure new databases meet your needs. It's easiest to change their design when they're still on paper.

Flat Files

A multiple-table database isn't the only possible way to store information. For example, a store could record sales data in a table, one row per sale. Some information would be duplicated, but there is less chance of error in a table with 50 rows, or even 500, than in a table with 50 million. Figure 5.5 shows such a *flat file* for hardware store sales records.

There's a lot of redundancy here. We're told three times that Alvin made a purchase on January 17 and lives in Santa Rosa. We are also told twice that hammers cost $15, that a pair of pliers costs $10, and that paint brushes sell for $5. More records would show more redundancy.

Flat files may suffice for small files. If you want to track your car's maintenance, recording what was done, when, where, and how much it cost, a flat file will do—and you can put it on a spreadsheet. If you need to track fleet maintenance for thousands of trucks, including information about the technicians who did the work, the time it took, and return visits to correct problems, you need more. Flat files are rare in business and are never used for important systems. Databases are near-universal. With a database, *normalization*, the subject of the next section, is important.

Date	Cust.	Address	Item	Quant.	Each	Tot. Sale	Unit Cost	Tot. Cost	Tot. Profit
Jan. 17	Alvin	Santa Rosa	Hammer	1	$15.00	$15.00	$9.00	$9.00	$6.00
Jan. 17	Alvin	Santa Rosa	Pliers	1	$10.00	$10.00	$7.00	$7.00	$3.00
Jan. 17	Alvin	Santa Rosa	Saw	2	$18.00	$36.00	$10.00	$20.00	$16.00
Jan. 18	Brenda	Napa	Nails	3	$3.00	$9.00	$1.50	$4.50	$4.50
Jan. 18	Brenda	Napa	Pliers	1	$10.00	$10.00	$7.00	$7.00	$3.00
Jan. 18	Brenda	Napa	Plane	1	$35.00	$35.00	$20.00	$20.00	$15.00
Jan. 18	Alvin	Santa Rosa	Light bulbs	2	$6.00	$12.00	$4.00	$8.00	$4.00
Jan. 18	Alvin	Santa Rosa	Sandpaper	3	$2.00	$6.00	$1.20	$3.60	$2.40
Jan. 18	Alvin	Santa Rosa	Paint	1	$21.00	$21.00	$13.00	$13.00	$8.00
Jan. 19	Carl	Novato	Scraper	2	$5.00	$10.00	$3.00	$6.00	$4.00
Jan. 19	Carl	Novato	Brush	5	$5.00	$25.00	$3.00	$15.00	$10.00
Jan. 19	Carl	Novato	Hammer	1	$15.00	$15.00	$9.00	$9.00	$6.00
Jan. 19	Brenda	Napa	Paint	2	$21.00	$42.00	$13.00	$26.00	$16.00
Jan. 19	Brenda	Napa	Hammer	1	$15.00	$15.00	$9.00	$9.00	$6.00
Jan. 19	Brenda	Napa	Light bulbs	3	$6.00	$18.00	$4.00	$12.00	$6.00
Jan. 20	Denise	Vallejo	Ladder	1	$75.00	$75.00	$40.00	$40.00	$35.00
Jan. 20	Denise	Vallejo	Brush	3	$5.00	$15.00	$3.00	$9.00	$6.00
Jan. 20	Denise	Vallejo	Roller	6	$6.00	$36.00	$3.50	$21.00	$15.00

FIGURE 5.5 Flat file sales example.

Normalization

Designing the MyVid database involves decisions that weren't mentioned above. For example, why not put each member's email address in the video table for each video a member uploads? It might be useful for contacting that member if there's a problem with the video. That approach, which has similarities to using a flat file, would result in a table like the one in Figure 5.6.

We don't, for the same reason we didn't store Laura's major in the volleyball team table earlier. Suppose Stuart Jones gets a new email address. If email addresses are only in one place, only that place must be updated. If it's in many places, all must be updated. Here, it would be in the member table plus twice in the video table. Updating all three takes longer, and is more likely to lead to errors, than one update. Therefore, we should only put the email address in the member table. When we need it, we can use the member ID from the video table to find it in the member table.

We should define entities at the lowest possible level, then push information up to higher levels. Each data item is stored only once. This is called *normalization*. There are rules for normalizing a database. A properly normalized database is more likely to function correctly than one that is not.

Another example of normalization: Suppose your school allows students to have multiple majors. Now the attribute *major* in a student database can have multiple values. Multiple values lead to difficulty in using a database. Instead, we add a new table that connects students to majors. Figure 5.7 shows this, with Kayla Coleman majoring in both French and History. (Go through the tables to be sure you understand how this information is stored.) The Majors table translates major codes to text descriptions. That allows us to change the title of the major, updating all applicable students' records automatically, and to store additional information about the major if we wish.

Video

VideoID	VideoTitle	VideoLength	VideoUpload	VideoViews	VideoVideo	VideoMembe	VidMemEmail
47	Birthday	4:37	7/20/2019	16	[video file]	1163	sjones@isp.com
66	Rose Parade	4:12	1/2/2020	46	[video file]	1163	sjones@isp.com
75	Auto Show	8:23	1/18/2020	207	[video file]	1764	mel@server.net
106	Disney World	15:06	1/5/2020	324	[video file]	2620	fashi@gmail.net
117	Trains	23:04	1/18/2020	125	[video file]	4705	danrose@photos.com

FIGURE 5.6 Unnormalized MyVid table example.

Students

StudentID	StudentName
2367	Kayla Coleman
3876	Henry James
4598	Ashok Puri
8432	Suzanne Wu

Majors

MajorID	MajorName
1	English
2	History
3	Psychology
4	French
5	Physics

StuMajor

SM_ID	SM_Student	SM_Major
1	2367	2
2	2367	4
3	3876	1
4	8432	3
5	4598	5

FIGURE 5.7 Student tables for multiple majors.

Does adding this table complicate the database? Yes. Does it slow it down for typical students, who only have one major—here James, Puri, and Wu? Yes. Is it worth it, in terms of reducing errors? Almost always, also yes.

You'll learn more about normalization if you go on to study database management. For now, it's enough to understand that it means giving a database a clean, correct structure.

Where you fit in: Normalization can slow down a database. Adequate performance then calls for faster computers. Database designers *denormalize* tables when the risk of errors is small, database performance is important, and potential speed gains are large. A decision to bypass normalization requires knowing how parts of the database will be used and what harm an error can do. (Giving the wrong major for an athlete can't do much damage. Sending a patient to the wrong operating room can.) Only database users know this. They must be involved in denormalization decisions.

Entity-Relationship Diagramming

In Figures 5.3 and 5.4, we illustrated the MyVid database by showing its tables with sample data. Database designers use *entity-relationship diagrams* (ERDs) to convey this information.

ERDs use symbols to show the entities a database stores information about, the attributes it stores for each of these entities, and the relationships among them.

There are several variations on the ERD theme. They differ much as English differs between New York and Sydney. Someone who speaks one will understand the other; it just sounds a bit strange. The ERD dialect you use on the job may differ from the one here, but you should understand it.

The ERD in Figure 5.8 shows the MyVid database as a database designer might sketch it. (Even if a designer can easily create a neat ERD on a computer, a sketch is often better. People are more willing to suggest changes to a sketch than to a computer-printed diagram, because the sketch looks less finished. Microsoft recognized this in 2019 when it added "sketchy" lines for shape outlines

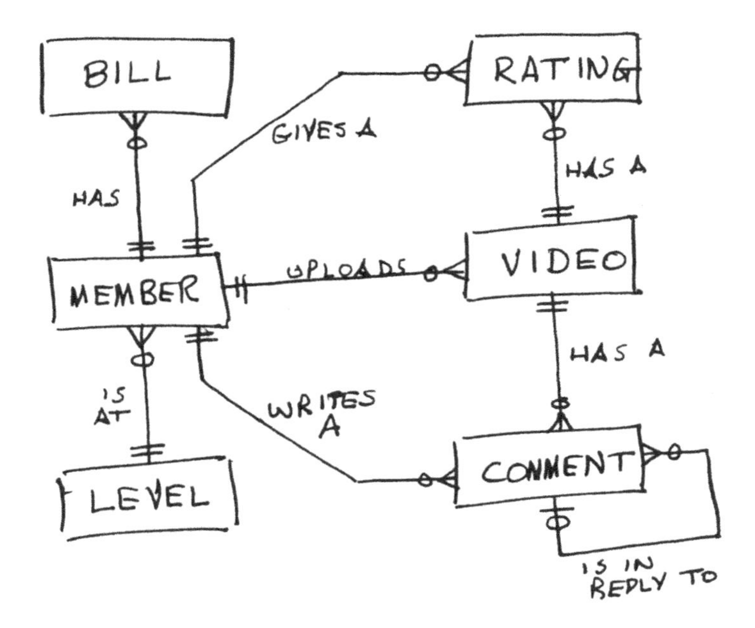

FIGURE 5.8 Sketch of MyVid database ERD.

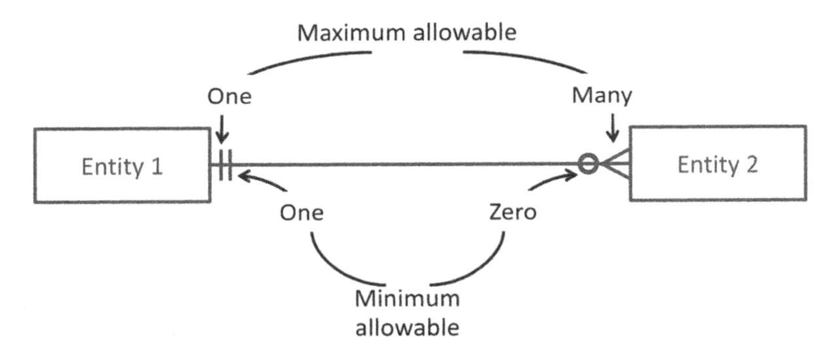

FIGURE 5.9 ERD symbols.

in its Office suite. Figure 12.10 in this book applies sketchy lines to a rectangle.) The ERD shows these ERD features:

- Database entities are shown as rectangles. Each entity corresponds to a table. There are six rectangles for the six tables: Videos, Members, Comments, Billing, Level, and Ratings.
- Lines show relationships among entities. The line between members and comments means that there is a relationship between them.
- The relationship is described in words next to the connecting line. Here, each comment is by a member. The relationship between comments and comments means that a comment may be in response to another comment. (Some versions of ERD notation don't have these words.)
- There are two symbols at each end of a relationship line, as shown in Figure 5.9:
  - The symbol nearer the entity gives the *maximum cardinality* of the relationship: Can there be one or many of it in the relationship? The members-comments relationship has a crow's foot at the comment end, to show that one member may make multiple comments. (This means the database allows them to, not that they must.) The line at the member end shows that each comment is by no more than one member.

 - The further symbol from the entity gives the *minimum cardinality* of the relationship: How many of that entity there must be as a minimum. If the minimum is one, there must always be an entity of that type in this relationship; if the minimum is zero, having one is optional. Using the same relationship as an example, the line across the relationship line at its member end means that there must be a member associated with each comment; there are no anonymous comments. The circle at the comment end means that a member may not have made any comments; comments are optional. (You can think of the circle as being a zero for the minimum cardinality, or the letter "O" for "optional.")

Putting the symbols together leads us to the three types of relationships shown in Figure 5.10:

- *Many-to-one* relationships are the most common. All MyVid relationships are of this type.
- *Many-to-many* relationships are also common. Students and courses are an example. Students can take more than one course; courses can have more than one student.

  Databases build many-to-many relationships from two many-to-one relationships. The student-course relationship has an entity in the middle, as shown in Figure 5.11. The entity in the middle stores that student's registration information, including his or her grade. The grade can't go in the student record, since a student can have several courses with different grades. It can't go in the course record, since a course can have several students with different grades. It needs its own place.

  Sometimes there is no data in the middle of a many-to-many relationship. A database still needs a table there to link the tables on the two sides. That table is called an *associative entity*. As a businessperson, you can trust professional database designers to know when to use associative entities. They'll be impressed that you've even heard of them.
- A *one-to-one* relationship might be the relationship between students and parking permits, if students can have only one parking permit each.

  One-to-one relationships are rare, since it is usually possible to combine both entities. They are usually used when both entities exist independently of the relationship. For example, consider restaurant pagers that tell waiting guests when their table is ready. A database might have a record for each pager with its ID number, the date it was purchased, from whom, for how much, and so on. When the pager is in use it is in a 1:1 relationship with a waiting guest. At another time it might be in a 1:1 relationship with a different guest or no guest at all.

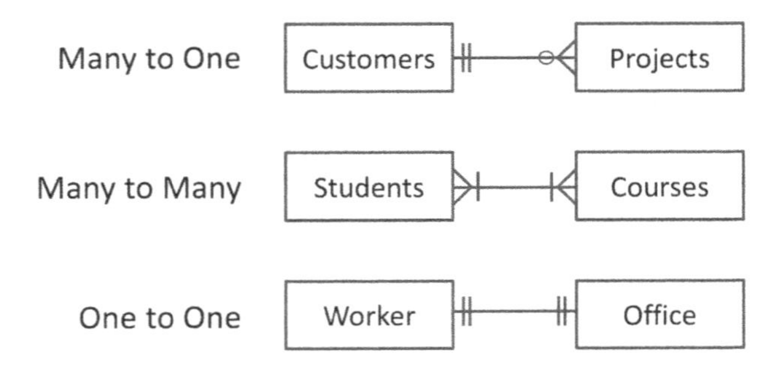

FIGURE 5.10 Three relationship types as shown by ERDs.

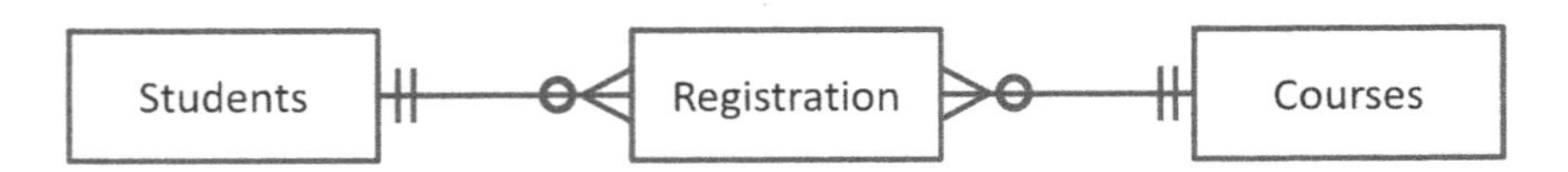

FIGURE 5.11 Student-grade-course relationships.

At a university, a student's parking permit number and vehicle information could be in the student table. Those fields would be blank for students who don't park on campus. If parking permits were purchased for a year and could be transferred from one student to another, it might be best to have them in their own table. They would then be in a 1:1 relationship with the student to whom they are assigned. That can be changed if the permit is transferred.

In addition, a finished ERD shows two more information items about the database:

- The attributes of each entity are listed in its rectangle. If there isn't enough room, attributes can be listed elsewhere on the diagram or on a separate page.
- The primary key of each entity is underlined. If an entity has a composite primary key (a combination of fields that might not be unique individually), all its components are underlined. Foreign keys are starred. (Some ERD notations identify keys differently.)

Businesspeople review ERDs to make sure they reflect business needs. For example, should comments on videos be by one member only? That's fine, but suppose an instructor uses the same approach to record students visiting her office and the topics they discussed. Is it reasonable for each office visit to be by only one student? What if students in a project group come together? A many-to-many relationship between students and office visits might be better.

Where you fit in: You'll use ERDs in two ways at work:

1. *Understanding the structure of a database is a good basis for developing it.* In developing personal databases for your own use, using a package such as Microsoft Access or FileMaker Pro, you should always start with a design. If you can't sketch an ERD for your database, you don't have a clear idea of it. If you can sketch one, you've thought it out. It still might not be what you need, but you've improved your odds.
2. *Database designers use ERDs to communicate when they design databases.* When they plan a database for your use, you'll have a chance to review their design. They'll use ERDs to show you what they came up with. They'll explain their ERDs, but the more you bring to that conversation, the more productive it will be—and the better the database you'll get.

Many database management packages can produce ERDs or similar diagrams for a database. Figure 5.12 shows MyVid's Relationship Map in Access. It uses the infinity symbol to indicate the "many" end of a relationship, and does not show minimum cardinality, but the concept of entities linked by lines is similar. The lines enter the boxes at the primary and foreign keys.

Figure 5.13 shows the diagram in FileMaker Pro. It uses the conventional crow's feet symbol for "many." The arrangement of the boxes is different, since each system arranges connecting lines differently. Both show two copies of the Comments table for the relationship from that table to itself, used when a comment replies to an earlier one. It only exists once in the database, though.

Distributed Databases

Databases aren't always stored on a single computer. A *distributed database* stores data in several places, while allowing users to access the data as a unit. This offers these advantages:

- Potentially higher performance, by spreading the database workload over several computers.

- Potentially higher reliability, since copies of the database may still be usable if some become unavailable due to hardware or network failures.
- A stronger sense of management control over local data. This is especially important when, prior to the use of a shared database, managers had full control of their unit's files.

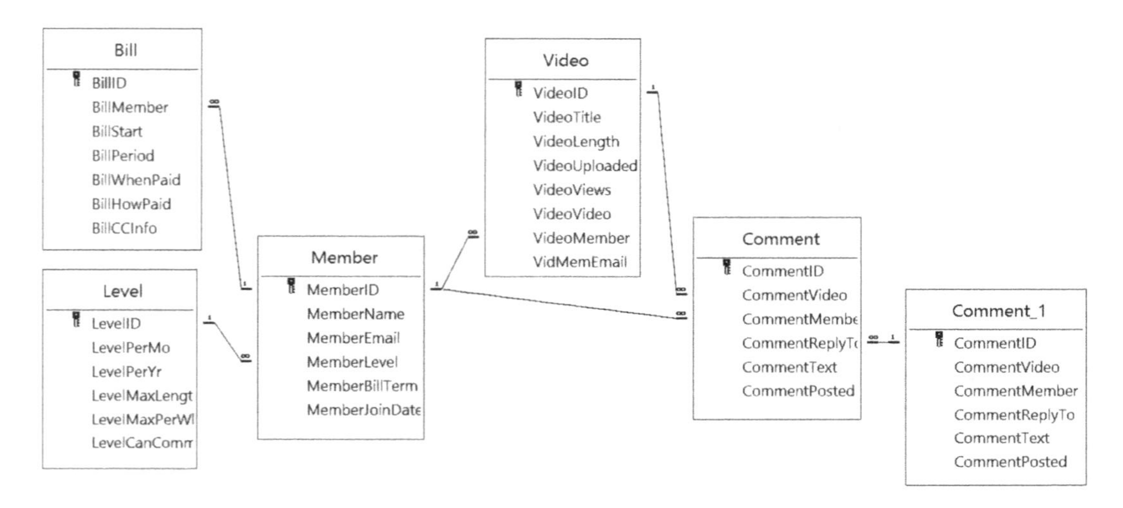

FIGURE 5.12 MyVid relationship graph in Access.

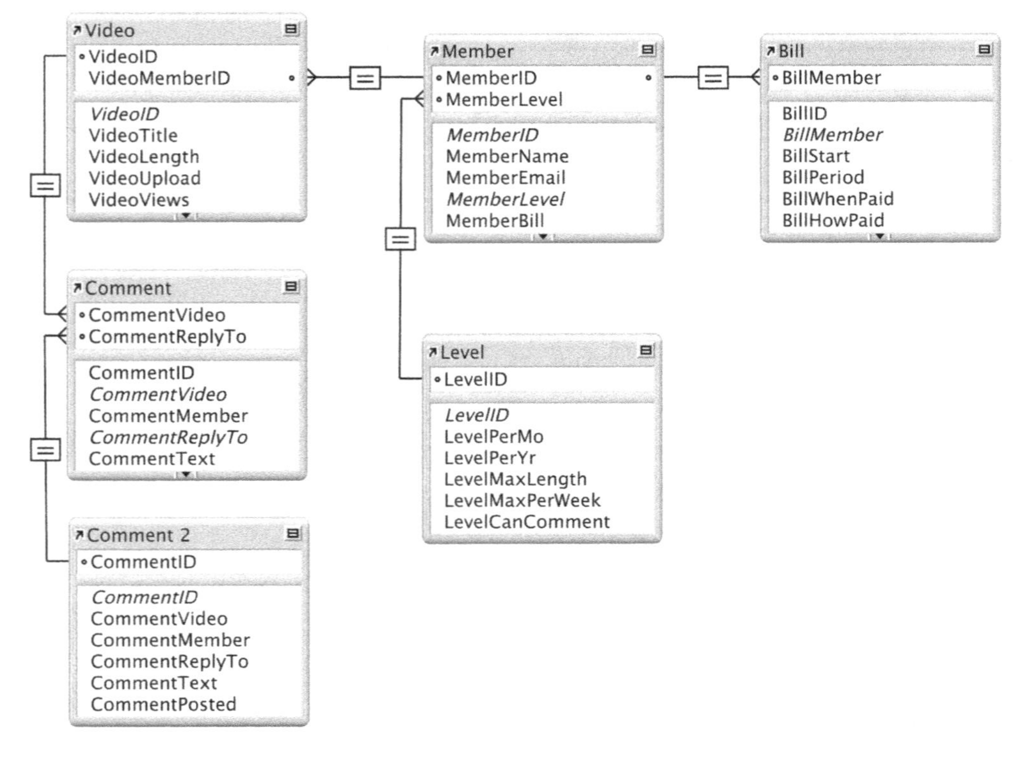

FIGURE 5.13 MyVid relationship graph in FileMaker Pro.

A distributed database can operate in several ways:

- A *partitioned* or *fragmented* database puts parts of the database in different partitions. The database is the sum of all those parts. There are two approaches to fragmentation:
  - With *horizontal fragmentation*, the table is split by rows. A company can store personnel data at each region's headquarters for employees who work in that region. (See Figure 5.14.) The whole database includes all employees from all regions.
  - With *vertical fragmentation*, the table is split by columns. A bank can keep checking account data in one place, mortgage data in another. (See Figure 5.15, where the table at the top is fragmented vertically into the two tables below it. Customer 1148 doesn't have a mortgage.) All fragments of a vertically fragmented row must include its primary key. The whole database includes all the columns in each customer record.
- A *replicated* database keeps a complete database in more than one location. All are updated when one is changed.
  - With *synchronous replication*, copies are updated immediately. This requires complex software and a great deal of computing power. An organization must decide what to do if some copies can't be updated right away for any reason, such as failure of the computers on which those copies reside or the network connections to them.
  - With *asynchronous replication*, copies may be updated later. If there is a master database, all the copies will eventually match it. If not, there must be a way to handle conflicting changes. If two people edit a paragraph at the same time in a document that can be edited by several people, the DBMS must figure out how to handle the

Emp. No.	Name	Job Title	Date of Birth	Date of Hire	Location	
471	Alice	President	12/17/1957	4/6/1981	Chicago	Stored in Chicago
675	Bob	Sales rep	7/13/1978	5/15/2006	Chicago	
378	Cynthia	HR assistant	10/15/1989	6/13/2011	Chicago	
993	David	Sales rep	1/26/1966	7/28/2008	Elgin	Stored in Elgin
234	Emily	Accountant	5/22/1991	8/4/2016	Elgin	
766	Frank	Custodian	9/23/1975	3/4/2013	Elgin	
116	Gina	Programmer	4/24/1985	2/23/2009	Rockford	Stored in Rockford
347	Henry	Trainer	8/6/1977	11/24/2003	Rockford	
985	Isabelle	Librarian	6/8/1982	12/11/2018	Rockford	

FIGURE 5.14 Example of horizontally fragmented database.

Cust. No	Ck Acct #	Balance	Last Activity	Mtgg Acct #	Balance	Rate	Last Activity
1763	100-1346	$157.12	2/4/2019	300-1966	$245,963.45	4.500%	2/3/2019
1983	100-1278	$3,145.24	1/29/2019	300-4229	$177,442.00	4.625%	1/30/2019
1148	100-1457	$786.43	2/6/2019				
1009	100-2084	$1,198.65	2/8/2019	300-2003	$533,983.26	3.875%	1/29/2019

Cust. No	Ck Acct #	Balance	Last Activity
1763	100-1346	$157.12	2/4/2019
1983	100-1278	$3,145.24	1/29/2019
1148	100-1457	$786.43	2/6/2019
1009	100-2084	$1,198.65	2/8/2019

Cust. No	Mtgg Acct	Balance	Rate	Last Activity
1763	300-1966	$245,963.45	4.500%	2/3/2019
1983	300-4229	$177,442.00	4.625%	1/30/2019
1009	300-2003	$533,983.26	3.875%	1/29/2019

FIGURE 5.15 Example of vertically fragmented database.

update. This may require human intervention, so this approach is not well suited to critical corporate data.

- A *federated* database consists of multiple databases, each accessible independently of the others, that provide controlled access to all their data as a whole. The component databases of a federated database may differ in their structure or in other ways. Federated databases are most common when multiple databases were combined, such as in a corporate merger.

Where you fit in: Decisions about database distribution must be based on how the database will be used. This is a business question, not a technical question.

5.3 DATABASES FOR DECISION MAKING

As you read earlier, data usage for decision making differs from data usage for daily operations. When a bank computer authorizes cash withdrawal at an ATM, it wants one customer's record. It will update that record with a lower balance and a newer "most recent access" date. When banks design new financial products, they ask broader questions: What is our customers' distribution of checks per month? How does it vary with balance, with whether or not a customer has a car loan with us, with customer ZIP (postal) code, or with how recent the most recent address change was?

Decision-making database access differs from transaction processing access in several ways:

1. Each access retrieves many records, not just one. It is difficult to optimize a database for both single-record and multiple-record access.
2. Analysis does not change the data.
3. Transaction processing requires up-to-the-second data. Decision making often does not. Up to an hour ago, or even up to last week, may be perfectly adequate.
4. Transaction data should often be organized by application for speed. Decision-making data should be organized by subject to provide users with an integrated view.

Decision-making application users do not access records by primary key. They use other criteria, such as "supermarkets in Ohio." They don't know these criteria in advance, since each analysis is different. Therefore, they need database structures that lend themselves to *ad hoc* retrieval using data-related selection criteria. Such databases may not use the row-column structure of relational databases, though it can be convenient to think of the data they contain in those terms.

A relational database can answer these questions. It will be slower than a database designed from the beginning for this type of information retrieval. This disadvantage can be overcome by buying enough fast hardware. That's costly, but setting up two databases for the same information and keeping them synchronized isn't cheap either.

Another drawback of relational databases in decision support applications is that a long query can tie them up for a considerable period of time, locking out transaction processing users until it is complete. The reason for locking them out is that changes to a database during a query can cause errors. Suppose one is going through an employee table to create an organization chart. Each row of the employee table gives the employee ID of that person's supervisor, as in Figure 5.16. Harry's supervisor is Hattie, and so on. (The figure shows only the first eight rows.)

To create the organization chart, the database management system must find each employee's supervisor. It sees that Harry's supervisor is employee 6. To get her name, it goes back to the table and looks for employee 6. What happens if, between the first step and the second, Hattie's row is removed from the table? Unlikely, but possible. Over time, unlikely things happen. To prevent such situations, the database must be locked until the query is complete. Since the database doesn't know what changes might lead to problems, it must block them all.

EmpID	Name	Title	SupervisorID
1	Alice	Sales Rep	3
2	Bobby	Sales Rep	3
3	Christine	Sales Manager	9
4	David	Programmer	10
5	Harry	A/R Clerk	6
6	Hattie	Accounting Manager	11
7	Havelock	Engineer	12
8	Howard	HR Director	11

FIGURE 5.16 Employee table indicating supervisor.

Where you fit in: The choice to use an existing database for decision support or create a new one is a business decision. Technical people can give you speeds and response times, the cost of extra hardware to speed up a relational database for decision support, and the cost of creating an analysis database. Businesspeople must take this information and interpret it in light of business factors such as how many people will use the system and how frequent different uses will be.

Dimensional Databases

Each transaction in a historical database can be described along several dimensions. Consider men's shirts. Every time Macy's sells a man's dress shirt, it creates a transaction that can be broken down according to the dimensions in Figure 5.17, plus more you can think of.

Sales data can be represented as a 20-dimensional cube. (We can't visualize more than three dimensions, but computers can.) Putting each sale into a cell leads to a database that can be "sliced and diced" along any mix of dimensions. If we want to analyze color preference trends, how method of payment varies with time of day, or if a shirt's color is related to the discount at which it was sold, we can find out by slicing this *data cube.*

Consider a pet store with four salespeople: Adams, Baker, Carstens, and Dorie. It sells cats, dogs, fish, and gerbils. It stores how many pets of each type each salesperson sold each week. Figure 5.18 shows part of its data cube. The face of the cube shows sales data for the week of April 27.

This data cube lets the pet store owner see the fraction of pets of each type that each salesperson sells. Carstens sells more gerbils than anyone else. Is that because he's an expert on small rodents and the others send him those customers? Or, is it because gerbils are easy to sell and he's lazy? The database can't answer that question, but it can lead the store owner to ask it.

Dimensional databases are well suited to *data warehouses.* A data warehouse is a historical database for decision making. Data warehouses take data from sources such as operational databases, standardize their data representation, and put it into a useful form for decision making. You'll learn about data warehouses in Chapter 9.

Neck size	Color	Date of sale	Original price
Sleeve length	Pattern	Time of day	Selling price
Trim fit?	2nd color, if any	Store no.	Wholesale cost
Collar style	Fabric	Salesperson no.	Coupon used?
Cuff style	Supplier	Meth. of payment	Rewards card?

FIGURE 5.17 List of sales dimensions for men's shirts.

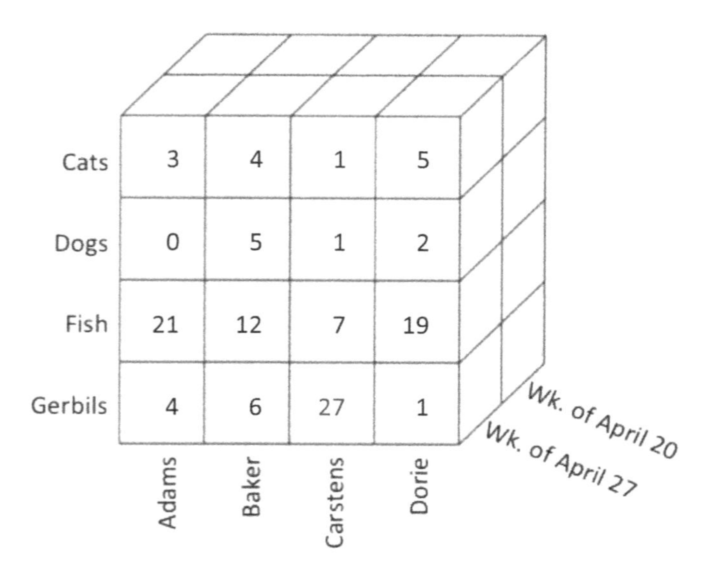

FIGURE 5.18 Dimensional database for pet store.

"Big Data"

IBM estimates that 90% of the data now stored in computers worldwide did not exist two years ago. Collectively, we generate new data at the rate of 2.5 quintillion bytes per day—enough to fill the disk drives of 20,000 new personal computers every week. Much of this data can be of value, but traditional approaches to organizing and analyzing data don't work with *big data*.

"Big data" has three characteristics. Only one of them is about size:

- ***Volume:*** Big data means a lot of data. Every day sees 12 terabytes of tweets, a petabyte every three months. How can any firm plow through this data to find out what people say about it, and compare that with what they say about its competitors? Walmart records a billion customer transactions in under two months. How can it possibly find useful nuggets there?

If size were the only issue existing methods could be extended to deal with it, but big data is also characterized by:

- ***Velocity:*** Data must be analyzed quickly. Credit card thieves use them within minutes, so fraud detection methods that take an hour or two are useless. If one investor detects a price mismatch between two securities in two seconds and another takes ten seconds to see it, the first will make money and the second will be left out.
- ***Variety:*** Big data includes structured data, unstructured data, audio, video, and more. The desired picture often emerges only when several data types are considered together.

These characteristics (which are sometimes given as Scale, Speed, and Scope) make using big data conceptually different from using traditional data.

Big data raises two challenges: storage and analysis. We'll focus on storage here. You'll read about analysis in Chapter 9.

Hadoop

Elephants, they say, never forget. When Doug Cutting needed to name the new file system he co-invented, he named it after his son's stuffed toy elephant. The name is apt, since Hadoop can handle far larger amounts of data than earlier approaches to data storage.

Hadoop, to quote the Apache foundation that is responsible for it, is "a framework for running applications on large clusters built of commodity hardware." In plain English, it lets an organization run an application on lots of ordinary computers.

With big data, the challenge is to get through all of it in a hurry. Putting it on many computers helps that happen. The second part of what makes Hadoop work is that, instead of giving each computer the entire database, it gives each computer a part of it. Each computer looks only at its part of the data, analyzes that part, and returns the result. With enough computers, the parts are small, and each computer finishes its task in a short time.

The third and final part of the Hadoop picture is an approach to computation called MapReduce. The idea behind MapReduce is that each computer first analyzes its part of the data. That's the Map step. Then, the Map results are combined in a way that eliminates duplication to produce the same result that a single computer would have reached. That's the Reduce part. This often allows analyses to be done in seconds rather than hours or minutes rather than weeks.

The Hadoop/MapReduce approach does not impose structure on data. Different parts of a data collection can have different structures. As long as the data a computer gets has a structure it can work with, it doesn't have to be the same as the structure of the parts that go to other computers.

The business implications of this are huge. Big data does not all have one structure. Tweets have one structure, spreadsheets another, customer purchase data a third. With Hadoop, one computer can run a program to analyze tweets while a second analyzes spreadsheets and a third handles purchase histories. As long as the three programs produce results that can be reduced to the same thing, all three data sets can be analyzed as a group. For example, a market planner can see if people tend to tweet about a product before or after they buy it.

Consistently with this, researchers at Aberdeen Group found in 2016 that 55% of companies that wanted to make use of unstructured or semistructured data were using Hadoop or planned to. Other business considerations, such as a desire for improved operational efficiency or supporting innovation, were not as strongly associated with Hadoop use.

In theory an organization could develop software to do this on its own, but few if any have the resources. Hadoop is supported by the open-source community. Because it is widely used, its users benefit from a support infrastructure of software, training, and expertise.

Where you fit in: Businesses are finding more and more value in big data as new technologies enable them to use it. Staying aware of opportunities means looking at all the data in the world around a business, not just data that people have traditionally associated with computers.

5.4 DATABASE MANAGEMENT SOFTWARE

By now you see that databases are complicated. Accessing data in a database doesn't have to be, though. Software packages can manage databases so you don't have to deal with the details. Such packages are called *database management systems.*

DBMS are a type of *system software.* They are above the operating system, below users and their applications. Applications access databases through a DBMS, as shown in Figure 5.19. It's based on Figure 4.3, which shows where software fits in the overall scheme of a system. The DBMS accesses data via the OS, which coordinates access to storage hardware as a shared resource.

Databases include *metadata* as well as data. Metadata means "data about data." DBMS access data via metadata. Metadata includes the name of a data element, so users and applications can refer to it; describes its format, such as a photo or a decimal number; and tells the DBMS where it's stored. Metadata may also include *access control* information: who may access a data item and what each such person may do with it.

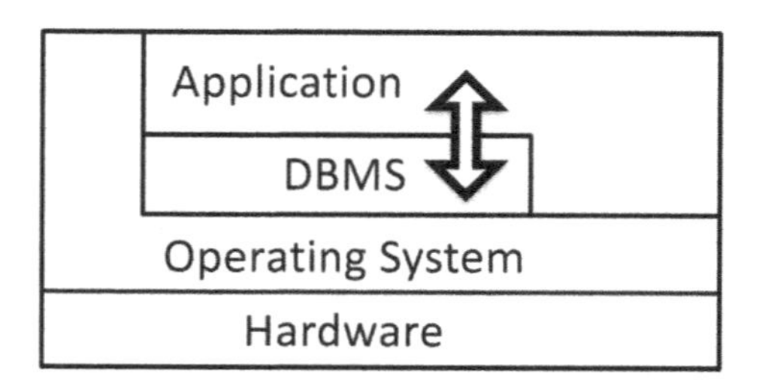

FIGURE 5.19 Application access to data via a DBMS.

An application that needs data sends a message to the DBMS saying what it requires, often using *Structured Query Language* (SQL, sometimes pronounced "sequel"). A SQL request for names and email addresses of Finance majors with cumulative grade point averages of at least 3.0, with the data in a table named STUDENT, could be written:

```
SELECT NAME, EMAIL FROM STUDENT WHERE MAJOR='FINANCE' AND GPA >=3
```

The DBMS sends back the (name, email) pairs that meet these criteria.

SQL can match keys to find information that is stored across more than one table. To get names and email addresses of MIS301 students, where the REGISTRATION table has student IDs but not their names, one could write:

```
SELECT NAME, EMAIL FROM STUDENT
WHERE REGISTRATION.COURSE='MIS301' AND STUDENT.ID=REGISTRATION.ID
```

Simple SQL as in these examples is not difficult to learn, but mastering it takes time and effort.

Figure 5.20 shows a user accessing a database. Most users access databases via applications, as you do when you check your grades. The application creates SQL messages and sends them to the database to retrieve your name, the list of your courses, and your grade in each. To you, the application and DBMS are one information system. You don't need to know SQL to use it.

Database management systems also provide tools by which a *database administrator* (DBA) can create a database, modify its structure, manage its security settings, and optimize its performance. Large organizations have full-time DBAs to ensure that their databases will provide high-quality information as needed.

DBMS that perform these tasks span a wide spectrum. Some are used by professionals. They access data from huge databases, with hundreds of tables having millions of rows each, used by thousands of people at a time. Such packages underlie airline reservation systems, insurance databases, and national tax records. The largest suppliers of proprietary DBMS are Oracle, IBM, Microsoft,

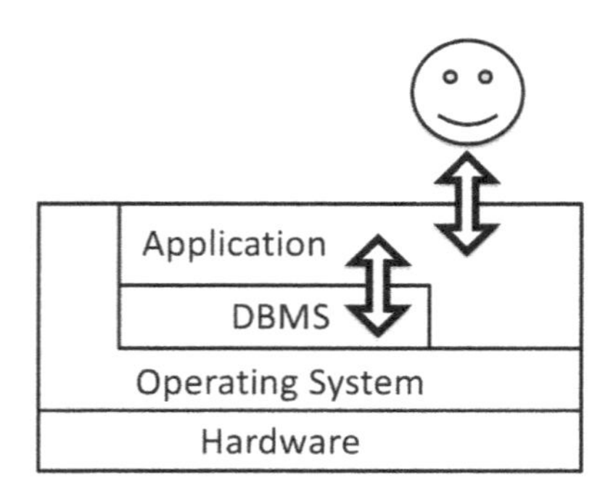

FIGURE 5.20 User accessing a database via a DBMS.

SAP, and Teradata. Several free packages are also widely used. Only an experienced professional can select the right DBMS at this level. This expert needs input from users (you).

Other DBMS support one user on personal hardware. They make it easy to set up and use a database. Personal DBMS let users enter data and query the database via simple forms. Their application development languages are easier to learn and use than languages designed for IS professionals. The leading packages in this market are Microsoft Access, part of Office for Windows; FileMaker Pro, which runs under both Windows and Mac OS; and Base, part of the open-source LibreOffice suite, which supports Windows, Mac OS, and Linux.

DBMS vendors offer query packages, report generation software, and data entry form creators as adjuncts to database management. These may be sold separately from multi-user databases, but are often packaged with single-user DBMS. That blurs the line between a DBMS and applications that use it. Most personal database users think of these applications as part of the DBMS, but they are separate internally. The structure of Figure 5.20 applies here also.

Where you fit in: You'll use databases at work, so you'll use database management systems. You'll work with database administrators to make sure you can get the data you need. The better you understand how DBMS work, the better you'll be able to take advantage of them to get your job done, and the better you'll be able to work with DBAs.

5.5 DATABASE SECURITY

Database management systems must protect databases from unauthorized access or modification.

(Database security is an aspect of information system security. Another is making sure that a system and its data will remain available for authorized uses. The general aspects of information system security are covered in Chapter 12.)

Unauthorized access to data can violate both an organization's privacy policies and applicable laws. U.S. laws, such as the Health Insurance Portability and Accountability Act (HIPAA) for healthcare providers and the Family Educational Rights and Privacy Act (FERPA) for schools, apply primarily to organizations that handle sensitive information. European Union protections are stronger: Its Data Protection Directive requires all organizations to protect information that can be associated with an identifiable individual. A company can't just move data from the EU to a more lenient area, either. *Trans-border data flow* regulations close that loophole.

Access to a database requires (a) that a person be *authorized* to use that database, and (b) that the information system *authenticate* that the person requesting access is who he or she claims to be.

Authorization is an administrative process. Personnel actions such as hiring or promotion trigger new authorizations or changes. Other changes can be initiated by an authorized manager. DBMS keep track of authorized users and their authorizations.

Once the database recognizes that a person has requested access as an authorized user, it must determine that the request is in fact from that person. It can use several methods:

- **Something you know.** The most common is a password or passphrase.

  Passwords can be stolen with no trace. Simple ones can be guessed: the most common* in 2018 were *123456, password, 123456789,* followed by a few more easy-to-type sequences, then *sunshine*, *querty*, and *iloveyou*. Users who must use complex passwords often write them down where others can see them. Since most breached passwords aren't guessed, but stolen, and a complex password can be stolen as easily as a simple one, the

* In English. Reportedly, the most common password—or at least one of the most common—in the world is *ji32k7au4a83*, Chinese for "my password" with its computer representation shown in the Latin alphabet.

value of hard-to-guess passwords is limited. (Microsoft's Alex Weinert says complex passwords are pointless as long as one avoids the "top 50" list.) The best personal protection is using different passwords for different sites, so stealing a password from one site won't help the thief penetrate others. *Password management software* can help keep track of multiple passwords.

- **Something you have.** This can be a physical device called a *dongle*, resembling a flash drive, that plugs into a computer's USB port. It can also be a clock-driven password generator that calculates a password for the time you log in. With such a device, a password copied at 2:10 will be useless at 2:15.
- **Something you are.** Your retina, iris and fingerprint patterns, and cardiac rhythm are unique. Many smartphones and some computers can recognize your fingerprint. These methods need a backup procedure, in case there's a cut or bandage on the finger it recognizes.
- **Something you do.** This includes voice print recognition, signature pressure patterns, and more. Again, backup procedures are needed to deal with a sore throat or a sprained wrist.

For any method, authorization information is entered into the user list. This list is encrypted to prevent people who obtain a copy from reading it.

Authorization can be limited further in several ways. The first three are shown in Figure 5.21:

1. **Column access control,** the right to access certain data elements. A manufacturing company might give quality engineers access to most supplier information, but not to their accounts payable. Advisors can see academic records, but not financial aid or medical information.
2. **Row access control.** You can read your own grades, but not your roommate's.
3. Type of access:
   3.1 Most users of any database have *read access*. You and your advisor can read your grades in your school's database, but can't change them.
   3.2 Some users can *create* new database records. The admissions office creates student records, populating them with information it collects from admission applications.
   3.3 Other users can only *modify* existing records. Your instructors can modify your registration records by entering or changing grades, but can't create or delete them.
4. A few users, typically database administrators, can **change the structure** of the database.

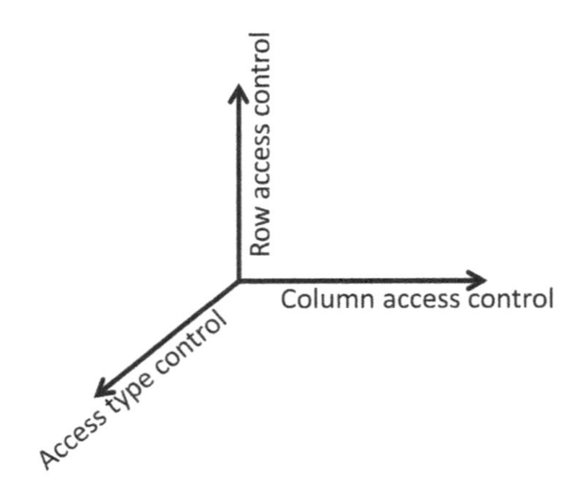

FIGURE 5.21 Three-way access control.

Access permissions can be determined in one of three ways:

- **Role-based access.** Your role in your school's database is **student**. All students have the same access rights. Faculty members have different rights. So do staff in the financial aid office, admissions office, and so on. Department chairs may have broader rights than other faculty members. The registrar may have broader access than clerks in the registrar's office. A person, such as a student who also works part-time at the information technology Help desk, can have more than one role. (What kind of relationship does this imply in the authorization database?)
- **User-based access.** Each user has individually assigned access rights. This was common before role-based access came into use. It has drawbacks compared to role-based access: Rights must be set up for each individual, people in the same role can have inconsistent rights, and people may keep access rights they no longer need after a job change. Today, when a person needs access rights that don't match his or her role, DBAs figure out what the person does that requires those access rights, find a corresponding role or define that as a new role if necessary, and put the person in it.
- **Context-based access.** Access rights can vary by time of day or by the activity in progress. A terminal on the factory floor may have access to inventory data during production hours, but not at night. Custodians are as honest as anyone, but they don't need inventory data. If they had access to it, a criminal could get inventory data by getting a job as a custodian.

There are two main approaches to protecting information resources after a user (or an application) has been authenticated as authorized. In the traditional *castle and moat* approach, once a person is inside the system (the moat), all its resources (the castle) are accessible. This is how your phone works and how your own computer probably works: You may need a password or a fingerprint to get in, but once you're in, there are no barriers to doing anything. This approach allowed Target's 2013 breach. Criminals entered through a poorly defended point-of-sale system. That system, and therefore the criminals, had free access to Target's databases.

The newer approach is called *zero-trust.* With this approach, a request to access any resource is checked to confirm that the user needs that resource to accomplish an authorized task. This would be analogous to setting up password protection for all the folders and files on your laptop, with different passwords for each of them. It is hard to retrofit zero-trust to existing systems, but new ones are increasingly being designed with this philosophy in mind.

The choice between these approaches is a business decision. Technology will make it work, but the tradeoff between the cost and benefits of added security must be assessed by businesspeople. This is discussed in more depth in Chapter 12.

Where you fit in: Using information effectively requires access to it. Access carries risk. This is a business tradeoff, not a technical one. Database designers can provide any controls a company wants, but they can't decide what those controls should be. Only users and their managers can weigh the benefits of access against its risks. Once again, that means *you*.

KEY POINT RECAP

Databases enable an organization to organize all its data in one place, with any item accessible directly or via its connections to other items.

As a businessperson, you use information. The better it is organized, the easier it is to do that. Having information in a well-organized database is the best way to achieve that goal.

Most operational databases use the relational approach today.
As a businessperson, you will almost certainly find your organization's databases structured in this way. The better you understand how they work, the better you will be able to use them.

Entity-relationship diagrams show how a relational database is organized.
As a businessperson, you'll discuss databases with their designers and may design small ones yourself. ERDs are the language that database designers use and a good way to draw up database designs before they go near a computer. Understanding them will help you in these situations.

Databases used for decision making may have different structures.
Because the use of decision-making databases differs from that of operational databases, different approaches to organizing them often work better. These methods, including ways to handle "big data," are evolving rapidly today. As a businessperson, you will benefit from keeping current on developments in this area.

Database management must include security considerations.
Having all an organization's data eggs in one basket places a premium on watching that basket. Security is important, but only businesspeople can decide what the appropriate level of security should be. Database administrators can then put those business decisions into practice.

KEY TERMS

Access control: Restricting access to a *database*, or to part of a database, to users who have been *authorized* and *authenticated.*

Asynchronous replication: An approach to *replicated databases* in which updates made at one location may be copied to other locations at a later time.

Attribute: A characteristic of a real-world concept or object about which a database stores data.

Authentication: Technical process by which a person requesting access to a database is verified to be who he or she claims to be.

Authorization: Administrative process through which a person or application gets permission to access a database.

Big data: Data characterized by high volume, high velocity, and high variety.

Blob: Short for Binary Large OBject, a database field that contains unstructured data based on multimedia such as sound, images, or video.

Cardinality: A statement of how many *entities* of one *entity class* must be associated with each entity of a related entity class if the database is to have a valid structure. See also *minimum cardinality, maximum cardinality.*

Castle and moat: An approach to security that verifies a user's identity on first accessing a system but does not limit that user's access to system resources thereafter. Contrast with *zero-trust.*

Column: A *field* (sense (b)) in a relational database.

Column access control: Restricting a user's database access to specific *columns* in that database.

Composite key: A *primary key* that is composed of more than one database *field.*

Context-based access: Database access based on the circumstances surrounding a request. Compare *user-based access, role-based access.*

Database: An organized collection of data about related real-world concepts or objects; the *tables* that describe a set of related *entities.*

Database administrator: A person responsible for organization, security, and maintenance of a *database.*

Database management system: Software to access databases on behalf of applications and users.

Data element: See *field*, sense (a).

Data warehouse: Database of historical data, used primarily for decision making.

Denormalization: Designing a database that is not *normalized* in order to improve performance.

Dimensional database: Database, used primarily for decision making, in which each *attribute* of an *entity* is represented as a separate dimension of a multidimensional cube.

Distributed database: A *database* whose data is stored on storage devices attached to more than one computer. See *federated database*, *fragmented (partitioned) database*, *replicated database.*

Dongle: Device that plugs into a computer port and authenticates the user who carries it.

Entity: A real-world concept or object about which a database stores information.

Entity class: All the real-world concepts or objects of a given type about which a database stores information.

Entity instance: One specific member of an *entity class*; described by the data in one *record.*

Entity-Relationship Diagram (ERD): A diagramming method that shows the *entities* in a *relational database*, their *attributes*, and the *relationships* among them.

Federated database: A *distributed database* in which parts of the database are independent of each other and may have different structures, but can be accessed as a unit.

Field: (a) A single data item of a specific type related to a particular real-world concept or object. (b) Conceptually, the place in a database structure where items of that type are stored.

File: (a) All the records that describe a particular *entity class.* (b) In non-database contexts, a named object in storage that the operating system treats as a unit.

Flat file: Single file or table that contains all the data about an entity.

Foreign key: A *field* in a *relational database* that corresponds to the *primary key* of another *table.* Used to connect information in those two tables.

Fragmented database (also partitioned database): A *distributed database* in which parts of the database are on different computers. See *horizontal fragmentation*, *vertical fragmentation.*

Horizontal fragmentation: An approach to *fragmented databases* in which different table rows are on different computers.

Many-to-many relationship: A *relationship* between two *entity classes* where the *maximum cardinality* of both entities is "many."

Many-to-one relationship: A *relationship* between two *entity classes* where the *maximum cardinality* of one entity is 1 and that of the other entity is "many."

Maximum cardinality: The maximum number of *entities* of one *entity class* that may be associated with another entity class. Values are usually 1 (there may be only one entity of this class) or "many" (there may be more than one entity of this class).

Metadata: Data maintained by a *database management system* that describes the form, structure, and other characteristics of the data stored by that system.

Minimum cardinality: The minimum number of *entities* of one *entity class* that may be associated with another entity class. Values are usually 0 (an entity of this class is not required) or 1 (an entity of this class is required).

Normalization: The process of ensuring that the design of a *relational database* meets rules for correct database structure, modifying its design to meet them if necessary.

One-to-one relationship: A *relationship* between two *entity classes* where the *maximum cardinality* of both entities is 1.

Operational database: A database designed to support the daily operation of an organization.

Partitioned database: See *fragmented database.*

Passphrase: Series of several words that functions like a *password.*

Password: Sequence of characters known only to one person, used to confirm that a person is who he or she claims to be.

Password management software: Application that helps users manage multiple passwords securely, making it unnecessary to memorize them or write them down.

Primary key: Unique identifier of a *row* (*record*) in a relational database.

Read access: Permission to read the content of a database, or part of a one, but not to modify it. Compare *write access.*

Record: The group of *fields* that, together, describe a single real-world concept or object.

Redundancy: Storing the same information more than once in a database.

Relational database: A database in which data is stored in the form of tables, with records in different tables connected by matching a *foreign key* in one table with the *primary key* of another.

Relationship: A statement of how one entity is related to another. This can include text and a statement of the cardinality of the relationship. See *cardinality, minimum cardinality, maximum cardinality, one-to-one relationship, many-to-one relationship, many-to-many relationship.*

Replicated database: A *distributed database* in which copies of the entire database are on different computers.

Role-based access: Database access based on a person's job. Compare *user-based access, context-based access.*

Row: A *record* in a relational database.

Row access control: Restricting a user's database access to specific *rows* in that database.

Structured Query Language (SQL): A language for manipulating *relational databases.*

Synchronous replication: An approach to *replicated databases* in which updates made at one location are copied immediately to all other locations.

Table: A *file* (sense (a)) in a relational database.

Trans-border data flow: Movement of data across international borders, e.g., from a division in one country to headquarters in another.

User-based access: Database access based on permissions associated with a user ID. Compare *role-based access, context-based access.*

Vertical fragmentation: An approach to *fragmented databases* in which different table columns are on different computers.

Write access: Permission to modify the content of a database, or part of a one. Compare *read access.*

Zero-trust: An approach to security that verifies a user's identity and authorization before accessing any system resource. Contrast with *castle and moat.*

REVIEW QUESTIONS

1. What are the three levels of data organization between a *byte* and a *database*?
2. What is a *primary key?* What two characteristics must a primary key have?
3. What is a *foreign key?* How does it work with a primary key?
4. Who uses entity-relationship diagrams, and when?
5. What does minimum cardinality mean? Maximum cardinality?
6. What three types of relationships can two entities have in a relational database?
7. What is an associative entity?
8. Differentiate among *fragmented*, *replicated*, and *federated* databases.
9. How do the database needs of decision makers differ from those of an operational database?
10. What are the characteristics of *big data*?
11. Explain, at the conceptual level without technical detail, how Hadoop works.
12. What is a *database management system*, and how is it related to databases?
13. What does a DBMS do with *metadata*?
14. Who might use SQL, and why?
15. What are the four categories of methods to authenticate a database user?
16. What are the three types of database access control?

DISCUSSION QUESTIONS

1. The Rolling Stones first recorded "You Can't Always Get What You Want" in 1968. Relate that song's title (not its content) to database management.
2. Susanna Clarke's book *Jonathan Strange and Mr. Norrell* is on the revival of English magic in the 19th century. (Assume that happened.) Its revival was slowed by lack of information about English magic in its heyday, centuries earlier. Many books were written

at that time about spells, materials needed, words to be said, timing, and effects. Most of those books had disappeared by 1800. Surviving copies of others were in private libraries, out of public view.

a. Discuss the value that a central database could have had to an aspiring English magician of 1810. (If we can assume English magic, we can assume databases in 1810.)
b. Sketch a rough design for that database, identifying tables, the most important columns of each, and the foreign keys that each one must have to connect with other tables.
c. What distribution approach (see the last part of Section 5.2) would you recommend for this database? Assume something like the Internet existed in 1810, too.

3. Refer to the Oracle database price list in Section 4.6. Your company uses Standard Edition and is experiencing slow response times. Upon investigation, you find that the four-processor system that hosts the database is overloaded and must be upgraded to at least six processors. (Assume those will equal six processor units.) This hardware upgrade will, in turn, require upgrading your software license to Enterprise Edition at considerable expense.

   One of your colleagues suggests partitioning the database for two parts of the company. Each partition would run Standard Edition on a four-processor system. Two Standard Edition licenses, while more expensive than the single license you now have, would be less expensive than a six-processor Enterprise Edition license. Calculate the three prices (now and for each of the two alternatives, using per-processor licensing). Discuss the pros and cons of this idea.

4. Consider the scenario of Chapter 3, Discussion Question 1 (Sir Noten and Kate use the Tube).
   a. Draw an entity-relationship diagram for the Oyster Card database. You must make some assumptions about what data to store. For example, when value is added to a card, should the database store the information about how much value was added, where, and how, or should it only store the new amount in the card? State all your assumptions.
   b. Draw sample database tables, with four rows of data in each, for this database.
   c. Assume there are 10 million Oyster Card holders, and the average holder uses it twice a week. (Regular commuters tap it 20 times or more. Occasional visitors to London use it a few times a month.) The system keeps data for a year. Estimate the size of its database. Make any assumptions you need and state them. (There is no one right answer to this question. Your approach is more important than the number you end up with.)
   d. Assume the system keeps data forever, the number of Oyster Card holders grows at 10% per year, and the average usage per cardholder remains stable. Estimate the size of its database after each of its first five years. (Part (c) above is the answer for Year 1.)

5. Review the MyVid example in Section 5.2. Using a spreadsheet program, compute the average rating of a video with the ratings in the table, and create a bar chart of those ratings.

Rating	Count
5	76
4	54
3	61
2	17
1	2

6. Answer the following questions for the sample data (Figure 5.3) in the MyVid database:
   a. Who (name, not member number) uploaded the Disney World video, and when?
   b. Who posted "Very Nice," and what video was this comment about?
   c. Who posted "Thank you," and what comment did that refer to?
   d. Which member paid for his or her membership with an American Express card?
   e. Which members are allowed to post a 12-minute video?
   f. On what date does Melissa Ohanian's Bronze membership expire, unless she renews it?
7. Draw an entity-relationship diagram for an eyeglass store database. Customers have prescriptions from doctors or opticians, buy frames that come from manufacturers, and are helped by a staff member. Think of at least two additional entities to include. Consider the many-to-one relationships here.
8. Draw an entity-relationship diagram for a medical records database. It includes multiple doctors and multiple patients. Doctors and patients are connected via appointments; a patient can have more than one appointment with the same or different doctors, at different times. At an appointment, a doctor must write a note summarizing the appointment, and can optionally (a) diagnose the patient with a condition; (b) write one or more prescriptions; and (c) order one or more lab tests. That isn't a complete medical records database, but it's the basic idea.
9. You work in market planning for Toyota Motor Sales USA. It chooses models to be imported to the U.S., options to be offered, and how they will be packaged. It sets prices, considering costs and market conditions (including competition). It wants to create a data warehouse with information about every Toyota sold in the U.S. Come up with a list of data elements that it would store for each sale, similar to Figure 5.17 for men's shirts.
10. Consider the example at the end of the Hadoop section about a marketing planner comparing Twitter activity with purchase activity for a product. What might this planner do if
    a. a spike in Twitter activity tends to precede a spike in purchases?
    b. a spike in Twitter activity tends to follow a spike in purchases?
    c. a spike in Twitter activity tends to follow an expensive TV commercial, but there is no corresponding spike in purchases?
11. Visit *www.getnymi.com* and at least one other site you find via a web search to find out about the Nymi device for authenticating users via their heartbeat voltage pattern. Do you think you would use such a device? How would you feel if your employer required you to use one? Compare it with password authentication in terms of security, acceptability, and two other factors of your choice.

KHOURY CANDY DISTRIBUTORS DATABASES

Visal Phan, KCD's database administrator, was returning from a lunchtime walk just as Isabella and Jake arrived at the building's front door. They were a few minutes early, but he ushered them into the now-familiar IS department conference room. He opened the conversation with "What did Lakshmi tell you about our databases last week?"

"She said that you use Oracle's database manager on your central system and on the servers in the distribution centers," Isabella answered. "Nothing more than that."

"Actually," Jake interjected, "she did say one other thing. She said Oracle's version numbers just jumped from 12c to 18c, but she didn't know why. We thought you might."

"It's not a secret," Visal replied, "though I can understand that she never had a reason to ask. Up through 12, they were numbered in sequence. Then Oracle decided to number each version for the year it came out. Version 18 came out in 2018. 18c is its third update. It's that simple.

"Anyhow," he continued, "she's right. Oracle is pretty much the only database software we use. A few people build small personal databases using Access or FileMaker Pro, but I don't get involved with those unless someone asks me for advice.

"As far as the Oracle database is concerned, what we have is fairly typical of the distribution business. We work with candy, but our database could be about almost anything. From a user point of view there are four kinds of data in the system, plus others behind the scenes that our users don't care about. We have data about customers, parts of KCD, suppliers, and transactions such as sales. Some kinds of data, like products, overlap these categories. That's what makes it a database and not just a collection of files.

"The problem is, databases are like onions. Peel one layer away, and you find another. Then you peel that one away and see the next one. It keeps going for way more layers than you'd think. For instance, take a customer file. Our customers are businesses, but there's a person at each one who's authorized to place orders. We can't put that person in the customer record, because they can change over time and we can't lose information about who placed previous orders. So, purchasing agents are another entity. Then we have to have the dates that a given purchasing agent started and ended. We can't put those in the purchasing agent record, because sometimes someone who was the agent in the past comes back again later, so we have another entity for the period that a person is a purchasing agent. And some customers authorize more than one person to place orders, so the database structure needs to allow for that."

"That's a lot more complicated than what we did in our database course," said Isabella. Jake nodded agreement and added, "We did a sales database exercise, but a customer was just a customer. Customers had customer numbers as primary keys, names, addresses, that sort of thing, but it didn't go past that."

"Right," Visal agreed, "but the concepts are the same. It's like reading. I learned to read English, after my family came here from Cambodia, on really simple stuff. Once I knew how to read, I could pick up a newspaper without having to learn a new way of reading. It took longer to figure out sentences, and I had to learn new slang to understand the sports pages, but reading was reading. The things I had learned were still useful. Your database course is like that. You have the concepts. You just have to apply them to more complicated situations."

"That's what our professor said," Jake agreed, "but it means more when you hear it from someone who works with real databases for a living."

"Your professor was right, at least this time," Visal laughed. "Anyhow, all the information we keep for our customers works the other way too, for our suppliers. We have data for each order we place with them, and for each item in each order because each item is a different product. When we record that an order arrives, we have to allow for split shipments—not just of the order, but of each item. We have to be able to record partial and full payments, credits that aren't payments, partial and full returns, and more. And then we have to transfer what we get into inventory, and keep track of that. We're getting ready to use tiny RFID transponders for inventory tracking. Never a dull moment!"

Questions

1. Draw an ERD of any five tables in KCD's database. One must be their product table. The others can be any four tables you wish as long as all the tables are connected somehow. Don't use the purchasing agent example that Visal discussed, but try to think the same way he did.
2. The best way to organize a database with 100 records won't handle 100 million records. As a database grows, it must be reorganized. If it isn't, response time will degrade. To the left of where the curves cross in Figure 5.22, organizing it for a larger size will slow it down. To the right of this point, performance calls for restructuring. This restructuring need not be done often, but when it is done, the database is unavailable for a few hours. Suggest two ways to minimize the business impact of this outage.
3. Chris Evans described KCD's distribution center databases in the Chapter 2 episode. What kind of distributed database are they? Do you think this type makes sense for KCD? Explain why or why not.

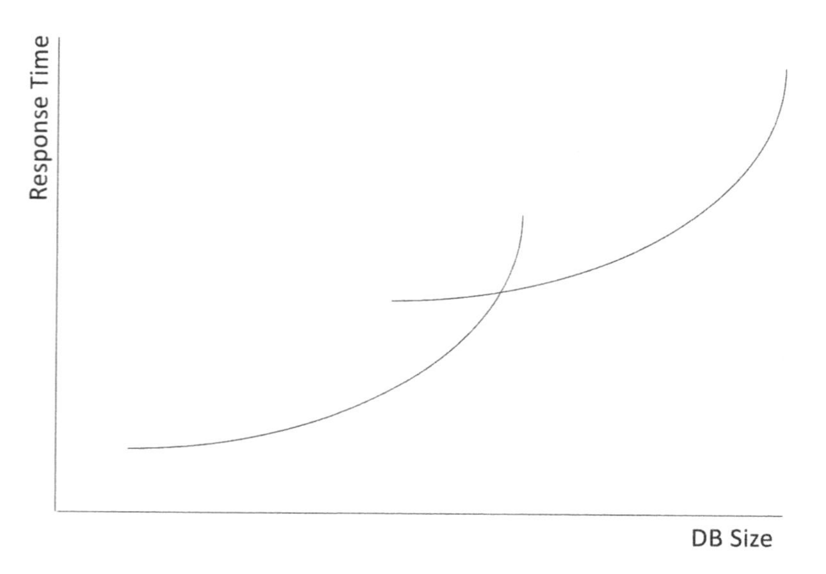

FIGURE 5.22 Illustration of reorganizing a database.

CASE 1: NETFLIX MOVES TO CASSANDRA

In 2018, Netflix accounted for 15% of the world's server-to-client Internet traffic, 19% in the U.S. That figure goes over 40% during peak hours. Amazon Prime and YouTube were in a virtual tie for distant second place with between 7% and 8%.

That's a lot of movies. Managing them all, sending the right part of the right movie at the right time to each of millions of customers, is a huge problem in managing data. Analyzing customer preferences, to stay ahead of the competition and develop new offerings, is another.

When Netflix began streaming movies to customers in 2007, it used an Oracle database. By January 2011 the volume of requests had grown to 37 times more than it had been in January 2010, thousands of times greater than when the service started. The original database couldn't keep up. Netflix needed to select a different database, and quickly.

Netflix chose the open-source Cassandra database as their DBMS for moving forward. They did not, however, simply download Cassandra from the Apache Software Foundation. (The Apache Software Foundation sponsors over 80 software projects. The best known is probably the web server for which it is named.) Instead, they obtained it from DataStax, as part of the DataStax Enterprise package. DataStax adds software for security, data analysis, workload management, and more to the basic Cassandra package; collects and provides updates as regular service packs; has Cassandra experts on call for customers; and makes the basic free package from Apache more like a full-scale commercial offering from a major vendor. Netflix needed that.

Netflix runs Cassandra on the Amazon Elastic Cloud 2 (EC2) public cloud because, as Netflix Cloud Architect Adrian Cockcroft puts it, "it could not build data centers fast enough." They see no reason to go to a private cloud, though some very large database users such as Facebook find private clouds more cost-effective and adaptable to their needs.

To prevent hardware failures from affecting service, Netflix has configured its EC2 database to store three copies of its data in every region that EC2 operates. (The eastern U.S., for example, is one region.) Normally, all three copies are updated. If there is a failure anywhere that prevents this, two are updated and the third will follow after the fault is corrected.

Scalability to accommodate growth was a Netflix requirement. Tests showed that the Cassandra database on EC2 scales linearly: If the number of computers supporting the database is doubled, the workload they can handle also doubles. This may sound obvious, but it is often not true. In many systems, the need to coordinate more and more computers prevents the system from scaling linearly—eventually reaching a point where adding more computers may even reduce the workload it can handle. Tests have confirmed that EC2 can handle many times the 2013 workload.

In addition to streaming movies to customers, Netflix uses data from its Cassandra database to understand customer preferences. This enabled them to picture the type of content that would be well received when it planned its successful "House of Cards," the first major show to bypass TV networks and cable operators. Matt Pfeil, co-founder and VP of Customer Solutions at DataStax, explained that this is the first TV program developed with the aid of big data analysis.

"Netflix has all these data points about movies getting watched, and they can look for things like, do people like actions or drama who are our highest returning customers? Who is the lead in most of those? What types of characteristics of films provide the most engaged watching experience? And then they can use that to go figure out which series they should potentially buy," Pfeil said.

"If you talk about this age as the data age, we're still in the teenage years," Pfeil continued. "The more data you have and the more you can do with it, the smarter [your] business decisions."

Questions

1. Much of the information from this case came from DataStax and Netflix. They cannot be objective about Cassandra versus Oracle. Many organizations are well satisfied with Oracle databases. Assuming Cassandra's advantages in scalability and availability are real, which may or may not be so, what advantages could Oracle have over Cassandra? You don't have to study those two DBMS. Just think of important factors that were not mentioned here.
2. Describe the relationship between DataStax and Cassandra. Your description should explain how DataStax can make money from free software.
3. Netflix's *replicated database* strategy, described midway through the case, means that the database one user sees might not be identical to the database another user sees at the same instant. Is this acceptable here? Justify your answer in business terms. Whichever way you answered, describe a different business situation for which you would answer the other way.
4. "House of Cards" was a Netflix drama based on U.S. politics, starring Kevin Spacey as a representative from South Carolina who became Vice President and then President. It was well received by critics and viewers, ending only after Spacey was credibly accused of improper behavior. (He did not appear in its final 2019 season.) Suppose it had, instead, flopped. What would this mean for the use of data analysis in planning television series?

CASE 2: HOMELESS IN LOS ANGELES

To moviegoers, the image of Los Angeles is Hollywood and Beverly Hills. Glamorous, rich people do glamorous, expensive (and, if movies are to be believed, occasionally illegal) things. Reality includes this, as it does in every large city around the world. However, also as in every large city around the world, reality has another side as well. No large city is immune to poverty and homelessness.

More than 57,000 men, women, and children are homeless on any given night in Los Angeles County. This number is increasing. According to the 2017 count, homelessness was up 23% from 2016 to 2017 despite many moving into permanent housing. The fastest-growing group of homeless are aged 18–24, followed by those under 18. Shelters have waiting lists and there is insufficient permanent low-cost housing. Most homeless live wherever they can.

Proposition HHH, approved by Los Angeles County voters in 2016, provides $1.2 billion in bond money to build 10,000 new affordable housing units over a decade and to provide support services. As this book is written about two-thirds of the money is committed, but to half that number of units. It appears that the numerical goal of the program will not be reached without additional funds, but even 7,000 units are significant. Measure H, passed by the county Board of Supervisors that same year, creates a sales tax of 0.25¢ cent per dollar toward the same goals. It is expected to raise over $350 million per year.

Implementing services effectively and building housing in the right places requires information: Where are the homeless? Where should programs be implemented?

Analysts from agencies across Los Angeles County use geographic information systems (GIS) to answer such questions. For example, Venice Beach homeless have different characteristics than those in Skid Row. Figure 5.23 shows types of sleeping accommodations by area. Fifty-seven percent of Skid Row homeless sleep in shelters, but in Venice only 12% do. The two areas' needs are different.

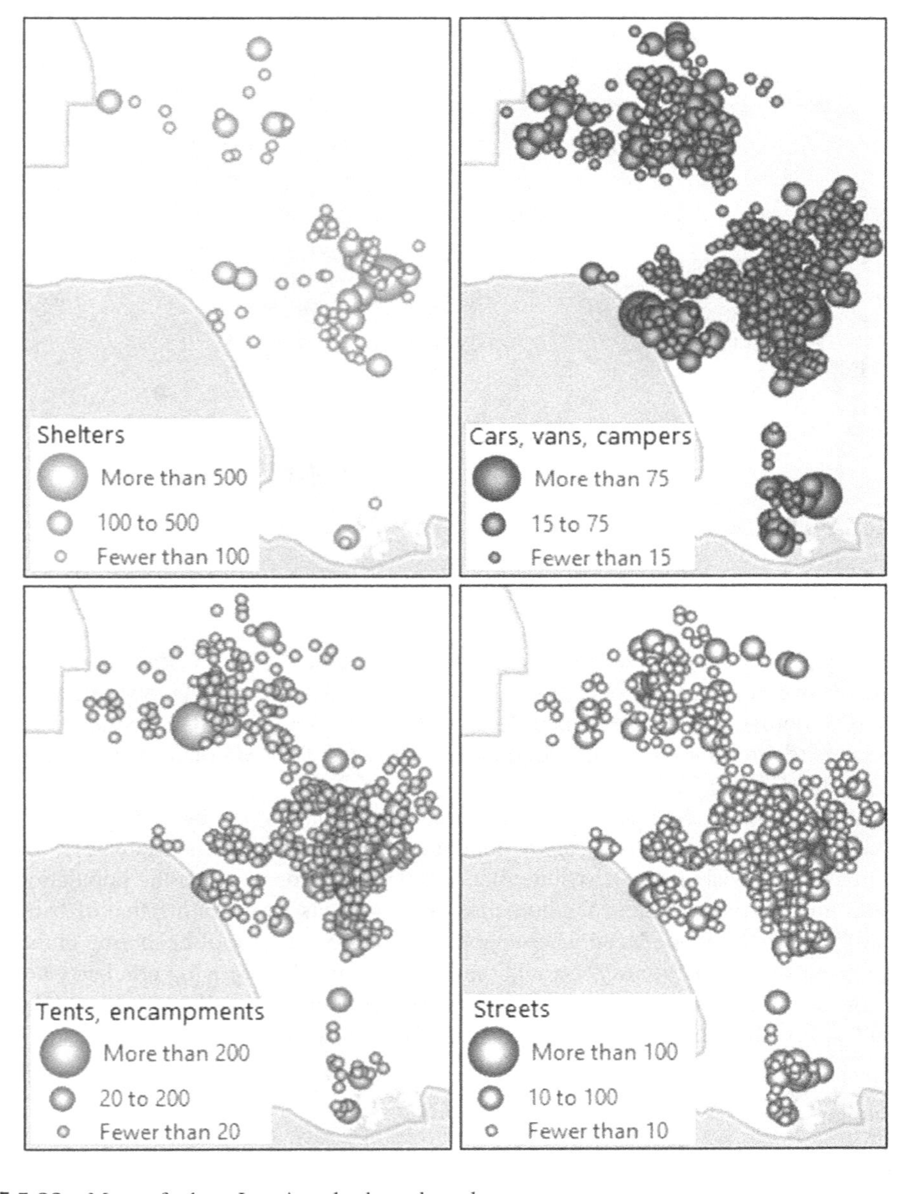

FIGURE 5.23 Maps of where Los Angeles homeless sleep.

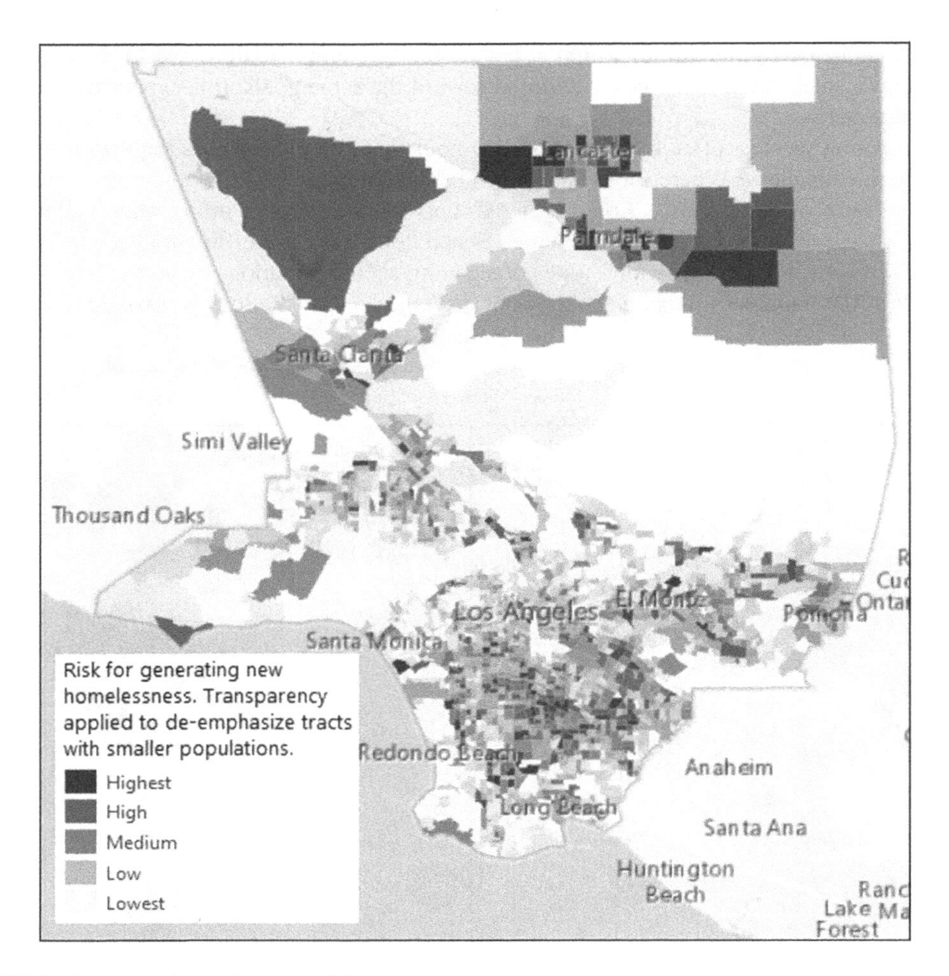

FIGURE 5.24 Map of homelessness risk.

Predicting where people will become homeless can be done via factors known to contribute to homelessness: poverty, unemployment, disabilities, public assistance, paying more than half of total income for rent, domestic violence, mental illness, and a few more. These can be mapped by census tract. A composite score for each tract can then be calculated. GIS can then show these composite scores by shading (Figure 5.24).

Using GIS can identify the likely impact of specific remediation programs in specific locations. This makes it possible to compare different approaches to providing for the homeless: Should services and housing be designed to distribute the homeless, in proportion to the population of each area? Should they be placed where the homeless are now, thus perpetuating that distribution (for good or bad)? Should they be placed where people are at greatest risk of becoming homeless, i.e., where the homeless lived previously? Should they be centralized, improving efficiency but encouraging the homeless to cluster around them? There are other options as well. Uninformed discussions are seldom helpful. GIS can provide a basis for rational comparisons.

Questions

1. Los Angeles prioritizes ending homelessness among military veterans. This requires asking homeless people about prior military service. You have a map that shows average

responses by neighborhood. Discuss: What do you think of the information on this map in terms of the information quality factors in Chapter 1? (Skip cost.)

2. A student registration database has four entities: students, faculty, courses, and sections. (A real one has hundreds.) Give one geographic data item that could be associated with each entity. (Your school's registration system might not include them.) How would you specify each location: street address, latitude/longitude, or something else (what?)
3. You plan to start a business producing replacement taillight lenses for antique vehicles by 3D printing. Describe three types of location-related information that your database could store.
4. Geographic databases store information in layers: a base layer for the terrain, a layer for man-made features such as roads, and additional layers for schools, movie theaters, etc. Layers can be shown or hidden to customize maps. The database has a table for layers, with data to say if the layer is hidden or visible. Every entity on the map has a foreign key identifying the layer it belongs to. How could this approach help manage what is displayed on a computer screen?
5. DigSafe (*www.digsafe.com*) is a program in the northeast U.S. that notifies utilities such as water, gas, and electricity providers of construction or other activities that involve digging. (Similar programs exist in many places.) Utility employees then mark underground facilities. To do this, they must know where those are. This was once done with marked paper maps. Today, computers are used. There are (at least) two feasible approaches:
   a. In a GIS, create a map of the area that includes utility pipes/lines/etc.
   b. In a GIS, create a map and store it as an image. Draw utilities on that image.

   Give two advantages of each approach over the other.

BIBLIOGRAPHY

Apache Software Foundation, "Apache Hadoop," hadoop.apache.org/, accessed March 12, 2019.

Apache Software Foundation, "Cassandra," cassandra.apache.org, accessed March 12, 2019.

Bowen, B., "Weighing and making the right decisions," *Sybase*, Winter 1995.

Cockcroft, A., "Migrating Netflix from datacenter oracle to global Cassandra," *Cassandra Summit Conference*, www.slideshare.net/adrianco/migrating-netflix-from-oracle-to-global-cassandra, accessed March 18, 2019.

Curtis, S., "Netflix foretells 'House of Cards' success with Cassandra big data engine," *Techworld*, news.techworld.com/applications/3437514/netflix-foretells-house-of-cards-success-with-cassandra-big-data-engine, accessed March 18, 2019.

DataStax web site, www.datastax.com, accessed March 18, 2019.

ESRI, "Combating homelessness in Los Angeles County," desktop.arcgis.com/en/analytics/case-studies/la-county-homelessness-1-overview.htm, accessed September 18, 2019.

Facebook, Company Info page, newsroom.fb.com/company-info, accessed March 11, 2019.

Foltyn, T., "The most popular passwords of 2018 revealed: Are yours on the list?" December 17, 2018, www.welivesecurity.com/2018/12/17/most-popular-passwords-2018-revealed, accessed March 14, 2019.

Lampitt, A., "Big movies, big data: Netflix embraces NoSQL in the cloud," *Javaworld*, May 2, 2013, www.javaworld.com/article/2078771/open-source-tools/big-movies--big-data--netflix-embraces-nosql-in-the-cloud.html, accessed September 17, 2019.

Lock, M., "The horsepower of Hadoop: Fast and flexible insight with results," *Aberdeen Group*, May 2016, hosteddocs.ittoolbox.com/horsepower-hadoop.pdf, accessed March 15, 2019.

Los Angeles Mission, "Measure H marks a big change for homelessness in LA," losangelesmission.org/measure-h-marks-a-big-change-for-homelessness-in-la, accessed September 18, 2019.

Oracle Corp., "Introducing Oracle Database 18c" white paper, July 2018, www.oracle.com/technetwork/database/oracledatabase18c-wp-4392576.pdf, accessed December 26, 2018.

Perry, J.S., "What is big data? More than volume, velocity and variety …," *developerWorks*, May 22, 2017, https://developer.ibm.com/dwblog/2017/what-is-big-data-insight/, accessed September 17, 2019.

Pratt, M., "What is zero trust? A model for more effective security," January 16, 2019, www.csoonline.com/article/3247848/what-is-zero-trust-a-model-for-more-effective-security.html, accessed March 20, 2019.

Sandvine, "The global internet phenomena report," October 2018, www.sandvine.com/hubfs/downloads/phenomena/2018-phenomena-report.pdf, accessed March 12, 2019.

Scribner, H., "Why 'ji32k7au4a83' is one of the world's most common passwords," *Deseret News*, March 7, 2019, www.deseretnews.com/article/900059031/why-ji32k7au4a83-is-one-of-the-worlds-most-common-passwords.html, accessed March 12, 2019.

Sengupta, S., "Machines made to know you, by touch, voice, even by heart," *The New York Times*, September 10, 2013, bits.blogs.nytimes.com/2013/09/10/beyond-passwords-new-tools-to-identify-humans, accessed September 17, 2019.

Sinha, A., "Use Hadoop to handle big data," *Open Source for U*, August 19, 2015, opensourceforu.com/2015/08/use-hadoop-to-handle-big-data, September 17, 2019.

Smith, D., "How close is L.A. to building 10,000 houses for homeless people? Here's a breakdown," *Los Angeles Times*, April 21, 2019, www.latimes.com/local/lanow/la-me-ln-hhh-spending-commitments-20190421-story.html, accessed September 18, 2019.

Weinert, A., "Your pa$$word doesn't matter," *Microsoft*, July 9, 2019, techcommunity.microsoft.com/t5/Azure-Active-Directory-Identity/Your-Pa-word-doesn-t-matter/ba-p/731984, accessed September 25, 2019.

6 Information Networks

CHAPTER OUTLINE

WHY THIS CHAPTER MATTERS

Those who supply data to an information system, those who use its output, and its computers are seldom all in the same place. When an information system serves people in different places, such as everyone who reserves a seat on a United Airlines flight, buys music at the iTunes Store, or follows a Twitter hashtag, they can't be. We need to move data from one place to another.

That's where data communication networks come in. Along with shared databases, networks are one of the two key *enabling technologies* that make modern information systems possible.

Purely technical decisions about networks are best left to experts, but many network-related decisions require understanding both business and how the technology works. After you study this chapter, you should be prepared for those decisions.

CHAPTER TAKE-AWAYS

As you read this chapter, focus on these key concepts to use on the job:

1. Having the right communications and networking is key to strategic applications.
2. Different communications and networking options involve different tradeoffs, and are best suited to different organizations or applications.
3. Long-distance communication is almost entirely over the Internet today, so most business choices involve local communication and connecting to the Internet.
4. Network security is a management responsibility, even when it is supported by technology.

6.1 COMMUNICATION LINKS AND NETWORKS

eBay, Amazon, and the iTunes store can't build direct communication links from their servers to the computers of every potential customer. They, and every other online user, share networks. In order to make business decisions involving those networks, we need to know a bit about how they work. They involve two basic concepts:

1. A *communication link* is a means of transmitting data from a device at one end of the link to a device at its other end.
2. A *network* is a collection of communication links, connected by devices that can route messages among devices attached to the network.

Networks are built from communication links and devices that connect them. The concepts are analogous to those of roads and a road network. A road network consists of roads, junctions to join two or more roads, and control devices such as traffic lights. Together, they enable vehicles to travel from one place to another.

6.2 COMMUNICATION LINKS AND DEVICES

A bewildering array of technologies can be used to transmit data. Some are seldom installed today, but you might encounter them in a system that was put in years ago and still gets the job done. They fall into two major categories: physical (wired) links and wireless links.

As you read about types of communication links, and as you learn of new ones in the future, keep in mind that everything sent over a link is managed by software. A link sends signals from point A to point B. Hardware at the ends of the links turns signals into 1s and 0s. What the 1s and 0s mean is up to the software that sends and receives them. The software at the two ends must agree on how to interpret them. Fortunately, there are standards for this. One standard, HTML, defines how to send web pages. A computer that sends a page sends it in HTML. Your browser (Chrome, Firefox, etc.) turns that HTML into a page on your screen. There are many others.

Software control, unfortunately, opens a door for criminals to send messages that seem to come from anyone or anywhere. This is the basis of "spoofing." You may get a message that seems to be from a friend, but is really from a criminal who wants you to open a file or click a link in order to access to your computer or personal information. More about this later on, in Section 6.6.

Software to operate communication links, and to route messages to and from applications, is built into all modern operating systems. It performs two OS functions: managing a shared resource (the communication connection) and providing shared services (sending and receiving data).

Physical Communication Links

Copper wires have been used to transmit signals since the 1800s. When computers came into use, the telephone network already connected every location of interest in the industrialized world. It was natural to adapt it for data.

Telephone lines were designed to carry *analog* signals: voltages that vary smoothly up and down, like sound waves (Figure 6.1). Computers use *digital* signals: 0 or 1, not 1/3 or 0.97. A sending device can turn digital signals into tones a phone line can carry: it *modulates* them. A receiving device can then *demodulate* these tones to interpret them as 0 or 1. A device that does both is a *modulator-demodulator*, universally shortened to *modem*. Devices that connect computers to communication links are still called modems, though today's links seldom use analog signals.

Sound waves on phone lines have speed limitations. Phone line modems topped out at 56,000 bits per second. That's fine for text emails, but not for web pages, photos, music, or video. *Fiber*

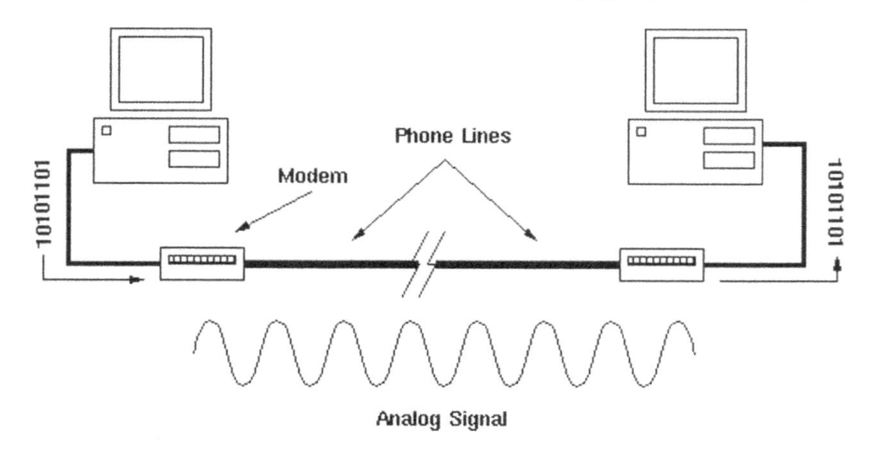

FIGURE 6.1 Diagram of communication link with modems

FIGURE 6.2 Fiber optic cable (Source: Wikimedia Commons).

optic links produced a big speedup. A light is switched on and off at one end of a glass cable, while a photocell at the other end detects its state. Fiber optics is inherently digital.

Today, fiber optic cables (Figure 6.2) connect switching centers to each other. Cables contain up to a thousand individual fibers and can transmit, in total, trillions of bits per second. Communication companies install cables with more capacity than they need, to avoid having to dig up the ground again when they need more capacity. They then activate fibers as needed.

Last mile technologies connect switching centers to homes and businesses. They use fiber optic cables where demand warrants. Elsewhere, copper wires are used, but those are not your grandparents' phone lines! Today's copper *digital subscriber line* (DSL) carries millions of bits per second, providing good response times for web users. Some DSL lines transmit at the same speed both to and from subscribers. *Asymmetric DSL* (ADSL) is faster to a subscriber than from it. ADSL is more common.

Cable television services are another last mile option. Their networks were originally designed to work primarily in one direction: from the central office to subscribers. Cable TV connections are therefore fast in that direction, but slow from subscribers to the central office (like ADSL). This suits the typical home user, who receives large files such as web pages, movies, and downloads, but sends very little in the other direction. It is not well suited to a business that sends out web pages, or to a software supplier that sends downloads to customers.

Where you fit in: You must understand your company's data usage patterns to choose the right technology, and hence the right supplier, for its Internet connection. This is a business issue.

Communications speeds keep rising. In early 2013, AT&T researchers sent a 400 GB/second (four billion bits per second) signal for 12,000 km (7,450 miles) over a single fiber optic cable. (Since optical signals deteriorate with distance, it's not enough to send a signal fast; it has to cover a long distance.) That's enough to send about 30 full-length movies in a second. But speed didn't stop there. In 2017, NICT Network System Research Institute of Japan and Fujikura Ltd. demonstrated a data rate of 159 TB/second—almost 400 times that fast—over 1,045 km (650 miles). By late 2018 they had increased this speed by another factor of 8, to 1.2 PB/second, over a shorter distance.

"Slow" transmission today would have been average five years ago, fast five years before that. A speed we think of as fast today will be average in five years, slow in ten.

Wireless Communication Links

Just as voice signals can be sent by radio and television, data can be sent over wireless links. Most long-distance wireless data transmission today involves satellites. They are of four types:

1. *Geosynchronous satellites* orbit about 22,200 miles (35,800 km) above the earth. At that altitude, they complete an orbit in 24 hours. Since the earth rotates once in 24 hours, they stay over the same meridian and are always in view from the same locations on earth (Figure 6.3).

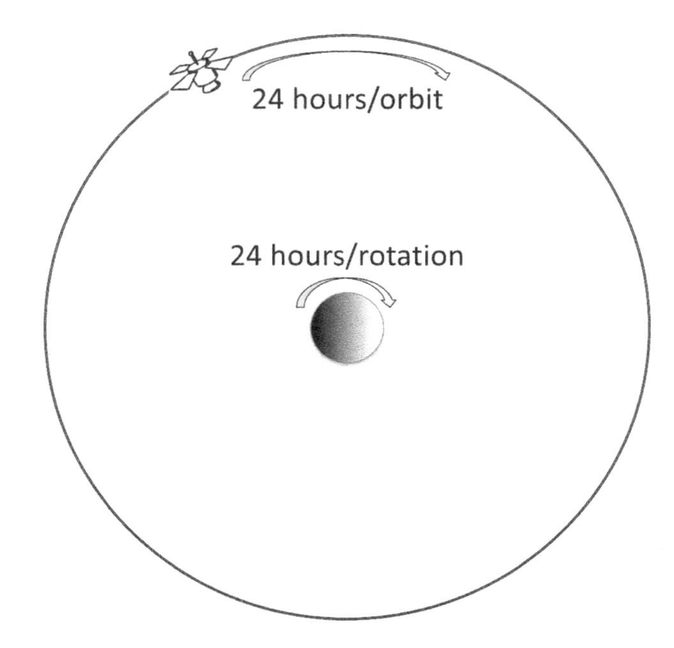

FIGURE 6.3 Diagram of geosynchronous satellite orbit

2. *Geostationary satellites* are geosynchronous satellites that orbit over the equator. They do not appear to move at all. They can work with a fixed antenna, such as a TV satellite dish. Most communication satellites are in geostationary orbits. If you have a satellite dish, it points toward the equator: south from Boston, north from Buenos Aires.
3. *Medium earth orbit* (MEO) satellites have altitudes from about 1,250 miles (2,000 km) up to geosynchronous altitude. Their main use is the GPS satellite network. Its satellites orbit at 12,550 miles (20,200 km), with a period of 12 hours. Communication satellites that cover the North and South Poles also use MEO. It is little used otherwise, as it offers neither the aiming benefits of geosynchronous orbits nor the launch economies of low earth orbits (LEOs).
4. *Low earth orbit* satellites are below 1,250 miles (2,000 km). The major use of LEO in communications is the Iridium telephone system. It covers the earth with 66 satellites at an altitude of 485 miles (780 km) with a period of about 100 minutes. At that altitude, satellite decay is a concern. Several Iridium satellites have fallen back into the atmosphere since the first one was launched in 1997. Ten replacements were sent into orbit in January 2019. Calls switch from one satellite to another as they pass over a caller. Iridium calls are costly, but they work where nothing else can. Starlink satellites (see below) will also be in LEO.

The mobile (cell) phone network can be used for data communications. Early mobile phone data rates were too low for effective use, but today's are more than adequate and today's phones take advantage of this capability. Laptop computers can use cell network adapters. Phones and computers with cellular connections can share this network connection by creating a *hot spot.*

Where cell towers are unavailable or not up to the demand, cell phones can still be used if a subscriber installs a *small cell.* There are three main types of small cells, listed in the following table. (Picocells and femtocells are named for the metric system prefixes for 10^{-12} and 10^{-15}, respectively.)

Type	Power	Max. Coverage	Capacity	Primary Use
Metrocell	5 W	1,000′/300 m	200 users	Outdoors
Picocell	1 W	750′/225 m	64 users	Indoors
Femtocell	0.1 W	60′/18 m	6 users	Indoors

Cell towers and communication satellites are not the only alternative to a physical connection. In June 2013, Google announced Project Loon to provide Internet access to disaster-stricken, poor, and rural areas of the globe. Loon uses helium balloons at 60–90,000 feet (18–27 km), well above aircraft (Figure 6.4). Balloons navigate by moving to an altitude at which the wind is in a desired direction, and can remain airborne for over 100 days. They provide Internet access at several hundred thousand bits/second to anyone within a radius of about 24 miles (40 km). In 2015 Loon brought the Internet to tens of thousands of people affected by flooding in Peru. In 2016 it provided connectivity to 200,000 Puerto Ricans after Hurricane Maria devastated that island. Its balloons were all launched in Nevada, in the southwestern U.S., and guided to their destinations.

Another venture to bring the Internet to underserved areas is Elon Musk's Starlink project. His SpaceX firm plans a network of 12,000 satellites orbiting the globe at 210, 340, and 750 miles (about 340, 550, and 1,200 km). Starlink is expected to cost about $10 billion in total. The first 60 satellites were launched in May 2019. As this is written, commercial operation is expected to begin in U.S./Canada latitudes in 2020 after six launches, with full coverage of the populated world after 24. SpaceX hopes that Starlink profits will support its plans for human travel to Mars.

There are other wireless technologies besides radio waves. Microwave signals were once popular, though they require connected locations to be visible to each other and can be disrupted by rain or snow. Infrared communication eliminates the need for wires within offices, using a line-of-sight connection to a ceiling-mounted central point. Neither of these is installed much today, though many existing installations remain in use.

The Internet does not use all these communication technologies. The GPS network is separate from the Internet. Mobile devices that get GPS data and send it to an Internet-based navigation application obtain this data via one network, process it, and then transmit it over a different one. Similarly, cell phone calls are separate from the Internet. You can get on the Internet via your phone, but that requires your phone carrier to know what kind of communication you want (not difficult) and then connect you to either the phone network or to the Internet.

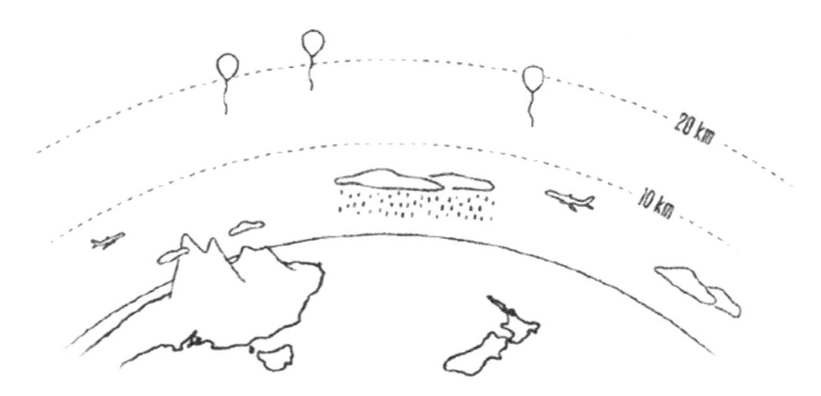

FIGURE 6.4 Drawing of Project Loon (Source: Google).

One-Way or Two-Way?

You should also know these terms and concepts:

- A *simplex link* communicates in only one direction. Lab sensors work this way. So do broadcast television, computer-projector cables, and garage door openers.
- A *full-duplex link* can communicate in both directions simultaneously.
- A *half-duplex link* can communicate in either direction, but only one way at a time. These links are common because equipment often operates faster in half-duplex mode than in full-duplex. Half-duplex devices mimic full duplex by alternating between sending and receiving.

Where you fit in: Individuals and businesses have many choices in the way they link devices on their premises and with the outside world. Their choices involve tradeoffs in cost, installation ease, speed, contract terms, security, and more. These are business tradeoffs. Businesspeople must understand what these tradeoffs mean from a technical standpoint to make them intelligently.

6.3 LOCAL AREA NETWORKS

As you read earlier, a network is a collection of communication links and equipment that enable several devices to communicate with each other. A *local area network* (LAN) is a network that covers a restricted area such as a building or a group of nearby buildings.

Why use a network that can only cover a limited area? Because of the tradeoffs in Figure 6.5. Such tradeoffs are common. One wants to achieve two objectives with a service or product. Cost control is usually a third goal. It may be impossible to accomplish all three.

Here, the options are:

1 Long distance, high speed. This can be done, if one accepts high cost.
2 Long distance, low cost. This is also practical, if one accepts low speeds.

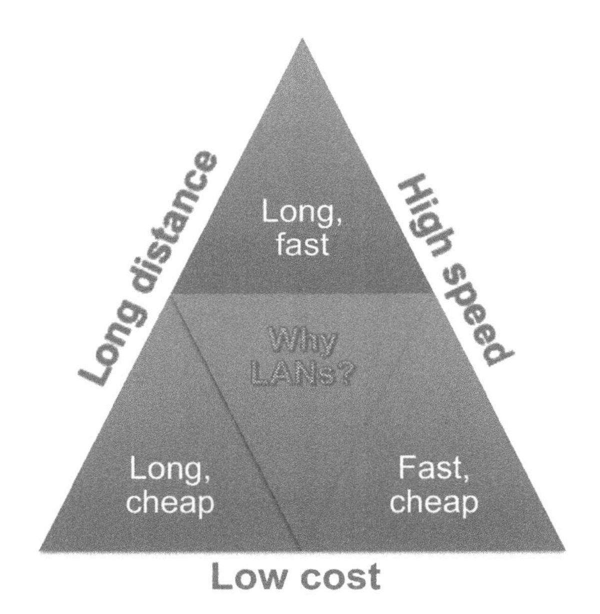

FIGURE 6.5 Data communications tradeoffs.

3 High speed, low cost. This is ideal if distance restrictions are acceptable. This choice leads to LANs. Few homes or businesses use any other kind of internal network.

Other tradeoffs are possible. For example, some wireless applications prioritize low power use to maximize battery life. These three, however, are those that make local area networks relevant.

LANs, like long-distance communication links, are of two types: wired and wireless.

Wired LANs

The dominant wired LAN technology in 2019 is Ethernet. It uses cables and connectors shown in Figure 6.6. Figure 6.7 shows an Ethernet network. Many computers today have Ethernet ports, though smaller laptops often omit them because wireless LANs are so widely available.

An Ethernet network has no central "traffic cop." When a device has a message to send, it listens to see if another transmission is in progress. If one is, it waits until that transmission ends. Then it starts to transmit.

What if another device was also waiting, and both send at the same time? When a device sends, it also listens. If two devices send at the same time, they notice a *collision*. They stop, wait a short random time, and try again. (This is what happens when you're with a group of friends and two of you start to talk at the same time.) Soon one message goes through. When that message is over, the other device sends. This takes microseconds. Users don't usually notice.

Today's state of the art is 100 Gbit Ethernet. It can move data at 100 billion bits per second (over ten billion bytes per second). The practical maximum is about a third of this figure. When the load exceeds this level, collisions happen so often that useful data transfer drops off.

Earlier Ethernet was slower: first 10 Mbits per second, then 100 Mbits, 1 Gbit around the year 2000, 10 Gbits in the early 2000s. 100 Gb Ethernet is still a small fraction of all Ethernet networks, but its share of market revenue grew from 8.1% in the second quarter of 2017 to 13.2% a year later. Ethernet is *backward compatible*: Devices built for a slower standard can participate in a newer network. They communicate at the fastest speed that they all support.

FIGURE 6.6 Photo of Ethernet cable with plug (Source: Wikimedia Commons).

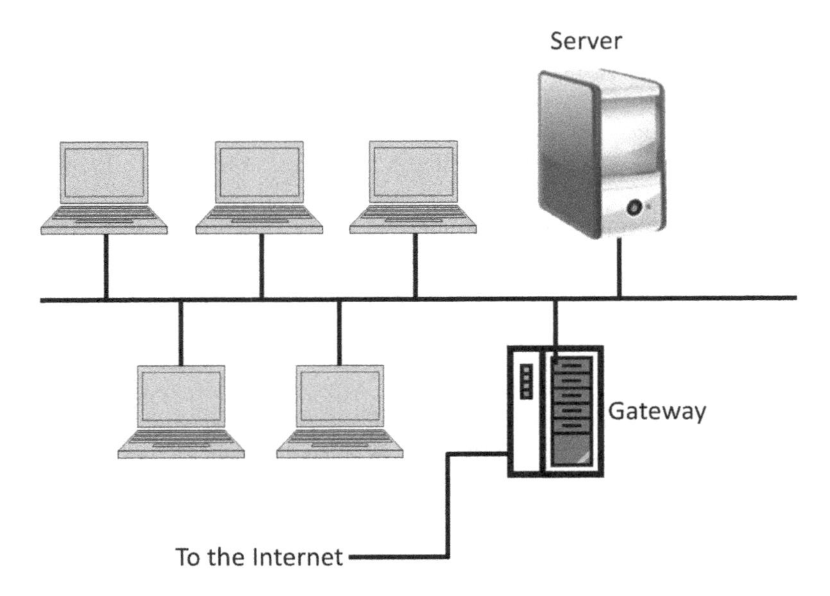

FIGURE 6.7 Diagram of local area network.

Wi-Fi

Most wireless LANs today use a technology called *Wi-Fi*, a pun on the term "hi-fi" that stands for *wireless fidelity*. Wi-Fi networks use routers which send and receive radio signals. Most routers connect to organizational networks and on to the Internet via wires, often fast Ethernet. Some also have Ethernet ports to connect nearby devices.

Wi-Fi versions are identified by numbers. Wi-Fi 5 replaced Wi-Fi 4 as the current version in 2013. You'll also see Wi-Fi versions referred to by the international 802.11 standard that they support—especially in products originally developed before 2020, charts and tables such as Figure 6.8 that originated before then, and so on. Wi-Fi 4 was originally known as 802.11n; Wi-Fi 5 as 802.11ac. This older notation will become less common over time.

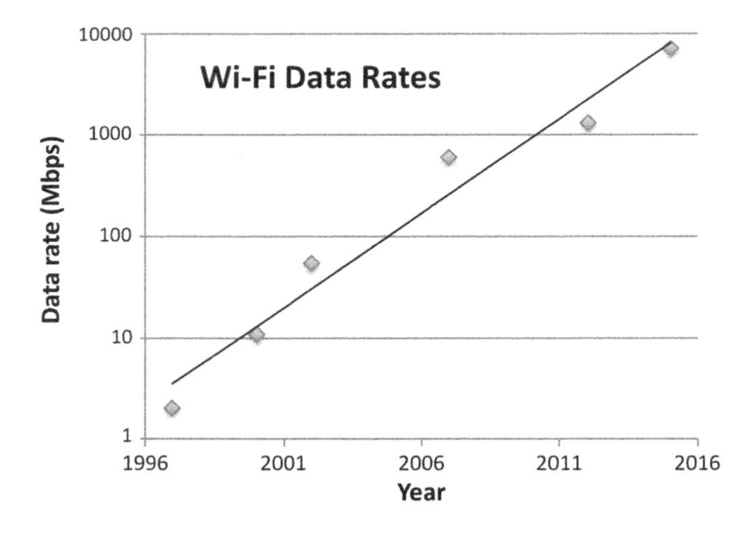

FIGURE 6.8 Chart of Wi-Fi speeds

Like Ethernet, Wi-Fi is backward compatible. If you have an old laptop that supports Wi-Fi 3 (802.11g) and use it with your home Wi-Fi 4 router, they will communicate at Wi-Fi 3 speeds. If you then replace that laptop with one that supports Wi-Fi 5, it will talk to your router at Wi-Fi 4 speeds. Earlier versions of the Wi-Fi standard were more than sufficient for web surfing, but uses such as streaming video require more speed, especially if more than one person is using the router. You can upgrade parts of a network that need speed while leaving other parts alone.

In 2019, Wi-Fi 6 (previously 802.11ax) is poised to replace Wi-Fi 5. It is expected to deliver data four times as quickly. Several semiconductor manufacturers have announced chips to support Wi-Fi 6, and Samsung has announced that its Galaxy S10 phone will support it.

The full potential of newer Wi-Fi versions is more than just moving bits faster. They can aim a beam to a specific device and combine multiple transmission streams into a single data stream. Higher speed means that portable devices can finish transmissions sooner, reducing battery drain. And their total speed can be split among several slower devices so they don't see the slowdown that they would if they had to share a slower link.

Wi-Fi Privacy and Security

Wi-Fi has security risks. They can be minimized or avoided at a cost in convenience. A common problem in many organizations is that people do not want to give up convenience for the invisible concept of security.

When a computer communicates wirelessly, it transmits and receives radio waves. Any receiver can pick them up. Most software is designed to receive only messages intended for it, so you have probably never received one meant for anyone else. However, it's easy to get software that will pick up all Wi-Fi traffic. (Such software has legitimate uses in network testing.) People communicate in public places. Anything they send or receive can be intercepted from the next table at a coffee shop, one row back at a conference, or two rooms down the hall at a hotel.

A Wi-Fi router can prevent this by encrypting transmissions. Instructions come with every router and are not hard to follow. You should do this, or, at work, make sure it's done. That doesn't help when you send a customer analysis from Starbucks or Sheraton. There, you must protect yourself by encrypting your data communication.

Where you fit in: Always encrypt transmissions when sending business information, even if the router doesn't do that automatically. Most email services offer this as an option. How to do this varies from one system to another. Once you're on the job, find out how—and do it.

Another risk of using Wi-Fi in public places: Suppose your laptop selects your home network automatically, as most do. When it starts up, it sends a signal to see if that network is in range. Say you're in a coffee shop across town. A device such as the Wi-Fi Pineapple, in the backpack of someone who looks like any other customer, can reply "Yes, I'm your home network." From then on your communications go through it. You won't notice a thing, but the crook will be able to collect your account numbers, passwords, and more.

You can avoid being victimized. When your computer selects a network automatically, check what network it thinks it's on before surfing. If it thinks it's on your home network but you're miles away, disconnect! Then select the coffee shop's network yourself.

Other Wireless LANs

Wi-Fi doesn't suit all local wireless connection needs. Many uses require less range and benefit from the ability to use less power or less expensive electronics. These tradeoffs and others have led to the development of other wireless LAN types:

Bluetooth

Typical range: 30′ (10 m)

Typical speed: 3 Mbits/second

Bluetooth is an alternative to wires for devices in close proximity: cameras to upload photos to computers, phones to headsets and car audio systems, computers to keyboards or printers. It is easy to connect Bluetooth devices, requiring only a simple and quick *pairing* process.

WHY "BLUETOOTH?"

This name comes from King Harald "Bluetooth" of Denmark. He unified the Danish islands into one kingdom around 960 AD (and had, even by Viking standards, bad teeth). Today's Bluetooth unifies devices into one network.

Radio Frequency ID

Typical range: 40′ (12 m), shorter or longer for specific applications

Typical speed: Slow, under 10,000 bits/second, but RFID messages tend to be short

RFID transfers information from a *tag* (or *transponder*) to a reader. Applications include product tracking for shipping and inventory management, identifying people via passports and animals via implants, highway toll collection, and more. The low cost of an RFID tag allows their use in library books, store merchandise, and individual medication doses. They can be small: One group of researchers put RFID tags about 0.1″ (less than 3 mm) square on ants to study their behavior.

RFID tags can be *active*, *passive*, or *battery-assisted passive*. Active RFID tags contain batteries and send out signals periodically whether or not there is a reader in their vicinity. Passive tags are powered by radio waves from a reader. Battery-assisted passive tags have batteries, but transmit only when interrogated by a reader.

Tags may be read-only, resembling an electronic bar code but with more information; writeable once; or updateable. A bar code on a camera box can identify it as a Nikon Z7; an RFID tag can give its serial number. Passive read-only tags cost as little as a few cents each in 2019.

U.S. passports issued since 2007, and most non-U.S. passports, contain an RFID chip. It is for use by immigration officers and others in related positions, but it can be read by anyone close enough. However, a U.S. chip passport contains no personal information. It identifies a record in the State Department database. Without that database, information in the chip is useless. Some countries' passport chips store personal information, but it is encrypted. So, RFID-blocking passport cases that travel sites and stores sell aren't needed. (If you want protection anyhow, wrap your passport in aluminum foil. It's just as effective, and a lot cheaper.)

Near Field Communication

Typical range: 4″ (10 cm)

Typical speed: 1, 2, or 400,000 bits/second

NFC is a variety of RFID that uses technologies that require the transmitter and receiver to be very close to each other. That improves security, since an eavesdropper is unlikely to get within a few inches of a device without being noticed, but sophisticated antennas make it possible to pick up NFC signals from a few yards/meters away. This is often far enough to escape notice.

Applications of NFC include contactless payment systems such as Apple Pay, social networking uses such as "bumping" two phones to exchange contact information, specialized devices for rapid transit or parking garage use, automatic setup of Bluetooth pairing, and many more. The common factor is that the two devices can get very close.

ZigBee

Typical range: 30–300′ (10–100 m), depending on obstructions and communication frequency

Typical speed: 250,000 bits/second

ZigBee is used for wireless control and monitoring: industrial control, home entertainment, medical data collection, hazard monitoring, building automation, and more. Each ZigBee network has a central *coordinator* and as many nodes as needed. Since ZigBee might be used to control devices that could be the targets of sabotage, it includes strong security features.

ZigBee supports *mesh networks.* In a mesh, nodes can communicate via other *router nodes* between them. A ZigBee node can communicate with its coordinator indirectly, via the nearest router node, as long as a path of router nodes exists to the desired endpoint.

Personal Area Networks

When the devices in a small LAN are carried by one person, such as a phone that sends Caller ID to a watch, the network can be called a *personal area network* (PAN). Its technology can be Bluetooth, any of several less well-known wireless technologies, or even a wired connection such as USB. A PAN that monitors its wearer's vital signs is called a *body area network* (BAN).

Where you fit in: You don't need to memorize the technical details of wireless protocols. They will change. You should know that there are different wireless protocols, each optimized for a different application area, and choosing the right one can affect a system's success. In a business situation, be prepared to ask questions of those who propose a network type for a new application.

6.4 WIDE AREA NETWORKS AND THE INTERNET

A *wide area network* (WAN) is designed to cover a geographic area, often anywhere on our planet. Most WANs today use either fiber optic or satellite communication links.

The key characteristic of a network, as opposed to a communication link, is that it includes switching nodes to control which links messages will use. Node types include:

Repeaters, which amplify signals so they can travel further but have no intelligence to control message routing. If a series of seven links numbered 1…7 is connected by repeaters, and a device in link 1 sends a message to a device in link 3, repeaters will pass the message to links 4–7 even though its destination is not on those links or reachable through them.

Bridges, which connect one link to another and reject messages that are not addressed to a device on their other side. In the same line of seven links, bridges between links 1–2 and 2–3 will pass the message, but the 3–4 bridge will not. This reduces traffic on links 4 through 7.

Switches, which connect several links at one point and behave like bridges between every pair. If the seven links were all connected to a switch, it would send the message only to link 3.

Routers, which look at the final destination of a message and select a link to use to route that message to that destination, often choosing from several links that all eventually lead there.

Gateways, access points from one network to another. Gateways between an organizational network and the Internet are common. Your school has one. Your future employer will, as well. At home, your Internet service provider functions as your gateway.

Where you fit in: It is not important for you to know the technical differences between a bridge and a router. People who design and install networks know what devices to use, where, and why. It's important to understand that having the right devices in a network is as important as having the right computers on users' desks or the right servers in a data center.

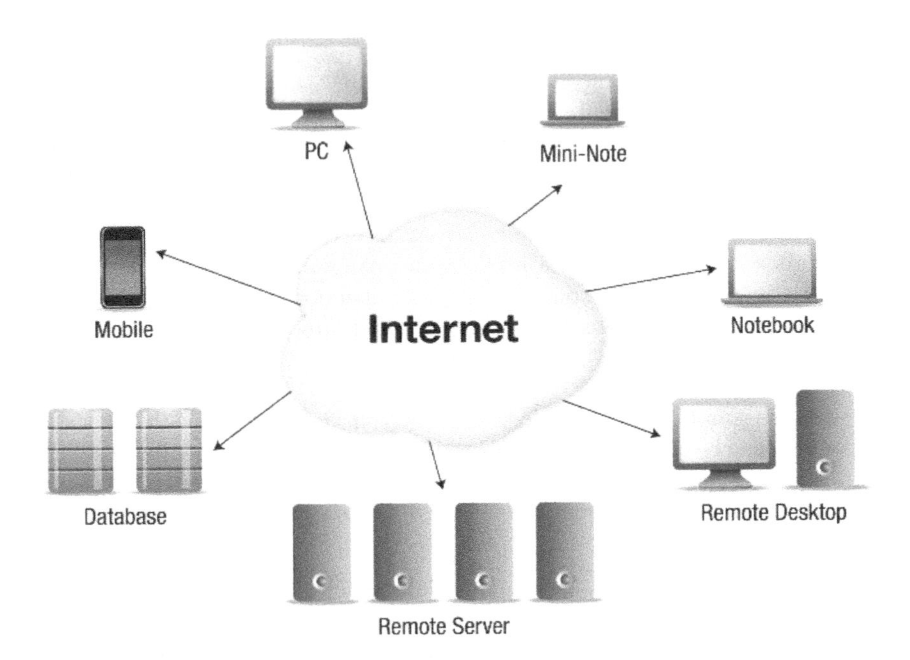

FIGURE 6.9 Diagram of the Internet.

The major worldwide WAN today is the *Internet* (Figure 6.9). Since its inception as a U.S. military project in 1969, it has grown to encompass people and organizations in every country, and beyond the earth to the International Space Station. Today's business and social life could not exist without it.

Internet users don't pay to use it. That has enabled it to grow rapidly. We all pay indirectly, in the U.S. through taxes on Internet connections and a portion of the National Science Foundation budget, but how much we pay doesn't depend on how we use it. This affects the economics of communication and has led to the establishment and growth of new businesses.

The *Internet backbone* (the high-speed links that connect major networks and metropolitan areas) is operated by telecommunications firms. It uses a variety of technologies, primarily fiber optics. As a businessperson, you can safely assume that the Internet backbone will continue to evolve as technology develops and needs demand.

Where you fit in: When your employer needs to communicate for business, it will almost always use the Internet. The better you understand how it works, the better you'll be able to use it. When Internet evolution requires changes in how your business connects to or uses it, the more you stay current on it, the more advance notice you'll have of those changes, and the better you'll be able to handle them in a planned and cost-effective fashion.

Packet Switching

A fundamental concept of the Internet, which explains some behavior that might otherwise seem mysterious, is *packet switching.*

In packet switching, messages are broken up into *packets* of about 1,500 bytes. Packets are numbered and sent individually. When they arrive, they are reassembled by number. When an entire message has been received, it is delivered to its destination.

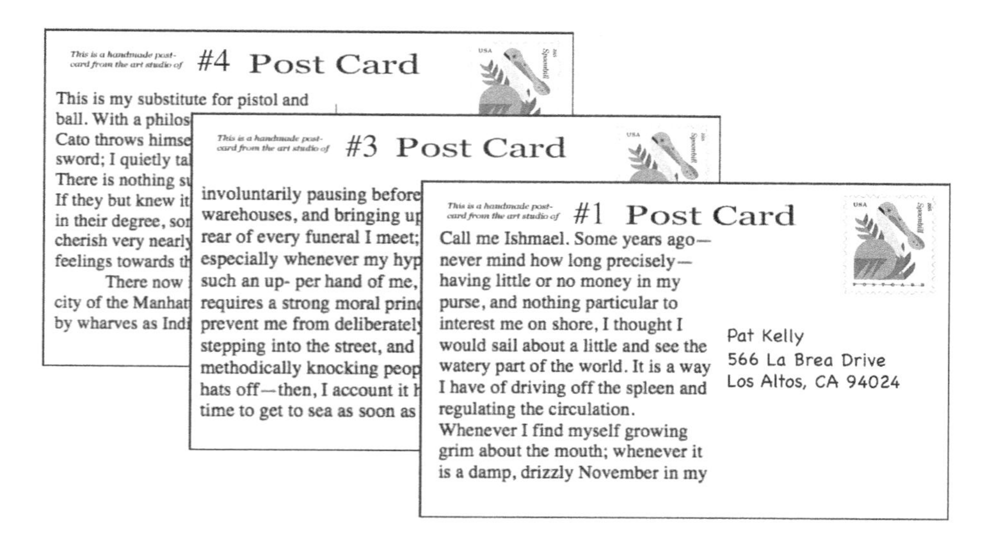

FIGURE 6.10 Packet switching on postcards.

This is like copying *Moby-Dick* onto 10,000 postcards, numbering them, and mailing them to a faraway friend (Figure 6.10). They take different routes. Some go by rail, others by air, with different transfer points. If a postal worker sees that a truck is full, other cards go on the next truck or a train. The cards arrive out of order, yet the book can be assembled from them after they arrive.

What if postcard 4,783 doesn't show up after a reasonable time? Your friend writes back, saying "Thanks, but I never got card 4,783." You make a new copy of that card and send it. You do the same if she writes, "I can't read card 1,815 because part of it was torn off." Then all is well.

Packet switching lets computers share a network efficiently. It lets the Internet bypass broken links. It helps balance the load on different links. Perhaps its most important benefit is that *packet switching creates reliable connections from imperfect physical links.*

No communication link is completely reliable. Modern links come close, 99.99% followed by several more 9s, but that's not 100. Even the best connection is occasionally disrupted by stray electrons or photons that turn a 1 into a 0 or vice versa.

Say you email a photograph to a friend. A typical photo is about 5 MB, or 40 million bits. If your photo traverses ten Internet links, 400 million bits move over links. If every link is 99.999999% reliable, there will be, on the average, one error per 100 million bits. On average, four bits will change in transit. Odds are that a few bits will change in any one photo: probably not exactly four, but a few. Software can't tell if a change is trivial or critical. All it knows is that there was an error. It asks the sender to retransmit.

Sending the photo the second time still requires 400 million bit transfers. The odds of success are no better. There are different random errors. This repeats until, in an hour or a week, fate smiles and the photo gets through. A ten-minute HD video would probably never arrive without error.

But what if your photo goes in 1,500-byte packets? Almost all arrive correctly. Your friend's computer asks yours to send packets 61, 904, and 2,789 again. It does. They arrive error-free. The photo is complete. The added delay is a fraction of a second. Your friend doesn't notice.

Alternative Switching Methods

You should also be familiar with the names of two other switching methods:

1 *Circuit switching*, where a communication path remains available for a conversation even if nothing is being sent over it at a given moment.

2 *Message switching*, where an entire message is sent as a unit and the communication path is then made available for other messages. Messages in a conversation may use different paths.

6.5 INTERNET APPLICATIONS

The Internet has, as you know, many uses. Here are the major ones. As a businessperson, you should be generally familiar with how they work—those you've used, and those you haven't.

The World Wide Web

Picture a researcher sitting reading a scientific paper on a computer screen 30 years ago. The paper says "As Wu found in 1985, …" with a bibliography entry for Jones's paper. The researcher needs to see it, but doesn't have a copy.

That researcher would go to a library that has Jones's paper and read it, photocopy it, or check it out. A team at the European Center for Nuclear Research (CERN, using the French word order) headed by Tim Berners-Lee thought it should be possible to click on the citation and have Wu's paper appear on the screen. The product of their work became available on August 23, 1991. None of those involved foresaw how their system would change our daily lives.

Web 1.0

A NOTE ON LANGUAGE

"Web 1.0" is a *retronym*: a word invented after other items of the same type come along. "Land line phone" is one: until mobile phones came out, they were just phones; there wasn't any other kind. Similarly, until Web 2.0 arrived, what we now think of as "Web 1.0" was just "the web."

The web is based on a few fundamental concepts:

- ***Uniform, universal access.*** Every page on the web has a Uniform Resource Locator (URL). Figure 6.11 shows a sample URL.
- ***Hyperlinks.*** Wherever you go on the web, you find clickable text or images. When you point to one, the pointer turns into a pointing hand. Click, and you're somewhere else.
- ***Discovery.*** Web developers have always been concerned with finding useful content. We use search engines such as Google, Bing, or Baidu for discovery.

The sample URL in Figure 6.11 has several parts:

`http://` tells the Internet to use *HyperText Transfer Protocol* (HTTP) to send this page. This is the standard protocol for web pages. Other Internet uses, such as email or file transfer, use other protocols. If you see *https://* the site encodes data for privacy.

`www` indicates the server at the site owner's data center that this request should go to. Using *www*, for World Wide Web, is common but not universal. Apple's store is at *store.*

`http://www.riccardis.com/new-bedford-restaurant.htm`

FIGURE 6.11 Sample URL

apple.com. If you have a blog at the popular blogging site *blogspot.com*, it's *blogname.blogspot.com*. (These servers are conceptual, not physical. Each *blogspot.com* server supports many blogs. Apple's store uses many computers.)

The last part of a domain name is its *top-level domain* (TLD). Originally there were few TLDs, listed in Figure 6.12. As the Internet grew, this caused problems. Should Hippo Shoes or Hippo Burgers be at *hippo.com?* The first to claim a domain name usually has it, as long as that buyer has a valid interest in the name and is not just grabbing it to force another business to pay for it. (That practice, common in the less-regulated early days of the Internet, is called *cybersquatting*.)

Here, `riccardis.com` is the domain name. Riccardi's is a two-restaurant group in southeast Massachusetts. A domain name covers all its owner's Internet activity, not just its web site. A Riccardi's employee might get email at `person@riccardis.com`. Someone at Riccardi's had to register this domain name with the Internet Corporation for Assigned Names and Numbers (ICANN), which manages Internet domain names, while hoping that Riccardi's Beef & Brew near Philadelphia didn't register it first.

To alleviate the problem of demand for similar domain names, ICANN has added new top-level domains since 2000. Now the shoe store can be at *hippo.com*, while the burger chain is at *hippo.biz*. The two unrelated Riccardi's restaurants can also be in different TLDs.

That doesn't entirely solve the problem, since many companies register domains in multiple TLDs to avoid visitor confusion. If you try to visit Weir's Furniture (four stores around Dallas, Texas) at *weirsfurniturevillage.biz*, you end up at *weirsfurniturevillage.com*. Other new TLDs include *.name* for individuals and *.pro* for professionals such as accountants and lawyers. There were 22 TLDs in December 2013. Over 100 more came in 2014, including *.music*, *.boston*, and *.bmw*, in what has been called the "dot anything expansion." By March 2019 there were several hundred. One in five new domain registrations in 2018 was in a new *generic TLD* (gTLD). Such registrations grew by 11% that year, versus 3.5% growth in registrations overall.

Will all these domain names be useful? Some people feel it will be easier to find sites of interest within a small "neighborhood" than on the entire web, or that the domain name *grg.consulting* is more descriptive than *grg.com*. Others feel they are confusing. A third group believes that, since most searches use search engines or lists of links, it doesn't matter much. In any case, these TLDs are here to stay. More will come.

Countries have two-letter TLDs. English speakers can figure many of them out: *.fr* is France, *.uk* is the United Kingdom, *.jp* is Japan. Some come from countries' names in their languages: *.de* is Germany (**De**utschland), *.es* is Spain (**Es**paña). Country TLDs may lend themselves to particular uses: television stations obtain domains in Tuvalu to get, for example, *wabi.tv*. Selling domain names provides significant foreign currency income to several small countries.

Larger countries often divide their TLDs into subdomains. For example, *oxford.ac.uk* is Oxford University, within the *.ac* (academic) subdomain of the *.uk* TLD. Some countries use U.S. TLDs as their subdomains; Australia's Monash University is at *monash.edu.au*. Others, such as the U.K. as the Oxford example shows, use different letters for their subdivisions.

TLD	Original Purpose
.gov	Non-military U.S. government agencies
.mil	Military U.S. government agencies
.com	Businesses and similar commercial organizations
.edu	Educational institutions such as schools
.net	Organizations involved in operating the Internet itself
.org	Organizations that don't fit any of the above categories, primarily non-profits

FIGURE 6.12 Original Internet TLDs.

Pages within a site are specified after the domain name, separated by a slash. The part of Figure 6.11 after the domain name, `new-bedford-restaurant.htm`, identifies a page on Riccardi's site. Pages can be structured within folders, several levels deep.

The Internet doesn't use domain names internally. *riccardis.com* must become 216.36.247.208 for devices to know where to send messages. Such an *Internet Protocol address* or *IP address* consists of four 8-bit numbers, each 0–255. Those allow about 4.3 billion combinations. This was once felt to be ample, but IP numbers are running out. Organizations are moving to *Internet Protocol version 6* (IPv6). It uses 128 bits for addresses. That's about 3.4 followed by 38 zeroes. If a billion IPv6 addresses had been assigned every second since the earth was formed, we would have used up less than one trillionth of them. The number of usable IPv6 addresses isn't as large as it seems, because addresses are often allocated in large blocks, but you won't have to worry about them running out during your lifetime.

Be careful with domain names!

Be careful when you click on a link whose source you're not sure of. Criminals register domain names that resemble real ones. Suppose you get an email asking you to visit Wells Fargo Bank and linking to *we11sfargo.com*. Did you notice that the third and fourth characters are digits? Many people wouldn't. If someone registers the altered domain name and puts up a site that looks like the real Wells Fargo Bank site, taking its graphics from that site, people might be fooled into entering account numbers and passwords. Most email programs and computer web browsers show where a click will take you before you click, but phones with small screens may not. You may have to take an extra step to check the target of a link. Check it, even if that's a hassle.

Where you fit in: Your employer will have a web site. Sites are often developed by technical experts who don't have a user point of view. These experts can give a site any desired content, appearance, and functionality, but they can't decide what those should be. That depends on the site's business objectives. A site can meet many objectives, but it can't meet them all at the same time. Your business focus and guidance will be vital. More on this in Chapter 8.

The content of most traditional web sites comes from the organizations that own them. In newer sites it may not be. That's where Web 2.0 comes in.

Web 2.0

Web 2.0, the Social Web, differs from the original World Wide Web in that most content comes from sites' users, not their owners.

The basic content of YouTube is user-submitted videos. Some show cats doing funny things, but many have business value. If you want to learn Microsoft Excel data bars, key "excel data bars" into YouTube's search box. You'll retrieve thousands of videos. Tutorials on data bars will be first. Then come videos that cover data bars along with, say, conditional formatting. You'll also see what people who viewed each video thought of it.

YouTube isn't a free-for-all. It is curated—monitored to ensure that content follows its rules—but its owner (Google) doesn't create its content. Craigslist and eBay don't create the products people sell or set their prices. Facebook doesn't decide who you friend or what your relationship status is. (It has been criticized for being too "hands-off" and not blocking content that, the critics feel, contributes to social ills.) Twitter doesn't tweet (except occasionally, about itself). And so on.

Any site can add Web 2.0 features. Amazon.com carries user reviews of products that it sells. Amazon and its partners still provide most site content, but user-written reviews enhance it.

Social networking sites are good ways to connect with customers and build customer loyalty. To encourage this, companies provide benefits to their social media friends, such as discount coupons in return for Liking a company on Facebook. Instagram reached one million advertisers in March 2017, evidence that businesses believe it can reach their target markets effectively.

Companies can learn from social media. Hotels and restaurants monitor Yelp and TripAdvisor, responding when it seems appropriate. (Some have been found to post fake reviews, and review sites have been faulted for not doing all they can to remove fakes.) Companies can follow tweets that include their names. Analyzing tweet activity following a TV commercial can yield insight into advertising effectiveness, though tweets don't necessarily result in sales.

> ***Where you fit in:*** Organizations of all types use social networking to their advantage. Companies have Facebook pages, blog about their activities, recruit employees via LinkedIn, tweet the latest corporate news, put training videos on YouTube and monitor what customers say on TripAdvisor. Doing these *right* is more than a technical task. It requires business understanding.

Streaming

Sending an email, photo, or spreadsheet is not time-critical. Nothing happens if it's delayed a second, or even a minute. We know that it will arrive eventually. For other purposes, seconds matter. There, *Internet streaming* is crucial.

Internet Telephony

International phone calls can be costly. In the 1990s, people realized that they could be cheaper:

- A phone call sends data from one point to another. The data may represent sound waves, but a network doesn't care.
- Sending data anywhere via the Internet is free.
- Local phone calls are free or inexpensive.

Overseas calls can cost less if we make local calls at each end to nearby switching centers, and send the data over the Internet between them. A provider must set up switching centers and get local lines in many cities, but once those are in place, it can offer cheap international calling.

This is *Internet telephony*. The technology is called *Voice over Internet Protocol*, *VoIP* for short. (Some people pronounce this as a word, "voyp." Others say the letters "V-O-I-P.")

A packet can carry 10–40 ms of voice data, so VoIP systems send 25–100 packets/second. Packets don't always arrive in order, but people need to hear them in order. VoIP systems handle this by:

- Delaying transmission by 100 or 200 ms so that all, or nearly all, packets arrive by the time they should be played. We don't notice such short delays.
- Reconstructing a missing packet using waveforms from the packets before and after it. When only one packet is missing, we usually can't tell this was done.
- As a last resort, garbling the transmission. The listener can then say "Sorry, I missed that." We're used to this. We don't mind, as long as it doesn't happen too often.

VoIP is popular. Microsoft's VoIP subsidiary Skype had 1.33 billion registered users at the end of 2017. That number is forecasted to reach 2.27 billion by the end of 2024, and Skype isn't the only VoIP provider.

Companies that make many overseas phone calls can use their own phone systems as switching centers. To call a U.K. customer from a U.S. office, one calls the company's London office (using the Internet) and keys in codes to place a domestic U.K. call from there. The concept is the same.

Where you fit in: Understanding the benefits and drawbacks of VoIP can increase your value to your employer, as well as potentially saving you money in your personal life.

Streaming Video

The Internet can transmit video in the same way that it can transmit voice. You can start watching a video before it's all been downloaded. That provides two advantages:

- To the person or company that has the video: They don't have to send the video to your computer. That reduces opportunities for piracy, though you can still record it as it plays.
- To the viewer: The video starts right away, not minutes (or hours) later.

Since video transmission is in one direction, streaming video can deal with out-of-order packets by not starting until a *buffer* of video is in hand. Once your computer has data for several seconds of video, the show starts. As the video plays, your computer receives more data, staying ahead of what's on the screen. If playback catches up with the end of the buffer, it will pause for a bit. You may see a "stand by" indicator or a *rebuffering*—refilling the buffer—message.

Teleconferencing

Combining VoIP and streaming video creates a video phone call. It's interactive, not one-way viewing of prerecorded content, so it doesn't tolerate delays well. Still, with fast links and allowing for an occasionally jerky picture, it can be the next-best thing to being there. You don't even need a computer: most phones' front-facing cameras support video conversations using a cellular connection or Wi-Fi.

Video-enabled conference calls can connect several people. Each screen shows as many of the other participants as a user wants to see at a time. (Users see *thumbnails*, small pictures, of all participants.) In a videoconference facility, each remote participant appears full-size on a monitor across the table. Such a *telepresence* facility comes close to the feel of a face-to-face meeting.

Several suppliers offer facilities for videoconferencing, such as the Teliris installation shown in Figure 6.13. Organizations that use them regularly can buy equipment for a conference room. Others can rent fully equipped rooms as needed.

FIGURE 6.13 Telepresence facility (Source: Wikimedia Commons).

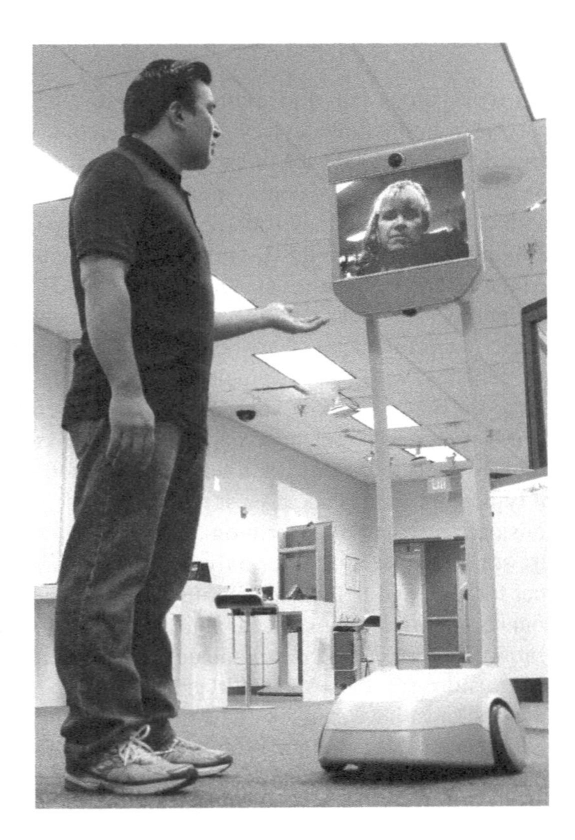

FIGURE 6.14 Telepresence robot (Source: Wikimedia Commons).

A high-quality telepresence conference requires high-speed, high-quality communication lines. Those can cost more than the equipment, so a decision to install such facilities—a conference requires at least two—should not be undertaken lightly. Despite the cost, this market is growing at about 20% per year and is expected to be a $2.6 billion annual business in 2022.

The fastest growth in telepresence is telepresence robots. These enable students to attend classes "in person" when they can't actually be there, doctors to be present in an operating room when they're thousands of miles away, and managers to monitor production lines or warehouses from their office. A telepresence robot (Figure 6.14) has a camera, a display, a microphone, a speaker, motorized wheels for mobility, and a computer to make it work. With some priced under $1,000 in 2019, there are many opportunities to use telepresence robots in business.

Net Neutrality

Streaming applications are affected by the concept of *net neutrality:* the principle that all Internet packets should be treated equally. This means that a packet in unsolicited email should be transmitted with the same priority as a packet in a videoconference.

Proponents of net neutrality do not dispute that it can degrade performance for applications such as VoIP, but are concerned that abandoning this principle can lead to favoring those who have more money and to invasion of privacy as carriers inspect packets to determine their priority.

Net neutrality is an open issue in the U.S. It was required in 2017, but on December 14 of that year the Federal Communications Commission ruled against it effective 180 days later. In 2019, this repeal is being challenged in court by states and technology companies. Legislation to reverse it is under discussion in Congress. The FCC, in turn, is challenging states that pass their own net neutrality requirements. Time will tell how this issue is ultimately resolved.

The Internet of Things

We've seen how devices of various types can send and receive data. We usually think of this in the context of a person sitting at one or both of the devices in a conversation: We're chatting with someone, posting on a social network page, ordering a T-shirt, or downloading a video. However, nothing says that humans have to be involved. When the communication involves devices that perform tasks not generally associated with computing, be they robotic home vacuums, self-driving cars, microwaves that get instructions by interrogating an RFID chip on a package of frozen food, or anything else, we have what is called the Internet of Things (IoT).

The IoT can multiply the number of Internet-connected devices manyfold. A car has dozens of computers, most of which have to communicate. Cars being designed today will communicate with each other, with emergency vehicles, with traffic signals, and more. If two cars are on a collision course, one or both will act to prevent a crash. If you don't stop for a red light, your car will. And if an ambulance is coming down a cross street, your car will stop—even if the light facing you is green. If you don't want your next car to do this, plan on getting a used car. By the time you're in your mid-30s, if you are of typical university age now, it will have to be a pretty old one.

As another example, consider an irrigation system on a farm. Today's computer-controlled irrigation systems can measure soil moisture and water only if it's dry. An IoT-enabled irrigation system could check the weather forecast and not water if enough rain is expected.

Sensors are key components of IoT systems. A car must have sensors that detect its position, velocity, and direction of motion before it can communicate those to other vehicles. A farm irrigation system must have soil moisture sensors. An intelligent parking garage, if it is to tell drivers where they can find open spaces, must have sensors that detect when a space is open.

In business, anything with a microprocessor, sensors, and actuators is a candidate for the IoT. This includes devices such as milling machines and conveyor belts that previously had specialized connections, if any. Just as the Internet is generally used for internal company emails, rather than the proprietary long-distance networks that companies once used for that purpose, the ease and generality of connecting to the Internet will soon make the IoT the technology of choice for all device interconnections. This raises security issues that will be discussed in Section 6.6.

Other Internet Applications

The Internet use we think of most often is probably the web, but there are others. Here are a few:

Email

Email existed before the Internet. Companies with large computers had applications where users could send text files to other users. Email came into its own when the Internet connected most of the people one would want to send a message to, not just those inside one's own firm.

Email goes from one server to another, and between servers and users' mailboxes, via a few standard protocols. *Simple Mail Transfer Protocol* (SMTP) moves messages from server to server, say from the university server that sends your message to Google's server for your friend on Gmail. *Post Office Protocol* (POP) and *Internet Message Access Protocol* (IMAP) send it from your laptop to your school's server and from Gmail's server to your friend's tablet.

"But I see email in my web browser," you may think. When you access email in a browser, your computer doesn't receive those messages. The server builds a web page with the content of the message and sends that page, not the message itself, to your browser. The effect is the same—you see the email on your screen—but what happens behind the scenes is different. Conversely, when you download messages into Outlook, Apple Mail, or Thunderbird, they're on your computer. You can read them when it's not online. When you delete them, they leave its storage but might stay on a server. When you access email via a browser, though, you're in your email provider's server. You can only read email when you're online. Messages aren't on your computer unless you choose to download them.

File Transfer

You go to the iTunes store. You find a song you like. You pay 99¢. The download begins.

You don't get this song over the web. You found it on the web, you paid for it on the web—but the song never became a web page. Instead, you get it over the Internet—not the web, but still the Internet—via *File Transfer Protocol* (FTP). FTP is used to send large data files. It isn't just for music. If you download course slides, a Firefox update from *mozilla.org*, an audio book from *audible.com* or an airline timetable, or upload a video to YouTube, you use FTP.

Modern browsers can receive files via FTP. Some can also send them, though people who upload a lot of files may prefer an *FTP client* program. Some browsers display common file types once they are downloaded. If you download a PDF file in Firefox or Chrome, it will display when it's received. However, it won't look or act like a web page.

Other

The Internet has more *application layer protocols* (to distinguish them from the protocols that move data over the network). They include SSH to control one computer from another, UDP to translate domain names such as *toyota.com* into Internet address numbers, SNMP to manage network devices, and more. They work behind the scenes. Such protocols are part of the Internet picture, but you don't have to be concerned with their details.

6.6 NETWORK SECURITY

People try to access information systems without permission for many reasons. Organizations that use networks try to protect their networks and computers from unauthorized access.

The access control concepts discussed in Section 5.5 apply to networks. Your network identity determines what you can do via the network, what applications or resources you can access. Your identity can be authenticated on the basis of something you know, have, are, or do. Organizations are increasingly moving to a *single sign-on* (SSO) approach, where a person signs on to access the network and is then also signed on to all applications that he or she is authorized to use. While this simplifies access from the user viewpoint, it puts a premium on the security of that sign-on.

Network security and database security differ in the methods people use to try to break in and in the defenses that can be used against break-ins.

A key element in defending against unauthorized network penetration is the *firewall*. Firewalls detect attempts at intrusion and keep them out of a network. A firewall can also keep those inside a network from performing unauthorized activities (such as visiting entertainment sites) outside it. Firewalls for large networks are separate devices, positioned to protect a network from damage (symbolically, fire), as shown in Figure 6.15. Firewall software exists for all popular computers.

Some network attacks attempt to obtain access for malicious purposes. Those attacks resemble those used against databases: impersonating authorized users by guessing or stealing passwords, using *social engineering* methods to be let in by a person who believes a story, and so on. Once inside a network a person must still access an application or database to cause harm, but these are often poorly protected on the assumption that network security will keep evil-doers out.

A different type of attack, designed to disable a network or the computers on it with no direct benefit to its perpetrator, is called a *denial of service* (DoS) attack. A DoS attack floods a network with huge numbers of requests in a short time frame. The idea is to overload its equipment so it is unable to handle legitimate traffic and may crash.

Firewalls can foil DoS attacks. If a firewall detects many messages from the same source, it can reject them. As a result, criminals developed *distributed denial of service* (DDoS) attacks. To carry out a DDoS attack an attacker first infects other systems with a DoS program, without their owners' knowledge. The infecting program lies dormant until it is activated. Then, those systems launch

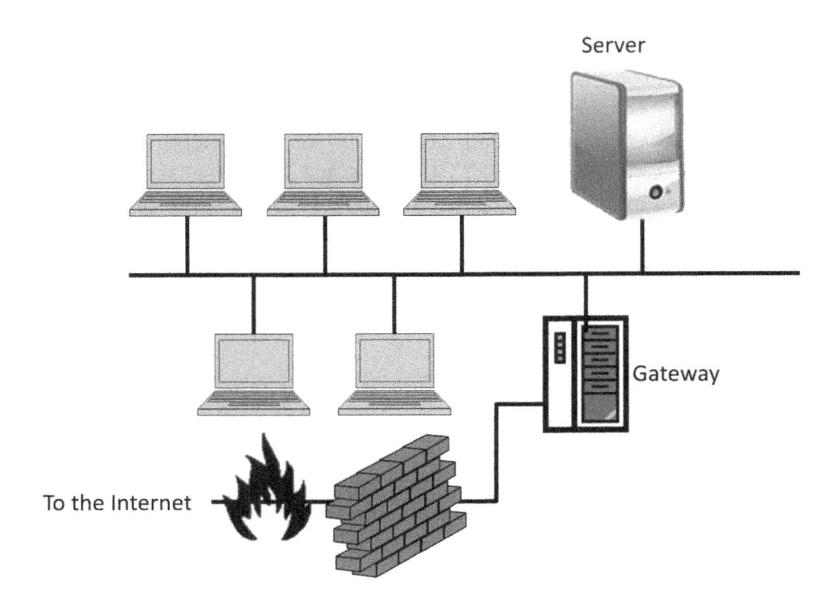

FIGURE 6.15 Diagram of network protected by firewall.

DoS attacks simultaneously. At this point users of infected computers may notice that their computers are working slowly or not at all, but it's too late to stop the attack.

A DDoS attack is harder to mount than an ordinary DoS attack, since the network of *bots* or *zombies* (often called a *botnet*) must first be infected by an attack program, usually through a *virus* or *trojan*. Only then can the attack be launched. Once it is launched, methods that defend against ordinary DoS attacks will not stop it.

GitHub, which Microsoft bought for $7.5 billion in 2018, is a hosting site for software and a vehicle through which many people can collaborate on a software project. In August 2019 it had about 40 million users. On February 28, 2018, it was the target of a massive DDoS attack. The attack originated from over a thousand different systems. At its peak, over 125 million packets per second were sent to GitHub, a data rate of 1.35 TB/second. GitHub detected the attack at 5:21 p.m. (all times UTC) and rerouted incoming traffic through Akamai's "scrubbing centers." Akamai's specialized servers kept DDoS traffic from reaching GitHub. GitHub was only offline until 5:26, though service was intermittent until 5:30. Under the circumstances, that is outstanding. It shows the importance of being prepared with contingency plans and watching network connections carefully.

A "zip bomb" is related to DoS attacks, though it is intended to disable processing and storage rather than communication links. It is possible to create a compressed file in the common *.zip* format that will expand to millions of times its original size. In one test, researchers created a 10 MB file that expanded to 281 TB, beyond the total storage capacity of most small- and medium-sized organizations. Users must be trained not to open compressed files of unknown origin.

Where you fit in: Businesspeople need not understand the details of network attacks or how to defend against them. (There are many more types.) They must be aware that they are varied and complex. As a manager, you must be sure your company's network security effort is adequately funded and supported. This can be difficult: Security doesn't increase revenue or profit, and if it works perfectly, nothing happens. Funding it is still necessary.

Human Aspects of Network Security

A chain is proverbially only as strong as its weakest link. The weakest link in network security is often the individual user. Many attempts to penetrate networks take advantage of human frailty.

Phishing

A *phishing* attack* is designed to fool a user, usually via email, into visiting a web site where that user will be asked to provide confidential information.

A common type of phishing attack tells people that their email access will be cut off unless they visit a web page and provide information. That page requests name, email address, and password. Many people, scared by the tone of this message, will visit that page and provide information—even if they never had an email account at that domain! Armed with this information, a thief can access emails that may include login information to other sites. Since many people use the same password for multiple sites, this may provide access to bank accounts, credit cards, and more.

Other phishing attacks claim that a person must requalify for online banking access. These emails often use graphics from the bank's web site, increasing their credibility. Again, recipients are told to visit a web page and enter their account number and online access password. Many people don't notice that the site they are sent to is not the bank's. This attack depends on large numbers: if a message that claims to be from TD Bank is sent to 10,000 people, most won't have accounts there, but a few hundred will. Some of those will fall into the trap. It doesn't take many victims to make the attack profitable.

A phishing attack could go to people at a particular company. Many names of employees at large companies are available in Internet discussion groups and other public sources. The format of corporate email addresses is easy to learn. An email asking "fellow employees" to renew their "internal network credentials" will probably find takers—all of whom will expose the company network to criminals.

The best protection against phishing is to be sure of the source of a request before providing confidential information. (It may not be what it seems.) Legitimate sites, including banks and credit card issuers, won't ask for it. Make sure the domain to which you are directed matches who the email claims to be from. Check its language: Many phishing emails use poor English. Don't click on links in the message; they may be to criminal sites whose domain names were chosen to resemble the real domain name to the eye. (Type the organization's URL directly into a browser address field.) Be especially wary of emails that are completely impersonal, that do not contain information such as the last four digits of your credit card number. Finally, confirm, by phone or email (not email Reply!), that it is real. The odds are at least 10,000 to 1 that it won't be.

If you get a suspicious email, wait a few minutes before replying. Stop and think. Most responses that disclose information come immediately after opening a message. Our human desire to cooperate, coupled with a tendency to obey authority, lead us to reply instinctively. Once we stop and think, we see things that alert us. You can also do a web search for some of the message text. You may find that thousands of others received the same email and that it is undoubtedly a scam.

Figure 6.16 shows a phishing email that was sent to the author. Notice these typical features:

- The U.S. Postal Service logo, taken from the USPS web site. Anyone can do this.
- A blatant grammar error in the subject line: "An package" should be "A package." The text above the button, "Delivery problems notification," is also not natural English.
- You can't tell in the screen shot, but the links, including the "View Details" button, all go to *http://www.hsm.co.th/wp-content/uploads/2019/saintlyfn.html*—a domain name in Thailand. (See the ".th" in the URL.) Most Thai sites are legitimate, but the USPS wouldn't have one.

* This term comes from *phreaks*, which once meant computer criminals. It changes *fishing* to *phishing* to parallel the change from *freaks* to *phreaks*. It is unrelated to the band *Phish*.

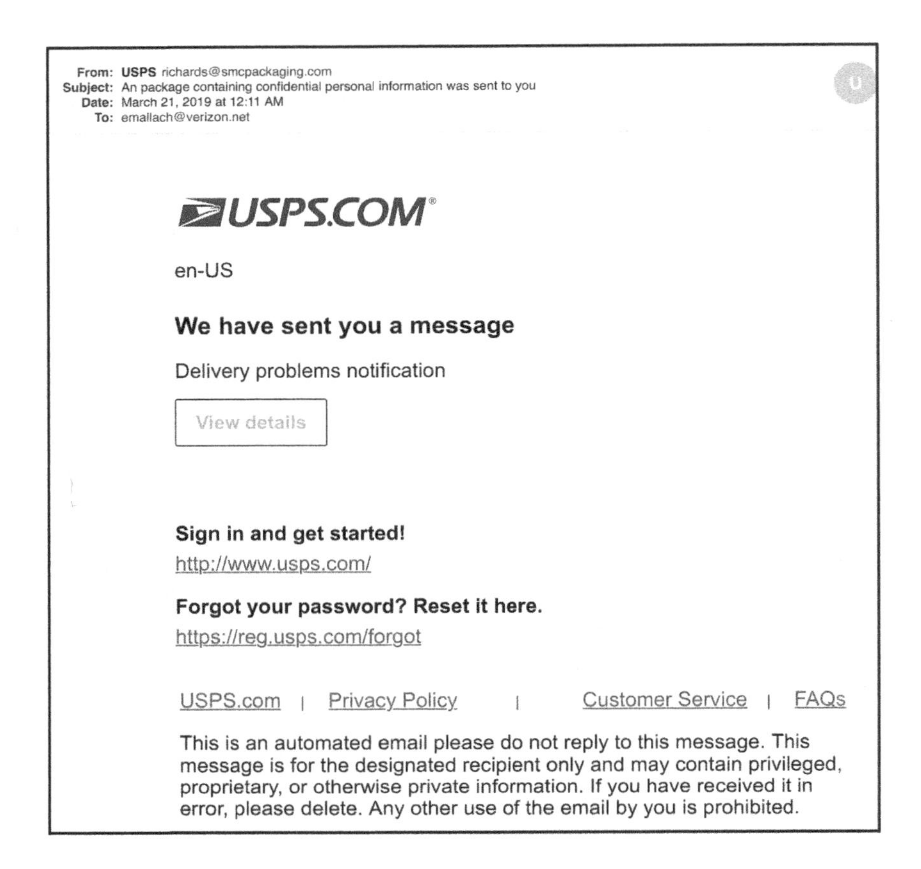

FIGURE 6.16 Example of phishing email, about here.

Other things to watch out for that often signal a phishing email:

- Use of an unusual TLD. A criminal who can't get *amazon.com* might get *amazon.name*. It is increasingly difficult for owners of legitimate sites to keep up with new TLDs.
- "Typosquatting," where a domain name resembles—but is not identical to—a legitimate domain name.
- "Lookalikes," which take advantage of character similarity. An inverted exclamation point resembles "i" in sans-serif fonts. The two letters "r n," with no space, often look like "m." Several Cyrillic letters look identical to Latin letters, but are encoded differently. (Some browsers flag URLs that mix Latin with Cyrillic to warn users of this threat, but not all.) Upper-case Greek letters can also cause confusion, but upper-case letters in URLs are rare.

A variation on phishing attacks, designed to evade these precautions, uses a QR code that goes to a criminal site when scanned. Someone who pulls out their phone to scan the QR code is likely, once at that site, to "sign in" with the login credentials the criminals ask for. This is especially likely if the sender spoofs the From address of the original message so it comes from the "HR director" and asks people to log onto a (fake) page and "agree to a corporate policy." Everyone must be aware that scanning a QR code in an email is as risky as opening an attachment!

Stolen Email Address Lists

Did you ever get an email from a friend saying "I'm stranded in Rome; I lost my plane ticket, I need $1,000 to get home?" Your friend is in his dorm, but his address list was stolen. You might say "Let him find his own way back," but a loving grandparent might fall for the trick.

Email lists can be stolen in a variety of ways. One is by introducing a *trojan*,* a program that appears to (and may actually) perform a useful function, while masking its real purpose: finding information on your computer and sending it back to an attacker. The data formats of popular email programs such as Microsoft Outlook are public. It is easy to write a program that will see if you are using Outlook, find your address book in its usual place, and send it off.

Trojans may look like a vacation photo, a newspaper article attached to an email reading "Do you think he has a chance to get elected?," a memo asking you to look at budget numbers, or anything else that will grab a reader's attention.

Targeted emails are sometimes called *spear phishing*, since they are a variation on phishing attacks that focus on something that is known about the target.

Ransomware

Copying email lists is not the only damage that a trojan can do. A recent threat is trojans that encrypt the recipient's files. That person then gets a ransom note: If money is sent to a given address by a certain time, the files will be unlocked. Otherwise, they will stay locked forever. The money must usually be paid in a cryptocurrency such as Bitcoin, which can't be traced.

Security experts recommend not paying. There is no guarantee that the criminals will keep their word, and every payment encourages them further. Experts urge people to keep regular backups, and to restore their files after using anti-malware software to remove the trojan. Still, many people and organizations pay the ransom. It is usually around $1,000, which most businesses can raise on short notice. In one survey about two-thirds of businesses that had not been attacked said they would never pay such a ransom, but two-thirds of businesses that *had* been attacked, paid it.

Where you fit in: When you're a manager, you can take three defensive steps: train people to recognize phishing attacks and not fall victim to them, use anti-malware software to detect most attacks and minimize their damage if they get in, and make sure files are backed up regularly.

Pharming

Pharming, from *farming* with the same change as *fishing* to *phishing*, is a broad-based attack on many people who want to visit a web site rather than one that targets specific people with "bait." Pharming often relies on changing Internet *domain name servers* so that, instead of converting "citibank.com" to the binary address of Citigroup's site, they send people to the criminals' site.

There is little individuals can do to protect against URLs being redirected by a pharming attack. We rely on Internet service providers and operators of the Internet backbone to keep domain name servers secure and to fix improper changes quickly. They generally do this well. However, you should earn the normal look and feel of web sites you visit regularly, and be alert to any changes. Also, be sure the prefix of a page's URL changes from *http://* to *https://*, indicating a secure connection, before entering sensitive data such as a credit card number.

Some sites protect users against pharming attacks with a two-step sign-in process.† First, users enter their user names—but not passwords. The site then displays a picture, chosen by each user from among many, and any security phrase a user chooses. Crooks with a fake site can't know a given user's picture and phrase. Only after seeing them is the password entered. If the phrase and picture don't match what a user expects to see, the user should not enter a password. This method

---

* Named for the Trojan Horse, which (in Homer's *Iliad*) appeared to be a gift from the Greeks to the city of Troy. The Trojans brought it inside their city walls. Greek soldiers emerged from it later and captured Troy.

† Not to be confused with *two-step verification*, also known as *two-factor authentication* and other names. That requires two independent forms of verification, such as possession of an ATM card and knowing the PIN associated with that card. Either is useless without the other. TFA can involve, for example, sending a code to your registered phone after you log in, to be entered before you access any services. Your password is of no use to anyone who doesn't also have your phone.

relies on user awareness, but a user who's used to seeing (say) a picture of a corgi and the phrase "My dog Blinky" will notice any change immediately.

There are variations on this process. Their common element: You don't enter any confidential information until you can confirm that the site you are accessing is the site you want to access.

Identity Theft

Identity theft isn't an attack, it's something that happens after an attack, but it's important enough to deserve discussion.

Criminals who obtain personal information can use it in many ways. Criminals who have access to your credit card account can change its billing address to theirs. They pay the first bill on time, so as not to arouse suspicion, and then ask for a replacement card for "you" at their address. They sign this card with your name in their handwriting. A printed bill to "you" at their address can be used to obtain other forms of identification. Over time, they'll build up "your" complete identity. They can then strip you of everything you own. This can ultimately pay off (to the crooks).

It is usually possible to recover the financial harm of identity theft. U.S. credit card holders are liable only for the first $50 of fraudulent charges. Most card issuers cover that amount as well. However, recovery involves time, effort, lost work or personal time, and other costs, to say nothing of emotional impact and damage to business relationships before fraud is detected.

There is little you can do to prevent someone who steals your information from trying to use it. You can reduce the likelihood of theft by following the guidelines in previous sections. You can't eliminate it; for example, you can't prevent rogue employees of credit bureaus from stealing your information. Your best defense is to watch your accounts carefully. Identity theft takes time. If you notice someone meddling in your affairs early, you can stop thieves before damage is done.

Privacy

Privacy isn't security—Chapter 12 discusses their relationship—but it can be important in a web context. Web sites can identify you in three ways without you doing anything:

IP Address

Your Internet Protocol address, the unique string of 32 or 128 bits that identify your computer's Internet location, is available to sites you visit. However, many computers return IP addresses to their Internet service provider's address pool when they go offline and get a new one from that pool when they reconnect. Meanwhile, the previous IP address may be reissued. This is known as a *dynamic IP address.* Since many IP addresses are dynamic, they are seldom used to identify a user from one time to the next. (An IP address that doesn't change is called *static.*)

MAC Address

A computer's MAC (Media Access Control, unrelated to Macintosh computers) address identifies that computer uniquely. It is also available to web sites you visit.

MAC addresses do not identify a person: They belong to a computer, not its user. However, they enable sites to track repeated visits from the same computer. Password-protected sites use them to identify people trying to use passwords that were reported stolen. Repeated attempts to use stolen passwords from the same computer can be detected, though the site may not be able to do much about it beyond sending a threatening-sounding message and denying entry.

A computer's MAC address is built into its hardware, but it is transmitted by software. Software can send another number instead. That is called *MAC address spoofing.* It has value in protecting privacy, but also has unethical or illegal uses.

Cookies

Cookies are small files that web sites ask your browser to save. When you return to a site, it can check for cookies it left previously. Cookies can hold information about when you visited a site, what pages you saw, what products you looked at, or what promotion code you used. If you click "don't show this again" on a store's offer of credit, a cookie will save your preference.

In contrast to IP and MAC addresses, you control cookies. You can tell your browser to refuse them. Some sites won't operate properly if you do this, since they use cookies to help navigate the site. In that case, you can tell your browser to refuse cookies from anyone but the site itself. After you leave a site, you can delete cookies you don't want to keep.

Each browser stores its own cookies. If you visit a site with Firefox and return with Edge, the site won't see any cookies it left with Firefox during your first visit.

Security and the Internet of Things

Closed networks are more secure than open ones. Closed networks don't ensure total security, but they help. Opening devices to the Internet via the IoT creates new security risks. Managers must stay on top of these risks and do what they can to minimize them. One checklist includes:

- Identify all IoT devices in the organization.
- Set security requirements for IoT devices. This may mean rejecting insecure devices, even if they would be useful. Those who ask for them would be the first to complain if a data breach occurs through them.
- Make sure IoT devices are included in security plans, policies, and training.
- Engage top management to support IoT security efforts actively.
- Add IoT to company liability insurance policies. A high premium may raise executive eyebrows, but it is an indication of high risk.

Where you fit in: Most of these steps will be taken by technical staff. Managers must make sure their technical staff knows this is part of their job. "Managers" will soon mean *you*.

A Totally Secure System?

A totally secure networked system is an impossible dream with the current or foreseeable state of technology. When absolute security is required, you must avoid network connections. Even that may not suffice: The Stuxnet malware that disabled Iran's Natanz nuclear facility came on a flash drive. This has implications:

1. Never relax your guard against penetration of your employer's network.
2. Assume that network *will* be penetrated—by the worst possible people, at the worst possible time. Take precautions, such as regular data backups and encryption, to minimize potential damage when—not *if*, but *when*—this happens.
3. As a manager, make sure that the people you supervise know this.

KEY POINT RECAP

Today's applications could not function without data communications and networks.
As a businessperson, you must be on the lookout for situations in which someone in one location could benefit from access to data that comes from, or is stored, somewhere else.

The tradeoffs in communications and network decisions are business decisions.
As a businessperson, you must understand enough of the technology not to be persuaded by one-sided ads and sales materials. You must understand the usage patterns and needs of your business in order to match them to the most cost-effective solutions.

The Internet is changing and will continue to change.

As a businessperson, you must be aware of new top-level domains, net neutrality, and other changes. They may affect your organization's use of the Internet in unexpected ways.

Internet applications will continue to evolve.

As a businessperson, you must be on the lookout for new and profitable ways to use it. You can't rely on your technical staff to find them before your competition does.

Internet security is a major management concern, as well as a personal one.

As a businessperson, you must make sure security has adequate funding, management support, and technical effort. (See Chapter 12 for more on this.)

KEY TERMS

Active RFID tag: *RFID tag* that contains a battery and broadcasts information about itself.

Analog signal: Electrical signal that can take on any value between two limits, with an infinite number of steps between them. Contrast with *digital signal.*

Application layer protocol: A standard way to treat data sent from one application to another, as contrasted with a standard way to send the data itself over the Internet.

Asynchronous Digital Subscriber Line (ADSL): *DSL* line in which transmission from switching center to subscriber is faster than transmission from subscriber to switching center.

Battery-assisted passive RFID tag: *RFID tag* that contains a battery for data transmission but is active only when interrogated by a reader.

Bluetooth: Short-range communication technology, alternative to cabling nearby devices.

Body Area Network (BAN): *Personal area network* that monitors its wearer's vital signs.

Bridge: Device that amplifies signals to extend the range of a communication link or network element, and which passes only messages addressed to devices on the other side of the bridge.

Browser: Computer program designed to display web pages.

Buffer: An amount of streaming audio or video content that a computer receives before starting to play that content, to be able to play continuously if transmission is interrupted briefly.

Circuit switching: Network approach that provides a dedicated path for the duration of a conversation. Contrast with *message switching, packet switching.*

Communication link: Means of transmitting data between two connected points.

Cookies: Small files that a web site has a browser save on its behalf, to retrieve information about a visit during a subsequent visit.

Cybersquatting: The (now illegal) practice of registering a domain name to force a business, with which that name is generally associated, to purchase it.

Demodulation: (a) The process of converting an analog signal to digital data. (b) By extension, the process of converting any communication signal to computer input.

Denial of Service (DoS) attack: Attempt to disable a network by overloading it with near-simultaneous requests.

Digital signal: Electrical signal that can take on only a finite number of values. When it has two possible values, they are usually represented as 0 and 1. Contrast with *analog signal.*

Digital Subscriber Line (DSL): Communication technology that transmits data at high speed over copper telephone lines.

Distributed Denial of Service (DDoS) attack: *Denial of service attack* launched from more than one computer.

Domain: Internet name assigned to a specific person or organization.

Dynamic IP address: Internet Protocol address that changes from session to session, usually taken from a pool of available IP addresses. Contrast with *static IP address.*

Email (or email): Transmission of messages to people over communication networks.

Ethernet: *Local area network* technology that allows connected devices to share a common wire by monitoring for simultaneous attempts to transmit, backing off, and retrying if they occur.

Femtocell: Type of *small cell* that can cover up to 60′/18 m and support up to six users.

Fiber optics: Communication link technology in which signals are transmitted by switching a light source on and off and guiding that light through a glass tube to its destination.

File Transfer Protocol (FTP): Method of sending large files in one direction over the Internet.

Firewall: Device designed to block unauthorized attempts to send data to or from a network.

Full-duplex link: Communication link that is capable of transmitting in both directions simultaneously. Contrast with *simplex link*, *half-duplex link*.

Gateway: Device that connects one network to another, typically an organizational network to the Internet.

Geostationary satellite: Satellite whose orbit keeps it over the same point on Earth.

Geosynchronous satellite: Satellite whose orbit keeps it over the same longitude on Earth.

Half-duplex link: Communication link that is capable of transmitting in both directions, but not simultaneously. Contrast with *full-duplex link*, *simplex link*.

Hot spot: Device that transmits Wi-Fi signals to other devices in its vicinity, passing their messages to the Internet.

Hyperlink: Content on a web page that, when clicked, loads a different page.

Hypertext Transfer Protocol (HTTP): Standardized means by which computers receive web pages.

Internet: Large network connecting all economically important areas on Earth.

Internet backbone: High-speed links that connect major organizations, networks, and metropolitan areas into the Internet.

Internet Message Access Protocol (IMAP): Standard for sending email messages from a server to a client.

Internet of Things (IoT): Extension of Internet use to devices that can communicate with each other without human intervention.

Internet Protocol address (IP address): Binary address that the Internet uses internally to identify domains.

Internet telephony: Telephone conversations that use the Internet.

Last mile: The communication links between carrier switching centers and subscribers.

Link: See *communication link*.

Local Area Network (LAN): Network covering a limited area, typically a building or a campus.

Low Earth Orbit (LEO) satellite: Satellite orbiting below 1,250 miles (2,000 km), with a typical orbital period of about two hours.

MAC address spoofing: Sending a false MAC (Media Access Control) address for a device.

Malware: Generic term for deliberately harmful software.

Media Access Control (MAC) address: A unique identifier assigned to the network interface of a device.

Medium Earth Orbit (MEO) satellite: Satellite orbiting between about 1,250 miles (2,000 km) and 22,200 miles (35,800 km), with an orbital period of more than two but less than 24 hours.

Message switching: Network approach that provides a dedicated path for each message in a conversation, but may use different paths for different messages. Contrast with *circuit switching*, *packet switching*.

Metrocell: Type of *small cell* that can cover up to 1,000′/300 m and support up to 200 users.

Modem: A device that performs *modulation* and *demodulation*. (From the first letters of *modulator-demodulator*.)

Modulation: (a) The process of converting digital data to an analog signal. (b) By extension, the process of converting data in a computer to any type of communication signal.

Net neutrality: The principle that all Internet packets should be given equal priority.

Network: Group of communication links with additional devices offering the flexibility to transmit data between two points on any of its communication links.

Packet: Part of a message to be transmitted, typically 1,000–2,000 bytes.

Packet switching: Transmitting packets independently of each other, reassembling them at their destination to create a complete message. Contrast with *circuit switching, message switching.*

Page: Document available on the *World Wide Web.*

Passive RFID tag: RFID tag that contains no battery but is powered by the signal from a reader.

Personal Area Network (PAN): Local area network connecting devices carried on or by one person.

Phishing: A message intended to lure users into providing confidential information.

Picocell: Type of *small cell* that can cover up to 750′/225 m and support up to 64 users.

Post Office Protocol (POP): Standard for sending email messages from a server to a client.

Radio Frequency ID (RFID): Short-range communication technology designed to share small amounts of information between a *tag* and a central computer.

Repeater: Device that amplifies signals to extend the range of a communication link or network element, but which does not analyze the messages it passes on.

Router: Network device that can select a path for a message based on its final destination.

Simple Mail Transfer Protocol (SMTP): Method of sending email messages from one server to another.

Simplex link: Communication link that is only capable of transmitting in one direction. Contrast with *full-duplex link, half-duplex link.*

Single Sign-On (SSO): An approach to information system security in which a person signs on to a network to obtain access to all resources available through it.

Small cell: Cellular telephone base station that covers a smaller area than a tower cell, often installed by individual users. See *femtocell, metrocell, picocell.*

Social engineering: Attempting to penetrate an information system by persuading an authorized user that an access request should be honored.

Spear phishing: Phishing attack targeting a specific person, using information about him or her.

Static IP address: Internet Protocol address that is the same each time a computer goes online. Contrast with *dynamic IP address.*

Streaming: Sending data continuously for continuous video or audio playback, without waiting for all of it to arrive.

Switch: Device that connects several communication links or network element, and analyzes messages to pass them on to only the correct link.

Tag (RFID): Small device containing stored information that can be attached to an object, then read (and perhaps written to) by an RFID reader.

Teleconference: Meeting that uses the Internet to transmit audio and video,

Telepresence: Dedicated teleconference system designed to approximate the effect of remote participants being physically present.

Thumbnail: Miniature version of a picture on a computer screen that can be enlarged if desired.

Top-Level Domain (TLD): Set of Internet domains that belong to a type of organization or to a country; last part of a URL before the first slash, if any.

Transponder: See *tag.*

Trojan: Harmful program designed to look like a useful program or document.

Uniform Resource Locator (URL): Unique identification of a page on the *World Wide Web.*

Voice over Internet Protocol (VoIP): The communication method used for *Internet telephony.*

Web 2.0: The "social web," characterized by user-contributed content.

Web site: Organized collection of pages at a single *domain* on the *World Wide Web.*

Wide Area Network (WAN): Network that covers a large area, up to the whole Earth.

Wi-Fi: Popular short-range communication technology for local area networks. (From *wireless fidelity.*) Versions of Wi-Fi are identified by numbers, through Wi-Fi 6 as of 2019.

Wireless communication: Data communication that does not use physical connections.
World Wide Web: The total set of pages that are available, or can be created, for access over the Internet by means of the *HTTP* protocol.
Zigbee: Wireless communication technology for control and monitoring applications.
Zip bomb: Attack that disables processing and storage by expanding a compressed file to an extremely large size.

REVIEW QUESTIONS

1. Compare, at the conceptual level, a communication link and a network.
2. What are the two main, broad categories of communication links?
3. What are the two main types of long-distance physical communication links?
4. What are the four major types of communication satellites? Give one application for each.
5. Why do local area networks (that communicate only over limited distances) exist?
6. What is backward compatibility (in the context of Ethernet and Wi-Fi)? Why is it important?
7. Describe two wireless local networking technologies that cover shorter distances than Wi-Fi.
8. What is the major wide area network? Who pays for it in the U.S., and how?
9. Give three advantages of *packet switching.*
10. What is the difference between the Internet and the World Wide Web?
11. What are the parts of an Internet domain name?
12. What is a top-level domain?
13. What is the fundamental difference between Web 1.0 and Web 2.0?
14. Can businesses benefit from social networking sites? If so, how?
15. Give two uses for Internet streaming. How does streaming differ from sending a photograph?
16. Describe two applications of the Internet other than the World Wide Web.
17. Why do organizations use firewalls?
18. Why can it be difficult to get sufficient funding for network security?
19. State three security risks associated with IoT.

DISCUSSION QUESTIONS

1. When might an individual user, at home, **not** want an asymmetric Internet connection such as ADSL or cable TV? Think of how people use the Internet.
2. In the U.S. many people obtain television service through satellite links, but few obtain their Internet service that way even if they already have a dish to receive TV signals. Why do you think this is? You may want to do research beyond this book to answer this.
3. Would you recommend a wired (Ethernet) or wireless (Wi-Fi) LAN in each of the following situations? Explain your choice in a few sentences for each.
   a. A fraternity house used by 60 students for meals and recreation, 30 of whom live there.
   b. An accounting practice employing four CPAs and four support staff.
   c. A classroom with 26 computers, one in the instructor podium and 25 for students.
   d. A hotel with 200 rooms in which many guests want to use their own computers.
4. A few organizations build their own wide area networks to obtain the utmost security. This requires devices such as those in Section 6.4. As a manager in such an organization, you have proposals from three device vendors. Your technical staff finds all three acceptable. How would you select one? Include factors you would consider and outside advice you might use.
5. The approximate location of anyone accessing the Internet can be determined (unless they hide it, which is usually not worth the effort). A wired connection can be located to the

network point of entry to the Internet. A mobile connection can be located to the nearest tower. GPS-enabled devices can be located precisely if they share location data. Explain how an airline and an automobile company could use such information about a user's location. State what they could do with this information that they could not do without it. Be as specific as possible about the business benefits of doing this.

6. Visit the site *www.nissan.com*. Using what you read there and any other information you find online, argue for either Nissan Computer's or Nissan Motors' claim to this domain name.
7. Internet names and addresses are assigned, and new TLDs are approved, by ICANN (Section 6.5). What do you think would happen if there were no central agency in charge of these?
8. You plan to open an event/studio photography business. Your customers will be primarily individuals and small businesses in your area. How could you use Web 2.0 and social media to connect with potential customers and keep in touch with existing ones?
9. Research and summarize the current state of net neutrality in your country. Are you in favor of net neutrality, or against it? Justify your position.
10. Find out what free or inexpensive firewall software is available for your personal computer. Do you use it, or do you plan to use it now? If so, why? If not, why not?
11. Visit "Find the Fake Social Profile" at *www.zerofox.com/find-the-fake*. This site, operated by a company whose business is protection from social media threats, shows seven pairs of social media pages claiming to belong to the same well-known person. You are asked to identify the fake, and are then told what clues should have led you to the answer. Take their test. At the end, continue by reading their blog post "What is the cost of a fraudulent account on social media?" (ZeroFOX is not unique; other firms can do the same.) Then, for a real company of your choice whose name starts with the same letter as your given name, discuss the damage that a fraudulent social media account could do. Estimate numbers wherever possible. Who might want to create such a page, and why? Also, discuss what you learned from the seven examples of fraudulent pages. (You don't have to tell anyone your score.)

KHOURY CANDY DISTRIBUTORS NETWORKS

The following week, Isabella and Jake returned to meet with KCD's network manager, Armand Rocher. Since the conference room was in use, they went to Armand's office a few doors away. There was, of course, a bowl of candy on his desk. Behind the desk, on the wall, was a large photo of something that looked like a ski jump emerging from a domed stadium.

Jake was the first to ask "What's that picture of?"

"That's the Montreal Tower, built into the stadium that was first used for the 1976 Olympics—and you're not the first to ask," Armand replied with a laugh. "It reminds me of home. Besides that, it's where I first got into computer networks. Today, the old Olympic Park is run by the National Sports Institute. It's a big organization with all the computing needs of any other organization. As you can imagine, the Olympic Park didn't include networks when it was built in the 1970s, but by the time INS—that's what we call the National Sports Institute, it's the word order in French—was formed in 1997, it had become essential. We had to create everything. A great education!" he concluded.

"How did you get from there to KCD, then?" Isabella asked him.

"It wasn't that complicated," Armand replied. "The North American Free Trade Agreement, at the time, allowed Canadian computer professionals with a degree, three years' experience, and a job offer to get a visa to work in the States. When KCD posted this opening on indeed.com I applied for it, and the rest is history. Bottom line is I like it here, it's interesting work with good colleagues, but my home will always be up north. I get some of the candy in that bowl, like Coffee Crisp, from our Ontario distribution center whenever I'm there. Try some. You can't buy it this side of the border!"

"What's the most interesting part of your job, to you?" she continued as she reached for the bowl.

"It has to be keeping up with technology to strike the right balance between not replacing things that work just fine, simply because there's something newer out, and falling so far behind the curve that we can't develop systems we need."

"That doesn't sound much like technology," Jake commented.

"No, it's not," agreed Armand. "It's the way Chris or Jason would think about networks, perhaps. I try to think the way they do. Still, I enjoy the technical side too, or I wouldn't want to be in this job. For me, it has the perfect balance."

"What's the best part of your job on the technical side?" Isabella continued to probe.

"Security. Definitely security," responded Armand. "If everything could be trusted to work as it should with no interference, my job would be a lot simpler. The security challenges make it interesting, because they keep changing.

"We've been lucky so far. We haven't been targeted—as far as we know! We don't sell to consumers, so we don't have credit cards or other useful data for identity thieves. Distributors don't get a lot of social media followers. We're not at the top of anyone's target list. Still, we can't be too careful. Bad guys may figure that we know we're not an attractive target, so we might get lazy and be an easy one.

"At one point we considered shielding our buildings so people couldn't get our Wi-Fi in the parking lot," he continued, "but decided not to because it would cost too much. We would have had to put metal-based tinting in all the windows, for example. But what really killed that idea is that it would also destroy cell phone reception inside the buildings. We might have been able to put small cells inside the buildings, but all in all the risk of someone trying to get into our Wi-Fi from outside didn't justify the expense."

"So, what are your major security precautions?" asked Jake.

"Number one is making sure that our people know they're the weakest link and do what they can to protect us. Phishing emails keep getting better. When I started out they all read like a Google Translate job on an original in Klingon. Now most of them use decent English. Then we have a firewall on our gateway, and we make sure all our users keep up to date with virus protection. Finally, we keep checking for network activity that doesn't fit our usual pattern. So far it's been because of new apps or changes to old ones, but being able to find those makes us think we'd be able to find criminal activity too.

"Other than security," Armand continued, "we spend a good deal of time reconfiguring and updating switches, routers and so on whenever software updates come out or the business changes. That includes the networking software in our servers, so we also get involved with them. Now we're updating our networks to support RFID tags for product tracking in distribution centers. We were involved in that decision because they needed to know how much it would cost to decide if it was worth doing."

"You keep coming back to business decisions that involve technology but aren't about technology," Jake observed.

"I guess I do," Armand mused. "Maybe that's the difference between a network manager and a network technician. Nothing against network technicians, but that's a different job description."

Questions

1. Think of three ways in which a national Olympic training organization such as INS Québec could use data networks. For each, suggest a network technology that would make sense for that use.
2. Consider the network diagram of KCD's distribution centers (Figure 3.27). It doesn't include any use of RFID for product tracking. On a copy of that figure (enlarge it) or your version of it, indicate where that would go.

3. Armand said "the risk of someone trying to get into our Wi-Fi from outside didn't justify the expense." How would he have determined this? (You don't have to analyze the issue. Just describe how you would. Think like a businessperson.) What differences from KCD's situation might lead to the other conclusion?

CASE 1: CONNECTING REMOTEST MICHIGAN

Three connection issues must be dealt with in order to access resources via the Internet. People often think about just two of them: the *Internet backbone*, which connects major metropolitan areas and switching centers; and the *last mile*, which connects individual users to "the Net."

A last mile connection takes user messages to a nearby telecommunications switching center. We assume this center has high-speed links to the rest of the world. In metropolitan areas, that's usually true. But what about elsewhere? That's the *middle-mile* issue.

Michigan's population density varies widely: from over 5,000 people per square mile in a few urban centers to under one person per square mile in much of its Upper Peninsula. (The state average is 175; that of Keweenaw County, which projects into Lake Superior, is 4.3.) Providing fast Internet access to people in low-population areas is hard for a business planner to justify. At the same time, such access is essential to keeping them in the economic mainstream.

To help provide it, Merit Network, Michigan's nonprofit research and education service provider to community anchor institutions,* received two Broadband Technology Opportunities Program grants through the federal American Recovery and Reinvestment Act. In total, Merit Network brought $103 million in federal funding to Michigan, plus $30 million from other sources.

With these funds, Merit began the Rural, Education, Anchor, Community and Health care—Michigan Middle Mile Collaborative (REACH-3MC) project. Merit engaged seven commercial Internet service providers as grant sub-recipients to create an infrastructure to serve all sectors of society, including homes, businesses, and CAIs.

Merit and sub-recipients constructed fiber optic laterals from the mainline to connect CAIs and businesses and to access cell towers and central office facilities. Merit and sub-recipients each own fiber strands over various portions of the network, ensuring competition at every interval.

The project hopes to solve the lack of infrastructure in Michigan's remote and rural areas. Residents there have had challenges accessing information. CAIs and businesses have had to contend with substandard levels of Internet access, putting those organizations at a disadvantage.

In all, REACH-3MC now provides 143 CAIs with 1 Gbps-dedicated connections to Merit, enabling collaboration with more than 230 other CAIs that are already connected. Michigan's public institutions have a mechanism to cut costs and provide more service to constituents. More than 900 additional CAIs have connected or will have the opportunity to do so.

As a middle-mile project, the aim of REACH-3MC isn't to connect every home and business in the network service area. Rather, it is to bring infrastructure to rural regions and give ISPs the opportunity to do that. It seems to be succeeding.

Questions

1. Read the Merit overview at *www.merit.edu/about*. Examine the map near the top of the page. It shows nodes in Ohio, Illinois, Wisconsin and Minnesota in the U.S., and two in Canada. What is the function of connections outside Michigan? Are they necessary? Why or why not?

---

* A *community anchor institution* provides public services to facilities such as a school, library, town hall, or fire station. Many public services depend on fast, reliable Internet access.

2. REACH-3MC is funded by a grant from the U.S. government. Do you agree that the Federal government should pay for this sort of project? If you don't, who do you think should? Justify your answers, stating what you believe the role of a national government in funding Internet access should be. Would your answer be different for a country with fewer economic resources than the U.S.?

CASE 2: HELPING MOTHERS HAVE HEALTHY BABIES IN GHANA

Over 130 million babies were born in 2018. In that same year, 79 of every thousand children in sub-Saharan Africa died before their fifth birthday (versus six in Europe and North America, four in Australia and New Zealand). Nearly 300,000 women died worldwide during pregnancy and childbirth in 2017, two-thirds of them in sub-Saharan Africa. Many deaths in both categories are due to a lack of trained healthcare professionals in rural areas and unawareness of good healthcare practices.

Ghana's CHPS (Community-based Health Planning and Services) program hopes to change that situation within Ghana. One of its projects is Mobile Technology for Community Health (MoTeCH). It uses mobile phone technology to increase the availability and quality of prenatal and neonatal care in rural Ghana.

MoTeCH has two components: Mobile Midwife for pregnant women, new mothers, and their families; and the Nurse Application for community health nurses to track patients, recommend care, and record care.

An issue with setting up the system was the question of handsets for nurses. Most nurses already had mobile phones that were capable of sending SMS messages to MoTeCH, but these were not always suitable. Many had old batteries in poor condition that would not stand up to the increased use. Entering messages that would be acceptable to a computer required templates; many phones could not handle enough templates for the app, and differences among phones required several versions of each. Trouble-shooting was difficult because handsets were not standardized, and the lack of Java support on many phones limited what the app could do.

As a result, it was decided to distribute MoTeCH phones to all nurses. As a side effect, this reduced costs. MoTeCH phones could send and receive messages using GPRS packet switching technology. A single SMS message cost about 3¢ (in U.S. currency) to send, while a megabyte of GPRS data can handle the equivalent of a thousand such messages for 11¢. The cost of a GPRS phone was recovered in less than six months through reduced transmission charges.

Mobile Midwife couldn't use messaging. Most mothers in rural Ghana are not literate in English, and local languages may have no written form. Messages had to be recorded in every local language. When it came to recording them, testing found that educated-sounding speakers were not accepted, as they were not seen as being able to relate to users' daily lives. Users also disliked voices with "deep in the village" accents, as they were not trusted to be knowledgeable.

Another challenge MoTeCH faced was identifying clients uniquely. MoTeCH assigned nine-digit client numbers, but since clients may not remember these correctly, another means of identifying them was necessary. Mobile phone numbers were not suitable: They change and may be shared. Names and dates of birth were not usable: Clients often go by several names, are not concerned with their consistent spelling in English, and may not know their date of birth. The solution was to triangulate from name, address, date of birth where possible, National Insurance number where the client has one, and relationships with other family members in the database.

Software for MoTeCH was developed jointly by two groups. The server side was handled by Columbia University and the University of Southern Maine, both in the U.S. Components that interfaced directly with mobile phones were done by software development firm DreamOval in Ghana. Due to a lack of experienced senior staff at DreamOval, some of its work had to be redone by USM for improved performance. A Grameen Foundation program manager was located in Ghana and coordinated the two development teams.

MoTeCH began in July 2010 in Ghana's Upper East region. By March 2013 it had served over 11,000 families and was put in the "Existing: Post-Pilot" category by the Center for Health Market Innovations. After it was seen to be successful in Ghana, it was implemented in Zambia and India as well. It was followed in 2014–2015 by the CHN-on-the-Go (Community Health Nurse) app, piloted with Concern Worldwide, John Snow International and the Ghana Health Service. The app gives nurses working in remote areas point of care tools to strengthen the quality of diagnosis and care; professional development and workplace support through up-to-date medical reference and training guides, work planning tools, and wellness aids.

MoTeCH appears well on the way to making a difference.

Questions

1. Consider the challenges faced in implementing MoTeCH. (Challenges listed in here are far from a complete list.) How many of those were technical, how many non-technical? What, if anything, can you infer from this about the challenges you would encounter in designing and implementing a customer-facing mobile phone application in an industrialized country?
2. MoTeCH could have chosen to identify clients uniquely via fingerprint scans. What are some advantages and disadvantages of this approach?
3. The decision to supply nurses with MoTeCH handsets was a business decision, though technical factors played a part. Suppose your university health service implemented a mobile application for nurses visiting ill students in their dorm rooms. Would the same factors apply? Would it be appropriate to supply those nurses with university handsets? Why or why not?
4. Facebook announced its Libra project in June 2019. Libra is intended to bring a form of electronic currency, analogous to Bitcoin, to millions of people with no access to traditional banking. How does Libra overlap MoTeCH, despite having an entirely different purpose?

BIBLIOGRAPHY

Awoonor-Williams, J.K., et al., "Lessons learned from scaling up a community-based health program in the Upper East Region of northern Ghana," *Global Health, Science and Practice*, March 21, 2013, www.ghspjournal.org/content/1/1/117.full, accessed March 16, 2019.

Bidle, T., "Confronting the rise of website, application, and DDoS attacks," *IT Toolbox*, September 12, 2019, https://it.toolbox.com/guest-article/confronting-the-rise-of-website-application-and-ddos-attacks-what-you-need-to-know-right-now, accessed September 18, 2019.

Bray, H., "Dot-specific domain names on way to the web," *The Boston Globe*, January 28, 2014.

Checkoway, S., et al., "Comprehensive experimental analyses of automotive attack surfaces," In D. Wagner, ed. *Proceedings of USENIX Security, 2011*, USENIX, August 2011, www.autosec.org/pubs/cars-usenixsec2011.pdf, accessed September 19, 2019.

Chickowski, E., "10 web-based attacks targeting your end users," *Dark Reading*, August 2013, twimgs.com/darkreading/drdigital/080713s/DarkReading_SUP_2013_08.pdf, accessed September 18, 2019.

Constine, J., "Facebook announces Libra cryptocurrency: All you need to know," *TechCrunch*, June 18, 2019, techcrunch.com/2019/06/18/facebook-libra, accessed June 29, 2019.

de Fremery, R., "What Elon Musk's Starlink mission means for global Internet coverage," *IT Toolbox*, July 12, 2019, it.toolbox.com/blogs/rosedefremery/what-elon-musks-starlink-mission-means-for-global-internet-coverage-071219, accessed July 17, 2019.

DMR, "26 amazing Skype statistics and facts," November 10, 2018, expandedramblings.com/index.php/skype-statistics, accessed March 16, 2019.

Fruhlinger, J., "What is ransomware? How these attacks work and how to recover from them," *CSO*, December 19, 2018, www.csoonline.com/article/3236183/what-is-ransomware-how-it-works-and-how-to-remove-it.html, accessed March 16, 2019.

Gold, J., "AT&T smashes distance record for 400Gbps data connection," *Computerworld*, March 12, 2013, www.computerworld.com/article/2495236/at-t-smashes-distance-record-for-400gbps-data-connection.html, accessed March 16, 2019.

Grameen Foundation, "Training health workers," grameenfoundation.org/what-we-do/health/training-health-workers, accessed September 19, 2019.

Greenberg, A., "Hackers reveal nasty new car attacks—With me behind the wheel," *Forbes*, August 12, 2013, www.forbes.com/sites/andygreenberg/2013/07/24/hackers-reveal-nasty-new-car-attacks-with-me-behind-the-wheel-video, accessed March 16, 2019.

Guarino, N., "This "Man in the Inbox" phishing attack highlights a concerning gap in perimeter technology defenses," *CoFense*, July 18, 2019, cofense.com/man-inbox-phishing-attack-highlights-concerning-gap-perimeter-technology-defenses, accessed September 25, 2019.

Guarino, N., "Under the radar – phishing using QR Codes to evade URL analysis," *CoFense*, June 28, 2019, cofense.com/author/nick-guarino, accessed September 25, 2019.

HakShop, "WiFi Pineapple Mark IV," hakshop.myshopify.com/products/wifi-pineapple, accessed March 16, 2019.

Hirota, S., "Record breaking fiber transmission speed reported," *phys.org*, April 16, 2018, phys.org/news/2018-04-fiber-transmission.html, accessed March 14, 2019.

Ingram, D., "Instagram says advertising base tops one million businesses," *Reuters*, March 22, 2017, www.reuters.com/article/us-instagram-advertising-idUSKBN16T1LK, accessed April 22, 2019.

Institute National du Sport du Québec web site, www.insquebec.org/linstitut-2/about-us/?lang=en, accessed May 3, 2019.

International Data Corporation, "IDC's worldwide quarterly Ethernet switch tracker shows solid growth in Q2 2018 while router market sees mixed results," September 6, 2018, www.idc.com/getdoc.jsp?containerId=prUS44262218, accessed March 14, 2019.

Kang, C., "Google to use balloons to provide free Internet access to remote or poor areas," *The Washington Post*, June 14, 2013, www.washingtonpost.com/business/technology/google-to-use-balloons-to-provide-internet-access-to-remote-areas/2013/06/14/f9d78196-d507-11e2-a73e-826d299ff459_story.html, accessed March 16, 2019.

Kelly, W., "Are connected devices out of control? Managing IoT risk," *IT Toolbox*, June 18, 2019, it.toolbox.com/blogs/willkelly/are-connected-devices-out-of-control-managing-iot-risk-061819, accessed June 29, 2019.

Kottler, S., "February 28th DDoS incident report," *GitHub*, github.blog/2018-03-01-ddos-incident-report, accessed September 18, 2019.

Lamble, L., "With 250 babies born each minute, how many people can the Earth sustain?" *The Guardian*, April 23, 2018, www.theguardian.com/global-development/2018/apr/23/population-how-many-people-can-the-earth-sustain-lucy-lamble, accessed September 19, 2019.

Legal Line (Canada), "Working in the USA under NAFTA (TN Visa)," www.legalline.ca/legal-answers/working-in-the-usa-under-nafta, accessed May 3, 2019.

Leverege, *IOT 101: An Introduction to the Internet of Things*, 2018, www.iotforall.com/free-intro-ebook-on-the-internet-of-things, accessed June 29, 2019.

Leyden, J., "Ancient zip bomb attack given new lease of life," *The Daily Swig*, August 23, 2019, portswigger.net/daily-swig/ancient-zip-bomb-attack-given-new-lease-of-life, accessed September 25, 2019.

Luis, R.S. et al., "1.2 Pb/s transmission Over a 160 μm cladding, 4-core, 3-mode fiber, using 368 C+L band PDM-256-QAM channels," *Proceedings of the 44th European Conference on Optical Communication (ECOC)*, September 2018, paper Th3B.3.

Markets and Markets, "Telepresence (videoconferencing) market," www.marketsandmarkets.com/Market-Reports/telepresence-videoconferencing-market-193932914.html, accessed March 16, 2019.

Matador, "Do you need RFID blocking wallets or bags for travel?" April 27, 2017, matadorup.com/blogs/news/do-you-need-rfid-blocking-wallets-or-bags-or-travel, accessed March 15, 2019 (URL has "or" as next-to-last word even though title has "for" there.)

Metz, C., "With its tiny PC, Google makes a play for your conference room," *Wired*, February 6, 2014, www.wired.com/wiredenterprise/2014/02/google-chromebox-meetings, accessed March 16, 2019.

Open Health News, "Mobile technology for community health (MOTECH) suite," www.openhealthnews.com/resources/entities/mobile-technology-community-health-motech-suite, accessed September 19, 2019.

Pratt, M., "Computerworld Honors 2013: Remote areas in Michigan get connected with broadband," *Computerworld*, June 3, 2013, www.computerworld.com/s/article/9239140/Computerworld_Honors_2013_Remote_areas_in_Michigan_get_connected_with_broadband, accessed March 16, 2019.

Project Loon, "Journey," loon.co/journey, accessed March 14, 2019.

Raffensperger, L., "Networked cars are coming, but their hacks are already here," *Discover*, July 30, 2013, blogs.discovermagazine.com/d-brief/?p=2409, accessed March 16, 2019.

Reardon, M., "Here's everything you need to know about net neutrality on the anniversary of its repeal," *Cnet*, December 4, 2018, www.cnet.com/news/the-net-neutrality-fight-isnt-over-heres-what-you-need-to-know, accessed March 16, 2019.

Reeves, S., "Pros and cons of using femtocells," *TechRepublic Data Center*, November 11, 2013, www.techrepublic.com/blog/data-center/pros-and-cons-of-using-femtocells, accessed March 16, 2019.

Repeater Store, "Femtocells, microcells, and metrocells: The complete guide to small cells," November 28, 2018, www.repeaterstore.com/pages/femtocell-and-microcell, accessed March 15, 2019.

Russell, J., "The world's largest DDoS attack took GitHub offline for fewer than 10 minutes," *TechCrunch*, March 2, 2018, techcrunch.com/2018/03/02/the-worlds-largest-ddos-attack-took-github-offline-for-less-than-tens-minutes, accessed September 18, 2019.

Sanders, J., "How fraudulent domain names are powering phishing attacks," *TechRepublic*, June 17, 2019, www.techrepublic.com/article/how-fraudulent-domain-names-are-powering-phishing-attacks, accessed June 24, 2019.

Starlink web site, www.starlink.com, accessed July 17, 2019.

State of Michigan, Department of Technology, Management and Budget, "Area and population density of Michigan counties, cities, and townships: 2000 and 2010," www.michigan.gov/documents/cgi/cgi_census_density_mcd10_380470_7.xls, accessed March 16, 2019.

Statista, "Number of estimated Skype users registered worldwide from 2009 to 2024 (in billions)," www.statista.com/statistics/820384/estimated-number-skype-users-worldwide, accessed April 22, 2019.

Stokel-Walker, C., "Forget dot com, 2019 will finally be the year of weird domain names," *Wired*, December 20, 2018, www.wired.co.uk/article/domain-names-future-of-internet, accessed April 21, 2019.

Terdiman, D., "Stuxnet delivered to Iranian nuclear plant on thumb drive," *CNET*, April 12, 2012, www.cnet.com/news/stuxnet-delivered-to-iranian-nuclear-plant-on-thumb-drive, accessed March 16, 2019.

University of Bristol, "How house-hunting ants choose the best home," April 22, 2009, http://www.bris.ac.uk/news/2009/6292.html, accessed March 16, 2019.

U.S. Department of Health and Human Services, www.HealthIT.gov, accessed March 16, 2019.

Willcox, M., et al., "Mobile Technology for Community Health (MOTECH) in Ghana: Is maternal messaging and provider use of technology cost effective in improving maternal and child health outcomes at scale?" *Journal of Medical Internet Research*, vol. 21, no. 2 (June 2018), www.researchgate.net/publication/327614508_Mobile_Technology_for_Community_Health_MOTECH_in_Ghana_is_maternal_messaging_and_provider_use_of_technology_cost_effective_in_improving_maternal_and_child_health_outcomes_at_scale_Preprint, accessed September 19, 2019.

World Health Organization, "Maternal mortality," www.who.int/news-room/fact-sheets/detail/maternal-mortality, accessed September 19, 2019.

7 Integrating the Organization

CHAPTER OUTLINE

WHY THIS CHAPTER MATTERS

This is the first of three chapters devoted to the key ways in which information systems benefit a company's strategic position and bottom line. As you read in Chapter 1, one is connecting parts of an organization so they function in a coordinated fashion. The others, in the next two chapters, are linking it with its customers and suppliers, and helping its people make better decisions.

The need to connect parts of an organization is ancient, but it is greater today than at any time in history. Many organizations have global reach. All are affected by global commerce. Fortunately, we can deal with this need in ways that our ancestors could only dream of. You must understand those tools to function as a 21st-century manager. This chapter will enable you to do that. It will also cover related topics that will stand you in good stead as you work for an organization that deploys these tools, that plans to, or that—perhaps for good reasons—has decided not to.

CHAPTER TAKE-AWAYS

As you read this chapter, focus on these key concepts to use on the job:

1. Enterprise information systems connect people and departments in an organization. That is one of the key ways that information systems benefit organizations.
2. Functional information systems can perform the individual tasks of enterprise information systems, but work independently of each other.
3. Enterprise information systems, such as ERP, tend to be complex and expensive, and may require major changes to how an organization works. Their benefits often justify these costs.
4. It may be possible to get most of the same benefits by connecting functional applications to each other, using the Enterprise Application Integration (EAI) approach.

7.1 ANCIENT HISTORY: INFORMATION SILOS

Early information systems served a single part of an organization. This was due to the limited power of early computers, which lacked database management systems and convenient remote access. Such *functional information systems*, so called because they supported an organizational function such as accounting or sales, are often called *information silos* because, like grain silos (Figure 7.1), they keep their contents apart. With grain, this is desirable: Wheat in one silo shouldn't mix with barley in the next. With information, it is not.

FIGURE 7.1 Photo of grain silos (Source: Wikimedia Commons).

Functional information systems were a step back from paper and pencil for sharing information. Sales managers can read paper production schedules; production planners can read paper sales forecasts. If production schedules and sales forecasts are in isolated information systems, they are harder to share.

Functional systems are still worth looking at in the 2020s for three reasons:

1 "Silos" haven't gone away. Functional information systems are used today, by:
   a. Organizations that want to keep using their *legacy systems* as long as they do the job.
   b. Small organizations that can't justify the cost of enterprise information systems or deal with their complexity.
   c. Organizations that don't want to get involved in complex enterprise information systems, perhaps because they don't see a strategic advantage in them. (They may be wrong, but people have the right to be wrong.)
   d. Organizations that link existing information systems through *enterprise application integration* (EAI), the subject of Section 7.4.
2 Functional information systems laid the groundwork for today's enterprise systems. We often understand things better if we know a bit of their history.
3 Some enterprise information system concepts are easier to understand if we look at them first in the simpler context of a functional information system.

Functional information systems are, therefore, the topic of the next section.

7.2 FUNCTIONAL INFORMATION SYSTEMS

Large organizations are structured into functional parts. Each is responsible for a set of related activities, such as sales or accounting. Figure 7.2 shows a typical organization. The specific functions depend on the organization—hotels, hospitals, and hamburger chains have different divisions—but the concept is universal.

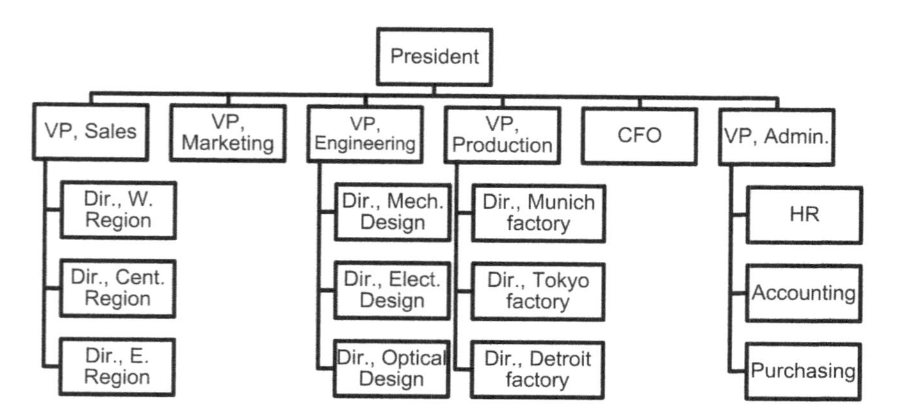

FIGURE 7.2 Typical organization chart.

Applications found in many organizations are discussed next. Few organizations have every one of these. Almost all have some that aren't listed.

Accounting

Organizational activities meet in the balance sheet and income statement. When anything is bought or sold, when an asset is depreciated, when an employee is paid, or when taxes are paid, its general ledger, income statement, and balance sheet are affected.

These tasks involve repetitive calculations. Pay varies from employee to employee, but the way it's recorded it doesn't. Debits and credits differ from sale to sale, but all sales are entered the same way. Early information systems automated these calculations.

The accounting modules of today's enterprise information systems get their input from other modules that record salary changes, sales, and other activities. They get other information from a shared database. That database is set up to alert the accounting module whenever certain fields change. When a production worker takes a part out of inventory, the value of raw materials inventory is reduced, balanced by an increase in the value of work-in-process inventory.

Finance

Accounting tracks the effects of past actions. Finance makes decisions for the future: how much to borrow and where, the best way to repay a loan, whether to use cash to buy back stock or pay dividends. The finance staff uses computers to assess the impact of such choices. Their software ranges from spreadsheets to specialized packages with complex calculations built in. They may also use accounting applications, but with assumptions about the future rather than historical data. *Pro forma* financial statements, which show income statements and balance sheets as they would be if an event such as a merger takes place, are an example of this.

Most computer applications in finance fit the "Making better decisions" theme of Chapter 9.

Sales

Sales information systems depend on a company's sales process. Stores improve checkout speed and accuracy while recording information about purchases. Construction firms calculate what it will cost to complete a project, so they can submit a bid that isn't too high (it won't win the job) or too low (they'd lose money on it). Airlines sell tickets for a bewildering array of flights, fare types, and connections. All forecast sales and keep track of customers.

Sales information systems fit the "Connecting with customers and suppliers" theme of the next chapter. We'll look at these applications in more detail there.

Marketing

Marketing figures out what a company ought to sell, how it should sell it, to whom, through what channels, and at what prices. Sales is where these plans are carried out.

Marketing professionals analyze market data to predict how well current products will sell, what other products customers will want, how the company should set its prices, and make every other decision that precedes selling. These include the "four Ps" from your marketing courses: Product (what to offer), Price (how much to charge), Place (where to sell, including the Internet), and Promotion (advertising and more, much of it online today).

Human Resources

Early uses of computers in human resources kept track of employee information: name and address, pay, benefit choices. Pay information went to a payroll application. That application created summaries of pay, taxes, and other deductions for accounting systems to use.

Today's HR managers use information systems for planning and decision making. Analyzing future needs in the context of today's staffing and anticipated turnover gives a company a head start in hiring. Information systems track training, to know which employees are qualified for certain positions and to plan future training needs; help develop and document succession plans, reducing confusion when a key employee is promoted or leaves; coordinate salary plans, so competent employees aren't penalized for having a stingy supervisor or personal characteristics that a manager doesn't like; and more. These smooth the operation of any organization.

Purchasing

Purchasing (or *procurement*) means buying what an organization needs. Purchasing professionals can do a better job than managers who buy one or two things a year. They know what questions to ask of a vendor, ensure consistency in a firm's approach to them, reduce favoritism, enable a firm to take advantage of better treatment (such as quantity discounts) based on its total purchase volume, and apply one division's experience with a supplier to purchases from other divisions.

These benefits require sharing information. Information systems with shared databases facilitate that sharing. They pass information from a manager whose department needs something to the purchasing agent responsible for buying it. They manage work flow: If a purchasing agent doesn't deal with a request in a timely fashion, it's routed to an alternate. Once a purchase order goes to a supplier, they check for responses and alert someone if there is none by a certain date. When a shipment is scheduled to arrive, they inform the receiving facility of any special actions to take.

Information systems can automate processes. When a production plan calls for materials that aren't in inventory, they can generate the necessary purchase orders. In a retail store, stock can be replenished automatically to reduce the probability of a stock-out, with warehouse space not wasted on items that may spoil, go out of season, or become obsolete before they're sold. A store's IS system can communicate directly with suppliers' systems, reducing delays and opportunities for human error. Human approval may be needed, to prevent standard actions from being taken when they shouldn't be, but these checks are designed not to delay the overall process.

Using computers in purchasing is part of *supply chain management*. It's in the "Connecting with customers and suppliers" theme of the next chapter.

Production

Manufacturing firms use information systems to support production. *Bill of materials processing* (BOMP) applications calculate parts and materials needed to produce products. Figure 7.3 shows an example. *Material Requirements Planning* (MRP) software compares needs with inventory to determine what is needed. *Manufacturing Resource Planning* (MRP-II*) also considers resources such as equipment and workers to figure out when products can be made, and may tie production into accounting systems. *Production planning* systems use this information to create a production schedule. These all led to today's *Enterprise Resource Planning* (ERP) systems.

Suppose a toy factory gets orders for 500 toy trucks and 2,000 toy cars. Each truck uses ten wheels; each car uses four. This comes out to 5,000 wheels for trucks and 8,000 for cars, 13,000 wheels in total. The firm has only 10,000 wheels in inventory.

Basic BOMP software would conclude that the factory can make the cars, and separately that it can make the trucks, with the wheels on hand. (It can't, of course.) Advanced BOMP software or MRP would realize that it needs 3,000 more wheels to make both. MRP-II would also "know" that cars and trucks are painted in the same booth, so they must be painted one after the other: If it takes two weeks to paint the cars and one week to paint the trucks, painting both takes three. Production planning systems could take the lead time for ordering wheels into account and defer either car or

Item No.	Qty.	Units	Description
1	84	each	Lumber, 2"x6"x96"
2	1	each	Lumber, 2"x6"x168"
3	8	each	Lumber, 4"x4"x144"
4	2	each	Kingpost, 6"x6"
5	24	each	Joist hangers
6	12	each	4'x8'plywood sheets, 1/2"
7	400	sq. ft.	Roofing shingles
8	400	sq. ft.	Roofing asphalt felt, thick
9	1	roll	Roof flashing
10	1	each	Roof center piece
11	6	each	1"x4"x96" decorative wood trim
12	28	each	1"x6"x96" decorative wood trim
13	40	each	Lumber, 2"x4"x96"
14	1	lb	Deck screws, 2 1/2"
15	1	lb	Deck screws, 3"
16	1	lb	Deck screws, 4"
17	2	lb	Nails, 16d
18	2	lb	Nails, 10d
19	2	lb	Nails, 8d
20	1	lb	Roofing nails
21	2	lb	Roofing staples
22	8	each	Adjustable post anchors
23	8	each	Anchor bolts
24	16	each	Roof bracing brackets

FIGURE 7.3 Bill of materials for gazebo.

* The Roman number II indicates the second use of those initials in information systems, differentiating Manufacturing Resource Planning from Material Requirements Planning.

truck production until more wheels arrive. ERP would draft a purchase order for 25,000 wheels (the predetermined economic order quantity) from their usual supplier, and submit it for approval. ERP is thus for more than just production. Here, it connects production and purchasing. It can do more. You'll read more about ERP later in this chapter.

Research and Development

Research and Development (R&D), also known as *Engineering* or *Product Design*, designs the products that a company will produce, market, and try to sell.

Engineers in R&D departments use *computer-aided design* (CAD) software to prepare drawings of parts and assemblies such as the one in Figure 7.4. This can take longer than drawing by hand the first time, but is more accurate, reduces errors, and saves time when changes are made. CAD files can be used as input to programs that calculate stresses to make sure a part is strong enough for its intended use, show an item from any direction as if made from any material in any color, and "print" three-dimensional parts. Architects, physicians producing hip implants, landscape designers, museum exhibit planners—almost anyone who designs anything—use CAD software.

Once a design is complete, CAD software allows it to be shared with others who will incorporate a part into a larger system. For example, the plastic lens of an automobile taillight must match the opening in the car's body. Their size and shape can vary, as long as both are the same. Once all parties agree, CAD software can create bills of materials and often cost estimates.

> ***Where you fit in:*** You will be in one or more of the above functions after you graduate. You will use information systems that your employer provides. Try to think of creative uses of information that tie into and supplement the applications you've been given. Your value to your employer will be greater if you appreciate what its information can do.

7.3 TRANSACTION PROCESSING

You may have learned in Accounting that a *transaction* is a business activity that affects the chart of accounts. Information systems use a broader definition: A transaction is a business activity that affects the organizational database.

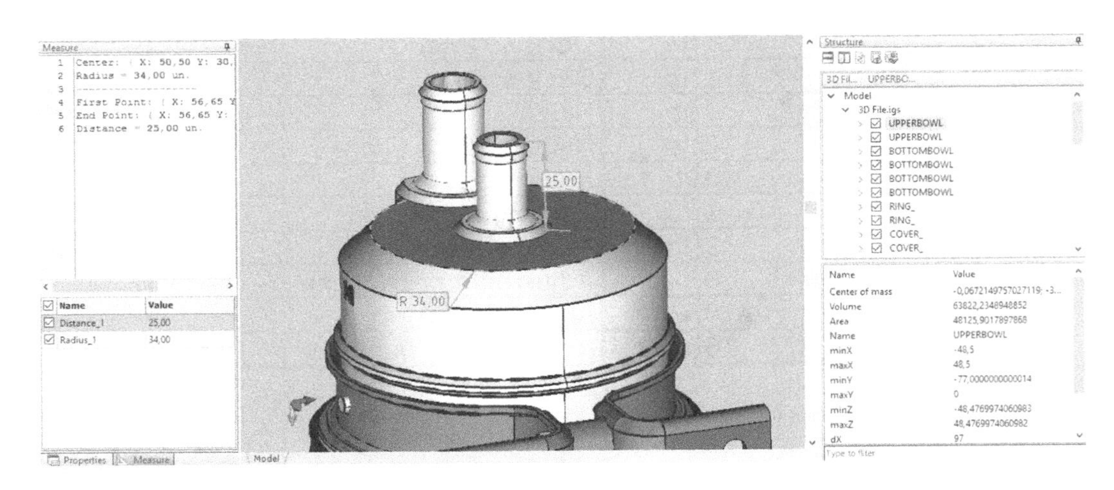

FIGURE 7.4 Output from CAD system (Source: Wikimedia Commons).

Suppose you switch from a 9 a.m. to a 2 p.m. section of Chinese History. Your tuition, fees, degree progress, and financial aid are unaffected. No money changes hands. A university's accountants may not call this a transaction, but to an information system, it is.

Every transaction involves the sequence of activities shown in Figure 7.5:

1 Input is obtained from an external source, often a person.
2 Additional information is read from a file or database, using the input to decide what to read.
3 Calculations are performed, using input data and data from the database.
4 The result of those calculations is stored in a file or database.

For your class change:

1 You are the external source. You enter your student ID number, your password, and what class sections you want to move out of and into.
2 The database provides your registration status, the capacity of the 2 p.m. section, and its current enrollment.
3 The system confirms that the 2 p.m. section isn't full and that it is not yet past the deadline for student-initiated class changes.
4 The system stores your new enrollment information in the database. It also creates an *audit trail* to record what was changed, when, and by whom.

In all but the smallest organizations, applications can be accessed by several people at the same time. Those people may be in different locations, unaware of each other's activity. Many system and communication elements may be involved in processing the transaction. This leads to two complications:

1 When multiple people update a record at the same time, there is a risk of losing updates.
2 When a system element fails, there is a risk of not completing a transaction.

Consider what can happen if you take $100 at an ATM at 10:00 a.m., and a $200 electronic deposit comes in at 10:01. To start the withdrawal, the bank's ATM software reads your balance: $500. It then subtracts the withdrawal from this balance, storing the new balance of $400. The deposit application also reads your balance—but the ATM software hasn't stored your new balance yet, so it too reads your original balance of $500. That application also computes your new balance, $500 + $200 = $700, and stores that.

What will your balance be after both transactions? Either $400 or $700, depending on which transaction finishes last. Neither is correct. It should be $600 (why?) but, because the transactions overlap in time, it won't be.

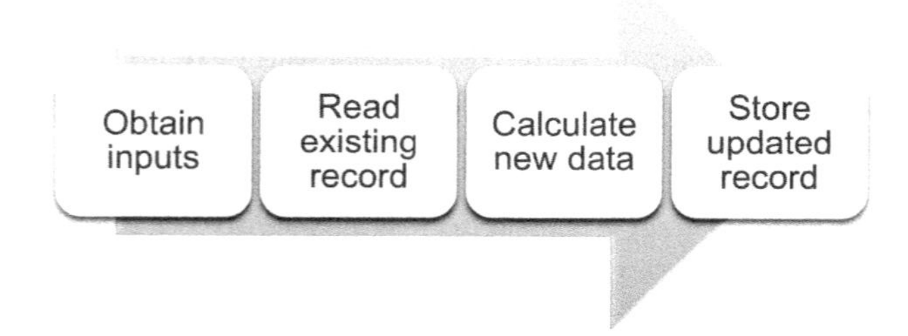

FIGURE 7.5 Diagram of transaction processing activity.

This error will be found when the bank's books don't balance or when you ask a manager where your deposit went, but it could cause other problems in the meantime. If the balance is too low the bank could bounce checks, charging you fees that will have to be reversed and perhaps triggering late fees from payees that will have to be reimbursed. If it's too high you could close the account, keep the money, and ignore (for a while) demands to repay it.

What are the chances of something like this happening to you on a specific day? Nearly zero. What are the chances of it happening once, in the tens of millions of transactions that a bank processes in a month, to one of its millions of customers? Nearly 100%. When an activity is repeated millions of times, unlikely things will eventually happen.

Another problem arises because many hardware components are involved in processing this transaction. Any of them can fail at any time. An ATM can fail after it dispenses cash, but before it tells the central computer that it has done so. If it does them in the other order, it could tell the central computer that it has dispensed cash, but fail before it actually does. Neither is desirable.

As with transactions overlapping in time, the chances of a failure in a given transaction are minuscule. We use ATMs without concern. Still, the chances of it happening once in the millions of transactions that a bank handles in a month approach certainty, so a bank must be concerned.

To avoid these problems and a host of others with large-scale transaction processing, an information system must ensure that transactions have the four *ACID properties:*

- **A**tomicity: A transaction is processed as a unit. If any part of it can't be processed, none of it will be. If it turns out later that the transaction shouldn't have happened, such as a salary change that wasn't properly approved, it can be reversed as a single unit.
- **C**onsistency: A transaction brings the database from one valid (internally consistent) state to another.
- **I**solation: Processing a transaction is isolated from processing other transactions.
- **D**urability. A transaction remains in the database indefinitely unless it is reversed.

Ensuring that transactions have the ACID properties requires complex software. Such software is expensive to create and to purchase, and requires high-performance hardware for acceptable response times. However, enterprise-level information processing needs can't be met without it. Personal computers won't do, no matter how many of them an organization is willing to buy.

Where you fit in: When you hear how much an enterprise information system costs, you need to understand that high cost is often a necessary consequence of high complexity. That doesn't mean you should accept any price without question, but you should know why the price levels you're used to with personal software don't transfer to enterprise software.

Transaction processing applications tend to be used at lower organizational levels. Managers and knowledge workers don't usually enter transactions, though first-level supervisors might when an employee doesn't know how to handle a complex procedure, to override a restriction, or to help with peak loads. Transaction data is summarized into reports for review by managers at higher levels. Managers query transaction databases to get answers to questions that come up in the course of their work. The results of a query might be a report, as in Figure 7.6, or a query response as in Figure 7.7. (It answers the question "Who bought an item that cost over $10, when, what was it, and how much did it cost?") A manager can easily learn to create

Customers in City: Boston

Customer	Order Date	Order Total
Adam		
	9/12/2014	$6.65
	5/5/2014	$2.48
Customer Total		$9.13
Belina		
	9/13/2014	$3.08
	9/7/2014	$7.15
	6/6/2014	$4.27
Customer Total		$14.50
Cameron		
	7/7/2014	$11.90
Customer Total		$11.90
Denise		
	9/15/2014	$1.19
	9/26/2014	$5.17

FIGURE 7.6 Report.

ProdPrice	ProdName	CustName	OrderDate
$11.90	Box of 12 Donuts	Cameron	7/7/2014
$11.90	Box of 12 Donuts	Eddie	9/8/2014
$11.90	Box of 12 Donuts	Cameron	1/14/2014

FIGURE 7.7 Query response.

these in Access, which was used for these figures, or another end-user-oriented DBMS such as FileMaker Pro.

Data Validation

Transactions begin with input from an external source. Before a transaction can be processed, its inputs should be validated to make sure they are correct.

It is impossible to validate all inputs. If a person checking the weather for Princeton, New Jersey, enters the state code as NE rather than NJ, the computer will happily provide the weather for Princeton, Nebraska. If southeastern Nebraska is expecting a blizzard while central New Jersey enjoys a warm winter, that person might take the wrong clothes on a trip. If someone orders 22 boxes of cereal online, a computer can't know that it should have been 2.

Source data automation, which you read about in Chapter 3, reduces the need for data validation. Supermarket cash register pricing is more accurate when bar codes are used to look up prices in a database than when clerks key in prices by hand.

However, nearly all data starts with manual entry. Bar codes may come electronically from a food supplier's database, but at some point someone at that food supplier decided that a 9 oz. can of tuna should have a particular code and entered it. Information systems should do what they can to reduce the chances of error here.

Information systems can incorporate checks to make sure data is of the right type and length. A U.S. Social Security number is nine digits long: not eight, not ten, no letters, no punctuation. Sometimes validation requires limiting data entry formats: Restricting names to letters makes it impossible to enter St. James, Gell-Mann, or O'Neill as people with those names often prefer.

Where you fit in: A system designer can program a system to accept telephone numbers in any format or in a wide range of formats, perhaps checking to see if they have the right number of digits. Is this worth it, though? That system designer doesn't know. He or she must be guided by the people for whom the system is being designed. That's you.

Some data can be validated against a list. Student IDs can be validated against a list of students. If a university with 10,000 students uses eight-digit ID numbers, a randomly chosen number has about one chance in ten thousand of matching a current student. (A criminal can improve the odds by studying their patterns.) If a supermarket checkout system can't find a bar code in its database, it beeps and asks the clerk or customer to rescan the item, key in its code, or enter a price.

Where the list of valid codes is too long to search, information systems use *check digits*. A check digit is a digit that is calculated from the other digits of a number. When the entire number, including the check digit, is entered, a computer repeats the calculation. If it comes up with the same check digit, processing can proceed. If it doesn't, it informs the user of a data entry error.

One might think random check digits will be correct one time in ten. However, human errors aren't random. Most errors involve pressing a key next to the correct one or transposing adjacent digits. Check digits can protect against 100% of those. Since other errors are rare, check digits are

FIGURE 7.8 Credit card with check digit.

effective overall. A credit card number always includes a check digit (circled in Figure 7.8). A computer will catch most data entry errors before trying to retrieve account information.

Reasonableness checks can help validate data. The interest rate on a given type of bank loan has lower and upper limits. The birth year of a kindergarten pupil should be four to six years before the current year. Manual overrides by authorized managers can handle occasional exceptions.

Where you fit in: As with format checks, system designers can program any validation check that one can dream up, but they seldom know what makes sense in a business situation. That comes from you and your colleagues. Understanding the value of such checks will help you provide designers with useful guidance. The end result will be higher-quality data in your system.

Batch versus Online Transaction Processing

When we pay bills, we wait until several pile up. The cable TV bill arrives in the mail; we set it aside. Our phone bill comes in an email later that day, a car payment reminder the next, and rent is due the following week. Once we have a few in hand we sit down at our computer, dig out our checkbook for the few we can't pay online, and take care of them all.

That is *batch processing*. Batch processing uses resources efficiently. That's why we don't rush into bill-paying mode to pay one bill that isn't due for two weeks. Instead, we wait until we have more. If a batch of computer transactions runs overnight when other computing demands are low, its hardware cost is effectively zero.

Batch processing doesn't provide timely information. If a shipment isn't entered when it arrives, an inventory check may not find parts for a production run, or a store may tell a customer, "Sorry, we don't have any" when it does. A dunning letter may go to a customer who just paid, causing embarrassment and possibly losing future business. If bad input is detected later, someone has to dig up the correct information—if it is still available.

Therefore, most applications use *online transaction processing* (OLTP) today. Each transaction is processed as it occurs. If you take the last seat in a class, the system will reject another student's request a second later. Immediate updating is vital to applications such as airline reservations.

7.4 ENTERPRISE RESOURCE PLANNING SYSTEMS

The parts of an organization are interconnected, as shown in Figure 7.9. (It is far from complete.) They must be if they are to work together toward a common goal. For example:

- When a salesperson sells an item, it must be produced (if it isn't in inventory) and shipped. This requires connecting sales, shipping, and perhaps production. Its production and sale must also be reflected in the chart of accounts.
- If that salesperson is paid on commission, information about the sale must go to payroll. It may also be tracked versus a sales quota or other incentive goals.
- When a company doesn't have materials to produce a product, they must be obtained. That means connecting production to purchasing. After the order for those materials is sent, the receiving dock must be alerted to expect them. When they arrive, the accounting department must create an accounts payable entry, and raw material inventory data must be updated.
- Business and human resource planning affect every part of the organization. A sales upturn may call for more production workers. New technology may require new types of engineers. Changes in the customer base may have implications for product design.

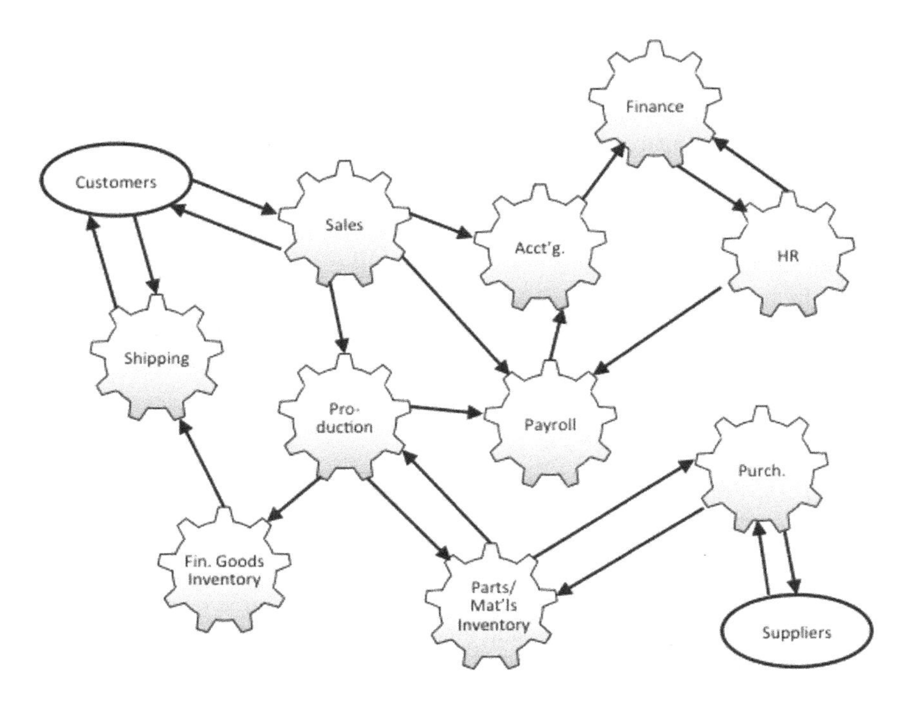

FIGURE 7.9 Diagram showing connections among parts of an enterprise.

Functional applications don't reflect this connectedness, but it is basic to how organizations work. That's why we have ERP.

Enterprise Resource Planning (ERP) Software

ERP software connects parts of an organization via a shared database. That database reflects all of the organization's operations. It has data from each function: general ledger for Accounting, order information from Sales, supplier data and purchase orders from Purchasing, personnel lists and payroll data from HR, bills of materials from Manufacturing, and so on. Departments share this database instead of passing data from one functional application to another.

The first ERP software grew out of MRP-II, as shown in Figure 7.10. It was designed for manufacturing organizations. Today, ERP packages are available for all major segments of the economy (often referred to as *verticals*, as you read in Chapter 4).

ERP software depends on the key enabling technologies for modern information systems: shared databases and networks. Its database is the focus of organizational coordination. Networks enable people in multiple locations to access it.

ERP systems have modules for all functional areas. Each one does what a functional application would do. The difference: They use the shared central database. These modules are designed with similar user interfaces, so people who work with several will find their look and feel consistent.

Business Processes

Consider what happens when a toy factory receives an order to make 500 toy trucks. That order puts several processes into motion, including (but not limited to!) these:

- The customer's credit is checked.
- The sales representative earns a commission, credit toward a sales quota, and perhaps credit toward another goal such as a weekly sales contest.

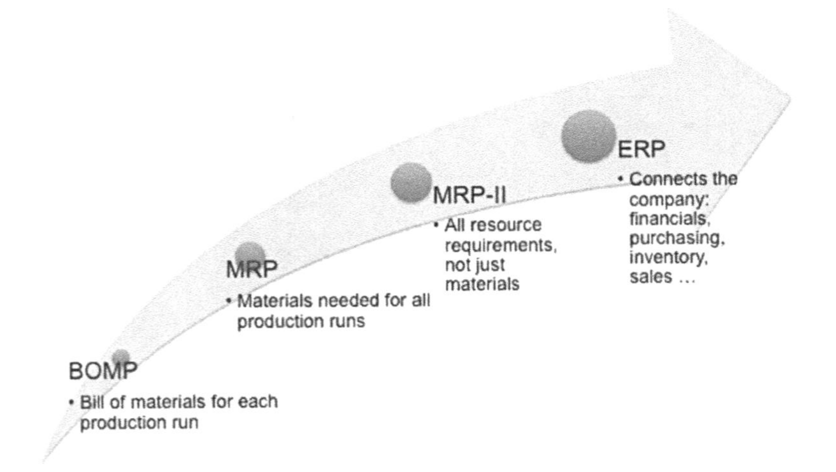

FIGURE 7.10 Evolution toward ERP.

- Inventory is checked for the required materials. Anything not on hand is ordered. Parts that are on hand are reserved for these trucks. If the trucks require parts that are reserved for something else, priorities of the two orders are compared. Those parts might, or might not, be reallocated to the trucks.
- Production is scheduled, taking into account lead time for parts to arrive and the availability of manufacturing resources that are also used to make other products.

Organizations handle processes differently. For example, what if a customer has poor credit? Should the firm reject its order, require a deposit, require payment in advance, ask for a bank reference, accept it anyhow, or something else? If it doesn't reject the order, should it wait until the credit situation is resolved before ordering parts? No answer is always right. Every business process raises such issues. Many issues are not obvious until one studies a process in detail.

ERP software developers study processes to identify industry *best practices*. These practices are then programmed into their package. Figure 7.11 shows a typical hiring process.

Businesses are not forced to use the processes designed into ERP packages. ERP software can be customized. Processes can be replaced by new ones. Coding them is not trivial, so ERP users try to minimize it. Organizations prefer packages whose built-in processes are largely acceptable.

Approaches to Using ERP

The approaches to computing that you read about in Chapters 3 and 4 apply to ERP:

- Traditionally, an organization licenses ERP software and uses it on its own computers.
- ERP can be used on demand as Software as a Service (SaaS). SaaS offers easier installation, with new features added as they become available rather than via installing an upgrade. It offers lower initial costs, and perhaps lower ongoing costs as well. However, it offers less flexibility for customization or for "mixing and matching" ERP modules with other software.
- A company can license ERP software to be operated by the ERP vendor or a third party.

In 2020 most organizations use the traditional approach, but many had no alternative when they first adopted ERP. About a third of all small companies that use ERP use SaaS: Small firms are more likely to have adopted ERP recently. Under 5% of larger firms, which are likely to have adopted ERP earlier, do. Aberdeen Group found that willingness to consider cloud-based ERP solutions

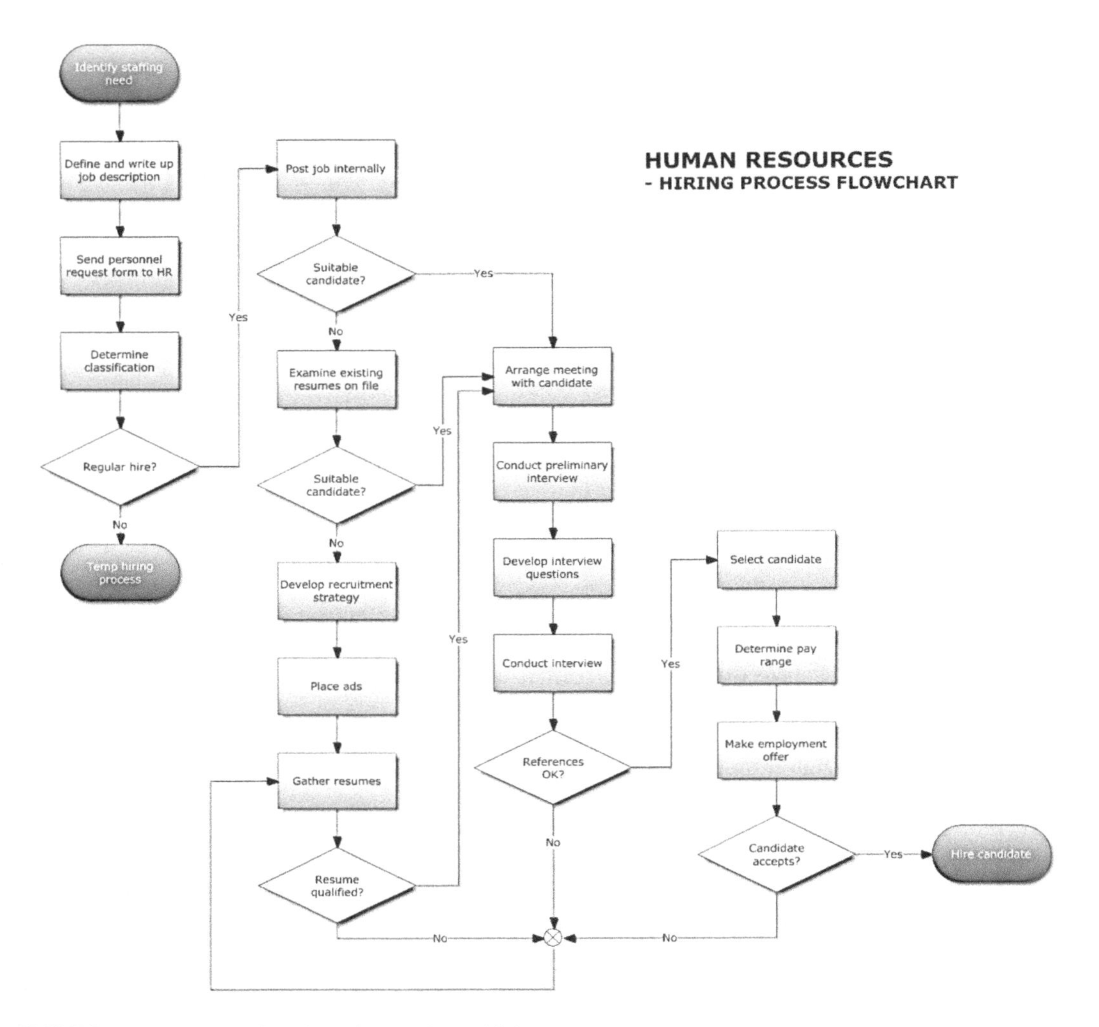

FIGURE 7.11 Process flowchart for employee hiring.

has increased steadily since they began tracking it in 2009. They expect use of cloud-based ERP to increase in the future.

Manufacturing ERP systems are often integrated with computer-controlled production equipment, an aspect of the Internet of Things (IoT) you read about in Chapter 6. ERP vendors are turning to *edge* and *fog* computing (see Chapter 4) to make that integration more responsive. This is an aspect of MES, Manufacturing Execution Systems.

MES are information systems that track what happens on the factory floor. Having direct input from production can reduce paperwork bottlenecks, improve planning, improve quality, and provide rapid response to problems that would otherwise cause lengthy delays as they are resolved. MES are sometimes thought of as an extension of ERP, sometimes as a separate application. However you think of them, they are part of today's manufacturing picture.

Where you fit in: Unless your employer is very small, it is likely to use ERP. In most cases the ERP system will run on their own computers. You should be aware that cloud-based options may make more sense in the future, and be prepared to participate in the firm's transition to the cloud.

Benefits of ERP

Companies go to the expense of acquiring, customizing, and installing ERP software because it offers valuable benefits:

- Manual information transfer from one department to another is reduced or eliminated.
- The central database provides a complete picture of all aspects of the business. Management has visibility into the overall operation, not just each piece individually.
- Sharing one database eliminates inconsistent views of the same data by different departments. Decisions are based on consistent data.
- Best practices reflected in built-in processes tend to be better than home-grown processes.
- Reporting is closer in time to the events that drive it. Information is available quickly.
- Having all data in one place can facilitate compliance with regulations such as Title 21, Part 11 of the U.S. Code of Federal Regulations. It specifies controls for companies that deal with the U.S. Food and Drug Administration. Many industries have comparable regulations.
- Inventory costs can be reduced with better forecasting.
- Eliminating duplicate data entry can save time and thus improve customer satisfaction.

Concerns with ERP

ERP is not a magic bullet. Getting its benefits is not always easy. ERP concerns include:

- ERP software can be expensive. Its initial cost for a large organization will probably be well into six figures (in U.S. dollars), plus annual support and update costs in five or six figures.
- Customizing an ERP package is complex. It may require new skills that will not be cheap. It may take months or years. It might not be successful, but without it, the ERP system may be unusable.
- Once customization is complete, implementation is also time-consuming and expensive. Customization and implementation for a large organization can cost over $1 million, perhaps several times that.
- The business processes in an ERP package, though excellent overall, may not suit a specific organization. It may not be practical to customize the package to be suitable, perhaps because an existing business process depends on information that isn't in the ERP database.
- Using a package's processes, even after customizing, may mean changing how people work. Changes can be upsetting, and sometimes require union contract renegotiation.
- Running ERP software may require expensive hardware upgrades.
- MIS staff, after being trained on an ERP system at their employer's expense, may use this knowledge to get higher-paying jobs elsewhere or demand exorbitant raises in order to stay.

Despite these concerns, most large organizations and an increasing number of smaller ones find ERP worthwhile. Once a company has moved to ERP, it is unlikely to go back—or to want to.

> ***Where you fit in:*** Moving to or changing ERP systems is a far-reaching business decision. It has high potential benefits, costs, and risk. IT specialists tend to favor it. Programmers tend to be optimists, and most would rather work on a modern ERP system than a bunch of old applications. Managers must temper this enthusiasm and make sure that ERP decisions consider negatives as well. They must then support the project. Without management support, ERP cannot succeed.

IS management must also work with human resource management to defuse the last risk above, the MIS staff taking their new skills to the highest bidder. Deal with this before a project starts.

Business Process Re-Engineering

Introducing ERP is an opportunity to re-examine business processes and change any that are no longer optimal. They can be replaced with a process that is programmed into the new ERP software or with something else. If the process change is radical, especially if it is enabled by new technology, it is referred to as *business process re-engineering* (BPR). It may be spelled without the hyphen, and is also known as *business process redesign* or *business transformation*.

While ERP may enable BPR, they are separate concepts. Either can exist without the other.

The BPR idea was first put forth by Michael Hammer in 1990. Instead of trying to optimize an existing process, he suggested looking for ways to eliminate it or perform it in a different, better, way. The enabling technologies you have studied, shared databases and communication networks, are fundamental to BPR. For example, the steps of a process may be performed in sequence because, when that process was first designed, paper had to be routed from place to place. Automating this process will send an electronic document through the same steps in the same order, just faster. Process redesign may recognize that its steps can be performed in parallel today.

BPR: A SUCCESS STORY

Ford Motor Company's North American accounts payable operation provides an example of business process re-engineering success.

The traditional method of paying vendors begins when they are issued a purchase order, on paper or electronically. When goods are delivered, they are accompanied by a packing slip to be signed by a receiving dock employee and returned to the vendor. The vendor's accounts receivable department then sends the customer an invoice. The customer's accounts payable department matches the purchase order, packing slip and invoice, and then pays the vendor.

Ford's insight was that packing slips and invoices are unnecessary in an age of databases and networks. Receiving dock personnel can confirm that arriving goods match a purchase order. Vendors who ship goods in response to a purchase order expect to be paid. The purchase order states how much to pay them. If ordered goods arrive, the vendor should be paid that amount.

This insight made it possible for Ford to eliminate most document-matching in vendor payment, reducing department headcount from 500 to 125 people.

The concept of process redesign isn't new. Assembly lines transformed manufacturing a century ago. Before that, steamships transformed port operations by removing the need to sail at high tide. Thousands of years earlier, farming redesigned food supply by making it unnecessary to gather wild plants. However, early advances that enabled process redesign were in one industry or a few related ones. Information systems can help redesign processes in any sector of the economy.

BPR has had failures. Failure can have several causes: misunderstanding strategy, and therefore trying a redesign that misses essential points; spending large amounts of money to redesign a process whose overall impact is small; initiating BPR with the goal of cutting staff, not improving a process; and ignoring human factors. It is estimated that over half of all BPR projects fail to achieve their goals, though some "failures" still yield worthwhile improvements.

Where you fit in: Business process re-engineering, while not a cure for all an organization's ills, offers tremendous potential benefits when applied properly. As a manager who understands the benefits of information technology, you should include BPR in your toolkit.

7.5 ENTERPRISE APPLICATION INTEGRATION (EAI)

An organization looking at ERP packages typically evaluates each module separately. It will probably find most modules in the most suitable package usable, perhaps with customizing. Some won't be practical to customize, but the organization can adapt to their built-in processes without undue difficulty. That may leave a few that fit this organization so poorly as to be unacceptable.

One solution is to "grin and bear it," saying "other organizations work that way; we can learn to." However, EAI makes this unnecessary.

With EAI, each application or group of related applications uses its own files or database. They read data from that database when needed. Their updates go into that database.

Updates made to one database in this scenario aren't in the other databases. However, data must be shared. If an employee gets a raise, but HR and payroll don't share a database, information about that employee's new pay rate must still reach the payroll system on a timely basis without human intervention. That is the role of *middleware* (Figure 7.12).

Middleware, as its name suggests, is in the middle of these applications. Updates come from one and go out to others. The middleware converts data as needed. Conversion may be as simple as extending a field from 10 to 15 characters, rounding numbers to two decimal places, or going from English to metric units. It may be as complex as restructuring a file into relational tables.

Middleware must be created, or customized if a commercial package is used, before it can do its job. A database administrator must define data sources, destinations, and conversions. Definitions must be updated whenever data structures or formats change on either side. Such changes usually coincide with software upgrades, though upgrades don't always change the database structure.

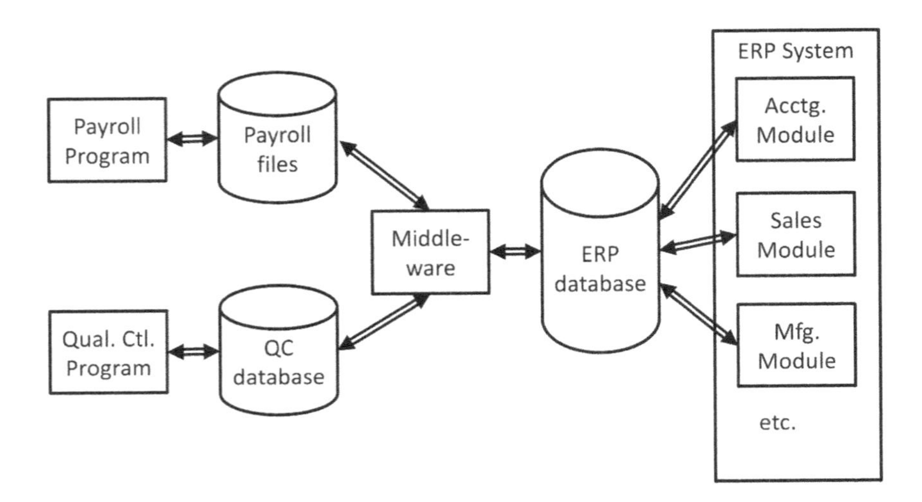

FIGURE 7.12 Diagram of enterprise information system with EAI.

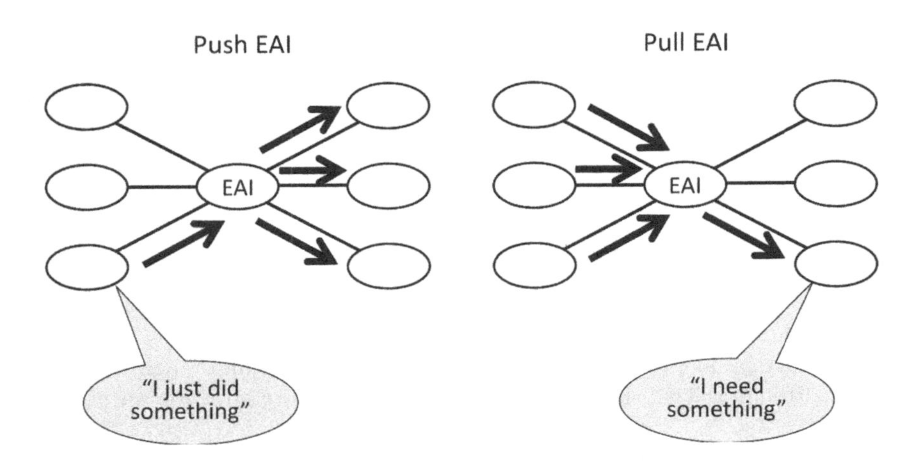

FIGURE 7.13 Diagram of EAI push versus pull.

Middleware can operate as *push* or *pull*:

- With *push*, on the left in Figure 7.13, changes to one database are pushed to other places as they happen. Some of this effort is wasted: New data might not be used before being replaced by newer data. However, when another application needs data, it's right there and up to date.
- With *pull*, data is moved when needed. This is more efficient. Many updates may take place before a transfer. However, there is a lag when an application needs data: Data elements must be retrieved, converted, and entered into the destination database before proceeding.

The choice—push or pull?—depends on business priorities. The value of immediate access to up-to-date data usually justifies the cost of added hardware needed to support pushing.

EAI is not limited to situations in which an ERP module can't be used. This approach can connect any set of applications, though if they are too old or too different it may be difficult to get them to share data. Most middleware packages are flexible in the data and file types they can work with, since their value depends on this flexibility, but that doesn't mean they accept everything.

ERP or EAI?

The choice between ERP and EAI is a business decision, though technical considerations are important in making it. Advantages of ERP over EAI include:

- The system comes from one place. Its vendor updates all of it at a time, not piece by piece on different schedules. When there's a problem, there is no question about where to turn to fix it.
- The best practices in ERP packages are better than the processes most organizations have.
- The organization's MIS staff deals with one system, not several plus middleware.
- Moving to a modern ERP system can push an organization to upgrade its technology. That benefits other areas as well.
- Having modern systems removes one reason that technical employees may want to leave.
- The common "look and feel" of all applications reduces user training requirements.
- Since there is no middleware, there is no need to customize it.
- Sharing one database eliminates the drawbacks of both the push and the pull approaches to middleware.
- It is easier to provide security for one modern database than for several, some of them old.

Conversely, advantages of EAI over ERP include:

- EAI lets an organization take a *best of breed* approach to choosing applications. It can get the best inventory management program from one source and the best purchasing program from another, rather than having to compromise on one.
- Each application can be chosen for its fit with the organization. Less customization is needed.
- An organization can continue to use existing applications, rather than changing them just because another department needs a new system for some other reason. This reduces implementation costs and training costs for current employees.
- Separate files and databases, none of which contain enough information to be worth stealing, may be less attractive to thieves than a single database would be.
- Technical skills may be easier to find within the organization, especially if some existing software is kept, and are less likely to cause people to depart for higher pay elsewhere.

As the cost of ERP drops, the technical skills required to customize and support it become more common, and people throughout the economy become more used to it, the trend is toward ERP. You may still find EAI in an organization you work for, though.

> ***Where you fit in:*** "ERP or EAI?" is a business decision even though technical issues affect it. Decision makers must understand the technical issues and maintain a business perspective in considering them. Being able to do this increases your value to your employer.

7.6 INTRANETS

Besides using shared databases to connect parts of an organization, often via ERP, organizations need networks to enable their users to connect to that database—and, through it, to each other.

Large organizations have LANs in most of their locations, linked through wiring within buildings and via the Internet beyond them. These provide connectivity, but a network by itself doesn't provide database access. That requires an application.

An internal network through which people in an organization can access its resources, and which uses Internet protocols, is called an *intranet*. Figure 7.14 shows a corporate intranet.

Users access intranets through browsers, just as they access public web pages. Intranet servers respond with web pages that look exactly as if they had come from a public site, but which contain

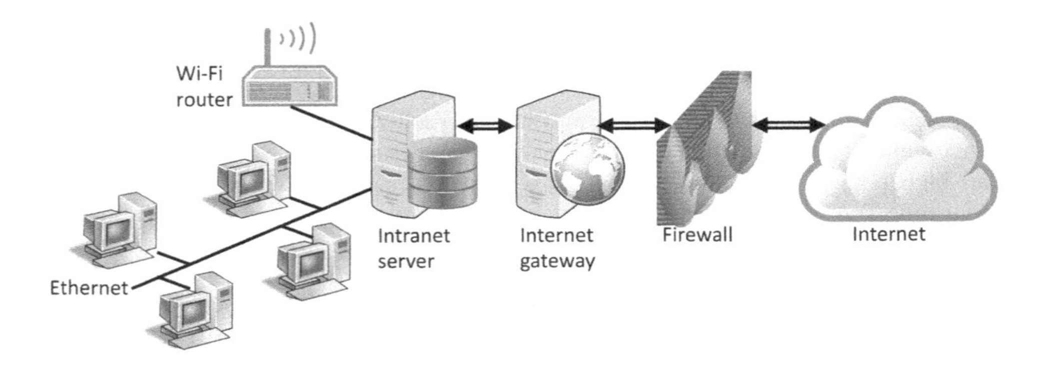

FIGURE 7.14 Diagram of typical intranet; illustration elements from openclipart.org.

internal information. This familiar interface makes it easier for people to adapt to new applications. Intranets also support email, file transfer, and other information-based activities.

Intranet access can be controlled in two ways. One is by physical connection: Computers in fixed locations such as offices often use wired connections and are always on the intranet. Access over wireless links and from other locations must be authorized, usually by user ID and password.

Once on the intranet, users have access to corporate databases and other information resources. This is an example of *single sign-on*. The user identification that they provided in logging onto the intranet determines their access privileges for those resources. Those who are automatically on the intranet via physical connections may have to log in to access controlled resources, since connections such as Ethernet jacks in conference rooms can be used by anyone who is authorized to be in the building (and perhaps by someone who isn't).

Most intranets, including the one shown in Figure 7.14, provide gateways to the Internet. Once on the intranet, a user can reach the web and other Internet-based resources as well. This access is often mediated by a firewall. It attempts to prevent harmful content from entering the internal network. It may also block users of the corporate network from visiting unauthorized sites.

Where you fit in: It may be annoying to have to log into an intranet before accessing company resources, but you should accept it as a necessary security precaution.

Intranets and the Internet of Things

Carlo Nizam of Airbus is generally credited with coining the term "Intranet of Things" for an intranet that connects non-human devices. The concept did not originate with Airbus, however. General Electric called it "the Industrial Internet of Things." Others gave it different names.

Just as an intranet is an internal network that uses Internet protocols, an "Intranet of Things" is an internal network that connects devices with sensors and actuators using the Internet protocols that many devices support. Using existing protocols saves a great deal of expensive, time-consuming work. Using them on an internal network removes some of the risk associated with exposing computer-controlled devices to anyone with an Internet connection, knowledge of how to manipulate it, and evil intent. (All three are common in 2019.)

Three examples of how the Intranet of Things is used in the real world:

- The Wegis and Young agricultural conglomerate began as a California family farm over a century ago. They farm over 10,000 acres (4,000 ha). To deal with severe California water shortages and water management legislation, the sensors in W&Y's system monitor well water levels and usage, water pumping, water quality, energy use (electricity, natural gas, diesel, propane), pump lubrication and vibration, and more. Sensors connect to a gateway via Bluetooth. That gateway links the Bluetooth intranet to a cloud computing facility.
- Offshore oil and natural gas platforms incorporate sensors and actuators on the platform and in pipelines connected to it. A platform is largely a self-contained information processing environment, since a satellite Internet connection is slow and expensive. Security demands are high, both because of the inherent danger of a breach and, in the U.S., the need to meet Maritime Transportation Security Act requirements. A platform can use a local network to connect flow meters, valves, and more to local servers. A firewall protects the slower connection between those servers and facilities "on the beach."

  The security of one such system was inadvertently tested when a contractor using an infected USB drive unwittingly exposed its intranet to a virus. Firewalls kept the virus contained and protected the safety systems of the platform. Strict rules against attaching foreign devices, instituted after this event, prevented a recurrence.

- Wind provides clean, renewable energy. One turbine incorporates hundreds of sensors and actuators: strain gages, bearing monitors, and more. The hundreds of turbines in a wind farm must balance power generation with load, and prevent damage during dangerous weather conditions, by adjusting blade pitch and generator settings, and must do this in difficult-to-access locations at sea. High data volumes call for an intranet to get data to local servers with minimal delay, with a gateway to pass summary data to more powerful land-based systems.

In manufacturing, MES (Manufacturing Execution Systems, see Section 7.4) increasingly take advantage of automatic production and process monitoring via the Intranet of Things. MES can be configured to accept manual data or data directly from machines, so factories can start with a manual system and evolve to a more automated one as they install machines that support it and the necessary networks.

7.7 INTEGRATING SOCIETY

Our largest organization, at least until we meet intelligent life elsewhere, is human society. Just as information systems can connect the smaller organizations we call businesses, universities, and so on, they can also connect elements of society at large. And, just as networks that connect the parts of an organization can help that organization run more smoothly, networks that connect parts of society can help society run more smoothly—as you have probably noticed in your own life.

We're familiar with social networks such as Facebook and Instagram that connect millions of people. (Facebook reached two billion members in June 2018, and had 2.41 billion who were active at least once a month in the second quarter of 2019. That's about 40% of all the people in the world who are old enough to join.) YouTube is close, with about two billion members, but is (as of 2019) about video. The next general social media platform is Instagram, which Facebook owns, with about one billion. There are tens of thousands of smaller social networks. Many of those focus on specific interests such as photography, BMW motorcycles, or gluten-free cooking.

One problem in connecting society is finding an economic interest in doing so. A company has an economic interest in improving operations by connecting its parts. A university has an economic interest in becoming attractive to applicants by showing them how connectedness will benefit them once they join its community. But who benefits from connecting members of society to each other? If they offer no direct economic benefit to anyone, how do social networks survive?

Some survive because an individual or group sponsors them as a public service. (Sponsors may see sponsorship as a form of advertising.) Some are provided as a benefit for an organization's members. Some charge for use. Some are supported by advertising: A focused social network is an attractive ad medium to companies whose customers match its focus. Thus, airlines advertise on *flyertalk.com*, camera companies buy ads on *dpreview.com* ("dp" is for *digital photography*). There are hybrids, such as a free site supported by ads, where paying a fee will remove them.

The way a company hopes to earn a profit is its *business model*. Finding a viable business model can be difficult for a social site. The business model must be reviewed regularly: one that works when a site is small may no longer work as it grows. A person who was willing to subsidize a site when it used an old surplus computer as its server may not be equally willing when it grows to require a rack of blade servers and a paid staff.

Social networks are subject to a phenomenon called *network effects*. With network effects, the value of being in a group increases as people join it.

Consider network effects and the telephone. A single phone is useless. Two are nearly so. Five hundred subscribers in a town would get value from their phones. Others would realize that they want to talk to some of those 500 people and businesses, and would also get phones. That would, in turn, bring some of their friends and relations into the telephone age, and so on until the whole town had phones—as in industrialized countries today.

A NOTE ON LANGUAGE

Despite the word, *network effects* are not about communication networks. Network effects don't require data communication. The usefulness of a language, for example, depends on how many people speak it. Those with more speakers are more useful. Since people study useful languages, popular languages acquire still more speakers. That's a network effect, too.

Network effects may not last. All educated Europeans once spoke Latin. Then England's role in the Industrial Revolution and its domination of shipping made English the international language. The economic power of the U.S. in the late 20th century reinforced that position, but few are sure that English will stay on top. Facebook defeated MySpace as a general social networking site. We don't know who will topple Facebook or when, but it would be folly to claim that nobody will.

Network effects show *positive feedback* to reinforce the position of a dominant firm. If you use Line for VoIP telephone calls, you urge your friends to use Line rather than WhatsApp or Skype. You do this for your convenience, but the effect is to help Line and hurt its competitors. *Negative feedback* works in the other direction: As a network grows, so can forces that limit or reduce its growth. This happens when a network becomes too large to function effectively.

Where you fit in: Using social networks can benefit any business. Some benefits can come from using existing networks such as Twitter and Instagram. Many firms can also gain by creating their own social sites or adding social features to existing sites. Be alert to these opportunities!

KEY POINT RECAP

Pulling parts of the organization together is one of the three key ways that information systems provide major strategic benefits.
This is one of the three areas in which you, as a businessperson, must look for creative ways to use information technology.

Transaction processing is fraught with complexities and potential problems.
These are a big part of why transaction processing systems are complicated and expensive. As a businessperson, you should understand where your company's money is being spent and why.

Integrated approaches such as ERP depend on a shared database and on the right processes.
As a businessperson, you are ultimately accountable for getting these right—even if you delegate the responsibility for them. ERP vendors provide good starts with database design and built-in business processes, but customization makes or breaks ERP systems. Customization isn't cheap.

Intranet
All but the smallest organizations can use an intranet. As a businessperson, you should look for ways to take advantage of the one in your company, if it has one, or for an opportunity to install one for everyone's benefit, if it doesn't.

KEY TERMS

ACID properties (of a transaction): *Atomicity, Consistency, Isolation*, and *Durability*.
Atomicity: Property of a *transaction* that it is processed either entirely or not at all.
Audit trail: Information as to how a database was changed, when, and by whom.

Batch processing: Accumulating a group of transactions to be processed together.
Best of breed: Approach to application selection in which the best application is chosen for each function, with less focus on their ability to share data.
Best practices: The ideal way to perform a business process, as determined by studying how successful companies perform it.
Bill of Materials Processing (BOMP): Information system that calculates the materials required to build a group of identical products.
Business model: The way an organization hopes to earn money from its activities.
Business Process Re-engineering (BPR): Making radical changes to business processes, to achieve significant technology-enabled improvements in efficiency and/or effectiveness. (Also *business process redesign*, *business transformation*.)
Check digit: Digit appended to a number that enables a computer to confirm, via a calculation on that number, that it was (with high probability) entered correctly. See *data validation*.
Computer-Aided Design (CAD): Information system that supports product design activities.
Consistency: Property of a *transaction* that maintains database validity after its processing.
Data validation: The process of ensuring, to the extent possible, that input data is correct.
Durability: Property of a *transaction* that it remains in the database indefinitely unless reversed.
Enterprise Application Integration (EAI): The process of sharing data across organizational functions by having software move data among them. See *middleware*.
Enterprise Resource Planning (ERP): Information system that ties parts of an organization together through a shared database, with its modules replacing functional information systems.
Functional information system: Information system that supports one organizational function independently of others.
Information silo: Informal term for a *functional information system* with little ability to share data among organizational functions.
Intranet: Internal organizational network that uses Internet protocols, especially those of the web, to bring information to users.
Isolation: Property of a transaction that its processing is isolated from that of other transactions.
Legacy system: An older information system that uses technologies that would not be chosen today, but which works and is still in use.
Material Requirements Planning (MRP): Information system that calculates the materials required to build a set of products and determines what must be ordered.
Manufacturing Resource Planning (MRP-II): Information system that calculates the materials, equipment, and labor required to build a set of products and determines when they can be built.
Middleware: System software package that moves data among applications that were not originally designed to exchange data. See *enterprise application integration*.
Negative feedback: Feedback that results in small firms becoming even smaller.
Network effects: The phenomenon via which the value of a group increases as it grows, or the value of a technology increases as more people use it.
On-Line Transaction Processing (OLTP): Processing each *transaction* immediately as it occurs.
Production planning: Information system that creates a feasible production schedule based on the output of other systems, order priorities, minimizing set-up times, and other factors.
Pull EAI: *Enterprise application integration* in which data items are moved from one application to another when the receiving application requires them.
Push EAI: *Enterprise application integration* in which data items are moved from one application to another when the sending application changes them.
Reasonableness check: Confirming that input data is plausible. See *data validation*.
Supply Chain Management (SCM): Using information systems to coordinate with suppliers.
Transaction: Any business activity that affects the enterprise database.

REVIEW QUESTIONS

1. What is a *functional information system*? What parts of the organization do functional information systems serve?
2. Explain the initials BOMP, MRP, MRP-II, and ERP.
3. Give representative tasks that functional information systems perform in your major field.
4. What is a *transaction?* How does the meaning of that term in information systems differ from its meaning in accounting?
5. List the activities that take place during a transaction, in the order in which they occur.
6. What are the four properties that ensure proper processing of all transactions?
7. What is *data validation*, and why is it useful?
8. Why are transactions usually processed online today, rather than in batches?
9. Explain how ERP differs from a group of functional applications.
10. Why must the processes built into an ERP system often be customized?
11. Give several advantages and disadvantages of ERP, compared to functional applications.
12. What is business process re-engineering?
13. Contrast ERP with EAI (enterprise application integration), giving two advantages of each.
14. Describe the role of *middleware* in EAI.
15. What is an *intranet*? Who can use one?
16. What are the two ways of controlling access to an intranet?
17. List some ways that social networks can cover their expenses.
18. Give an example, other than telephones, of a phenomenon that is subject to network effects.

DISCUSSION QUESTIONS

1. You have four related functional information systems: accounts payable (AP), accounts receivable (AR), payroll, and general ledger (GL). The first three create output files in a format that the GL system can read. At the end of each day, the three files are input to the GL program and an updated GL file is produced. This GL can be used the following day. Think of one business for which this mode of operation would be acceptable and another for which it would not. Explain why, for both. You must describe both businesses in enough detail for it to be clear why each explanation applies to one business but not the other.
2. Describe three different types of transactions that could take place in a hotel information system. For each, explain what could potentially go wrong if two transactions of the same type take place at the same time and interact in an unlikely, but possible, way.
3. Indicate what moves along each of the arrows in Figure 7.9.
4. A friend, who works at a manufacturing firm, has been asked to manage her employer's ERP project. Write her an email warning her about possible concerns.
5. Look up cloud ERP offerings from Plex (*www.plex.com*), NetSuite (*www.netsuite.com*), and Acumatica (*www.acumatica.com*). Summarize the advantages that these firms claim over on-premise ERP software (which a company installs and runs on its own computers). They can't be objective on this point. Why not? If you were considering a cloud ERP solution, how would you deal with their potential lack of objectivity?
6. Research the *hype cycle*. This concept, which describes the stages of enthusiasm of a new idea, originated at technology advisory firm Gartner Group and has been applied to many innovations, though it doesn't fit all. Don't rely solely on Wikipedia: Anyone can edit an entry to suit their bias, so you can't be sure it's objective. Describe it in a few paragraphs, citing your sources. Then, discuss how this concept applies to business process re-engineering.
7. An old quote, often attributed to Henry Ford, is "If I had asked people what they wanted, they would have said 'faster horses.'" Relate this statement to business process re-engineering.

8. A process design expert noted that many members of an orchestra play the same notes on the same instruments. This expert recommended that the cello section be replaced by one cellist plus an amplifier, and so on for the rest. Is this a good idea or a bad one? Why?
9. You work for a company that sells middleware. One of your sales prospects makes custom racing bicycles, employs about 100, and has annual revenue around $25 million. They use a collection of functional information systems, developed or acquired over years. They feel that better integration would enable them to serve customers better and save money. You are competing with ERP vendors, but not with other middleware firms, so if you can sell them on EAI, they will buy from you. Create a presentation of eight to ten slides to their managers that explains, in non-technical terms for a business audience, the benefits of the EAI approach.
10. Sports teams are often formed by choosing the best available player for each position and then training them to work together. This is analogous to the EAI approach to enterprise information systems. How would you form a team using the ERP approach?
11. Sketch the factory and offices of the bicycle manufacturer in Question 9. Draw an intranet on your sketch, showing which areas it will connect and who works there. Show a central server and a connection to the Internet. List five types of data that would be stored in the central server, and that people in more than one area on your sketch will want to access.
12. "A social network is an intranet for anyone who wants to be on it." Do you agree with this statement? Explain why or why not.
13. Choose three social networks or other Web 2.0 sites. Compare their approaches to making money from their activities. Of all the approaches, which will bring in the most revenue? Lead to the fastest growth? Have the most, and the least, potential to upset users? If you were starting a new social site, which would you choose—or would you choose a different one?

KHOURY CANDY DISTRIBUTORS CHOOSES AN ERP SYSTEM

The following week, Isabella and Jake took seats at the back of the conference room. They saw several familiar faces plus half a dozen or so that they didn't recognize. One person handed out printouts of the slides Chris Evans planned to use. His title slide was already on the screen at the front of the room.

Jason Khoury was seated near them in back, content to let Chris run the meeting that would decide if KCD should stay with on-premises ERP running on their IBM computer or move to ERP in the cloud.

"As you all know," Chris opened, "our current ERP system is from SAP. Like many SAP customers, we use an Oracle database. Oracle is trying to persuade us to drop SAP and move to their NetSuite. They say we'd save money, and they have numbers to back that up. We're also coming up on a new SAP release, so we'd have to do some converting either way. Still, it's not a slam-dunk. Lakshmi and her people have put some pros and cons of both approaches together." He clicked a remote and the next slide appeared.

"The specifics are on the slides. You have copies, so you can read them. What it comes down to is, if we were starting from scratch today, and we didn't have anything else on our database, we'd probably go for a cloud solution. NetSuite is up there with the best of those. We'd save money. We'd get ERP that's easier to use out of the box, and easier to customize where it needs customizing. Problem is, that's not where we are. We aren't starting from scratch. We have investments in our database, in training, and in our customized SAP modules. We'd have to redo the customizing with NetSuite, though it would probably be easier the second time around.

"Then we did a weighted decision analysis," he continued. "That's on this slide. You can see the factors we considered down the left, how much each factor counted out of 100 points in the next column, and each system's score on these factors in the next two. When we multiplied everything out and added the points up, NetSuite came out ahead by a small margin but not by enough to make a strong case.

"The bottom line," Chris continued as he clicked the remote again, "is that we figure to save about $25,000 a year with NetSuite, but it would take us six to eight years to make up the switching cost—and that's without taking the cost of money into account. Once you factor in the value of money in hand now versus the expectation of having money later, we probably never would.

"Because of that, Lakshmi's team recommended staying where we are. They don't think the benefits of switching justify the cost. I'm leaning that way. Before we make that call, I want to know if anyone can think of anything we've overlooked, that we counted too heavily, or not heavily enough. If anyone has questions, we'll take them now too. If I can't answer them, someone who can is probably in this room."

A hand across the table went up. A person Jake and Isabella didn't recognize said, "We're growing. In a few years we could have more distribution centers, sell in more states and provinces, and for all I know, carry snack foods and who knows what else. Did you look at how these systems can handle growth?"

"I'll take that one, if I may," said Lakshmi. Chris nodded to her, so she continued, "That's an important factor, so, yes, we did. SAP happens to be a bit better that way. It has a lot of features that we haven't tapped into yet. That's not knocking NetSuite, but to some extent it's a case of getting what you pay for."

"Did you consider one of SAP's cloud offerings as an option?" asked another participant.

This time Lakshmi answered without asking for Chris's approval. "Yes, early on. The one we'd go to is different enough from our SAP ERP that the conversion would be as big as moving to NetSuite. Between those, we felt that NetSuite met our needs better and didn't have any significant downsides. We already have relationships with both vendors, so that wasn't a factor either, and NetSuite was less expensive. So, we didn't include that in our final comparison, since we knew it would come in behind NetSuite."

"That's it, then," said Chris as no more hands went up. "We'll stick with SAP.* I'll break the news to our Oracle rep. I suspect that renewing our database contract will do a lot to cushion his disappointment!"

As the KCD participants filed out, Lakshmi stopped by where Jake and Isabella sat. "What did you two think?" she asked.

Isabella was the first to reply. "From what we've read, the cloud seems to be the way to go from a tech standpoint," she said. "Yet most of what we heard today wasn't technical. It was about what it would take to get from here to there and whether saving money was worth it."

"Exactly," Lakshmi agreed. "If you noticed a couple of sad faces, those were our lead programmers. They're fantastic at what they do, I'd hate to lose them, but they sometimes forget that we use technology to move the company forward. If we can't make a case for the bottom line being better off with a given bit of tech, we won't get it. They buy into the concept, but it can be hard for them to feel it in their gut."

"I'll bet they didn't major in business," Jake joked.

"You'd win that bet, they didn't," Lakshmi said. "Still, it takes all kinds. We wouldn't be here if Jason hadn't gone to B-school, but we also wouldn't be here without people like those two!"

Questions

1. Early in the meeting, Chris showed slides with pros and cons of NetSuite and SAP. As an experienced presenter who knew his audience could read, he didn't bore them by reading his slides to them—so this meeting report doesn't include their content. Create two of his

* A fictional preference for one product over another in a fictional situation should not be taken as an endorsement.

slides, one with the advantages of each system. You don't need to get into technical specifics, but you may want to do some additional research.
2. This episode omits the work that precedes a meeting such as this. (You'll read about it in Chapter 10.) Much of that work consists of gathering information online. Find five web pages with useful information for this comparison: information about one of the two candidates, a comparison of the two, case studies documenting user experience with one or both, or almost anything else. Don't use Wikipedia: Anyone can edit it to suit their bias. Your instructor may compile, or ask students to compile, a composite list of all the sources you found.
3. If it would take KCD seven years to break even on the conversion without considering cost of capital, but they would never break even if cost of capital is taken into account, what is the internal rate of return that KCD uses in its cost of capital calculations?

CASE 1: CITY OF EDINBURGH MODERNIZES

Edinburgh, which received a royal charter in 1125, has been the capital of Scotland since 1437. With a population of nearly 500,000 (one of every 11 Scots) and an area of over 100 square miles (264 km^2), governing it is no small task. This task falls to the City of Edinburgh Council.

Within Council government, the Corporate Governance department is responsible for information systems and technology—plus human resources; managing the Council's museums, galleries, and monuments; purchasing; tax collection; and more. It pays more than 45,000 city employees and recipients of social benefits. It would not be practical for a department with such a broad range of responsibilities to administer the Council's 18,000 computers spread over 120 locations. As many organizations do in this situation, they turned to a specialist.

Back in 2001, Edinburgh partnered with BT (originally British Telecom, now BT except in legal documents) to create the "Smart City" to replace old systems on mainframe computers. That partnership, originally for ten years, was extended through 2016. The city then contracted with CGI to continue the program and enhance it through 2023. Its accomplishments include:

- Functional applications have been modernized. One payroll system replaced a hodgepodge of departmental ones. This new system is accessed by over 200 Council staff members from departments, including payroll, internal audit, human resources, and accounting, improving departmental communication. A new city-wide finance system replaced earlier general ledger and accounts payable systems. Other new systems now provide electronic procurement.
- Citizen service was improved through a public portal. Benefit claim processing time has been reduced by two-thirds. With fewer inquiries, staff can handle the ones they get more quickly.
- New functional applications were added. A new geographic information system provides location-based information to planners. This system was used to analyze data from the 2011 census, creating maps such as the one in Figure 7.15, and is to be used in 2021. Knowing how population is shifting from one part of the city to another enables planners to anticipate needs for increased or decreased services, rather than reacting to those needs after they arrive.

Shared databases are essential to the Smart City. The citizen portal could not function without a central database from which it can pull information. A geographic information system that does not also store non-geographic information is of limited usefulness.

Smart City won a Most Effective IT Partnership award from *Information Age* magazine as well as other awards. Its real benefit, however, is not in these awards, but in the improvements in civic efficiency and effectiveness that led to them.

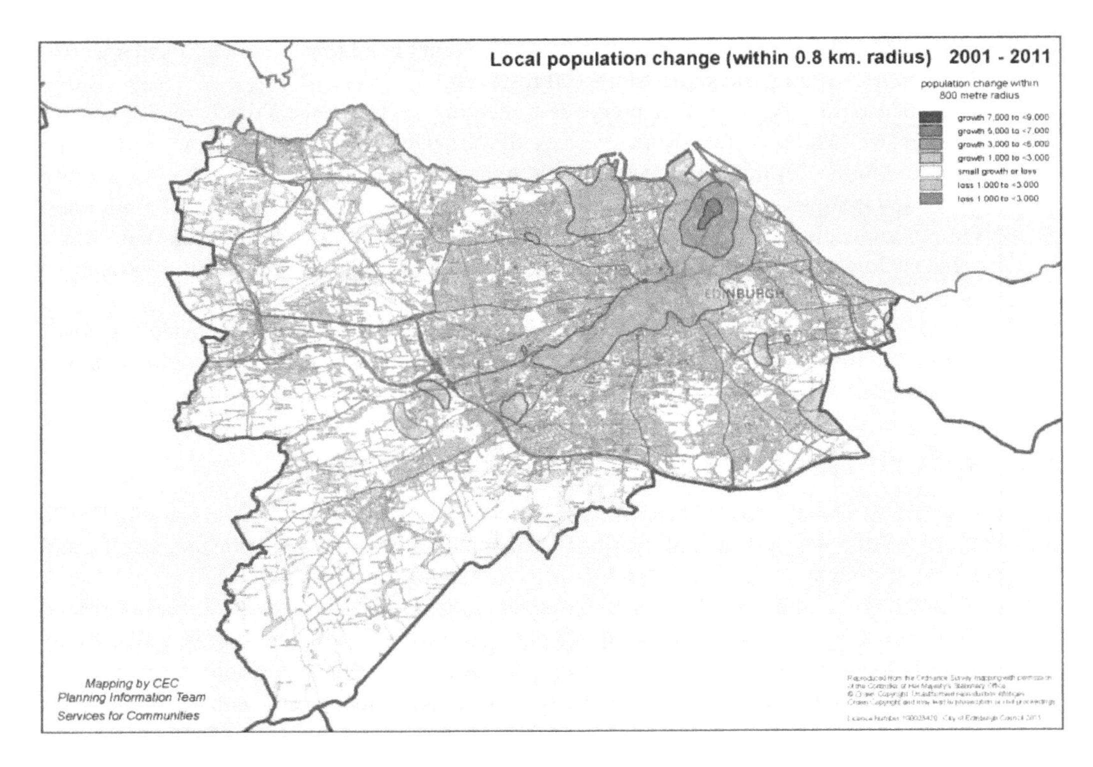

FIGURE 7.15 Map of Edinburgh showing population changes.

The City of Edinburgh is not all about traditional information processing systems. It is fully into social media. As one example, in late 2019, it was preparing City Plan 2030 to "set out policies and proposals for development in Edinburgh between 2020 and 2030." Citizens can keep up to date with the plan via its blog at *planningedinburgh.com*, with an option to be notified of new posts by email; can follow it on Twitter *@planningedin*; can download a document to learn how to get involved in the planning process; and can subscribe to an e-newsletter at *cityplan2030@edinburgh.gov.uk*. Edinburgh has come a long way in nearly 900 years!

Questions

1. Find a city or town near you that has a *public portal:* a web site through which citizens can find information and access municipal services. Using this portal, try to find out:
   - What are the name and title of the head of its government? How is this person chosen, and when does the term of the present one end?
   - What are the hours of the public library? (Its main branch, if there are more than one.)
   - How does one obtain permission to open a restaurant? What fees are required?
   - How does one appeal a parking ticket that one feels was issued incorrectly?
   - What facilities, if any, does the city or town offer for recycling paper, glass, and plastic? Based on the ease or difficulty with which you found these items (or your inability to find some of them at all), make suggestions for improving this portal.
2. The theme of this chapter is *linking parts of an organization*. Give three examples of how a system such as this could link parts of city government. (They need not involve Edinburgh, and need not be real.) For each example, give two items that would be in a shared database, which one of the departments would input to the database, and which the other would use.

3. Identify three departments in a typical city that need geographic information (information that ties a database entity to a geographic area or location, such as an address). Describe two ways in which two or all three of them could share information from this database. Be specific as to what data items would be used by two or all three of these departments.

CASE 2: SERENIC ACCOUNTING SOFTWARE*

Before developing software to connect parts of an organization, one must ask "What parts have to be connected, and why?" The answer for a manufacturer might not be right for someone else.

When it comes to nonprofit organizations and charities, the major piece to be connected to others is accounting. It connects to fund-raising on the inbound side and to awards (what they do with the money they raise) on the outbound side. A system that can handle these, without getting bogged down with features that other organizations need, may be exactly what they need.

That is what Serenic set out to develop. Their system now includes an accounting base, a grant management application, a budgeting application, and a human resources/payroll application. Figure 7.16 shows a Serenic grant management screen. Note the use of two currencies: LCY (Local CurrencY), U.S. dollars, and ACY (Additional CurrencY), Canadian dollars.

Nonprofits must respect their donors' intent. Donors may be passionate about an organization's mission, but may also express intentions for spending their money. If they find out it was spent contrary to those intentions, future funding from that donor would be in jeopardy and there is a potential public relations crisis.

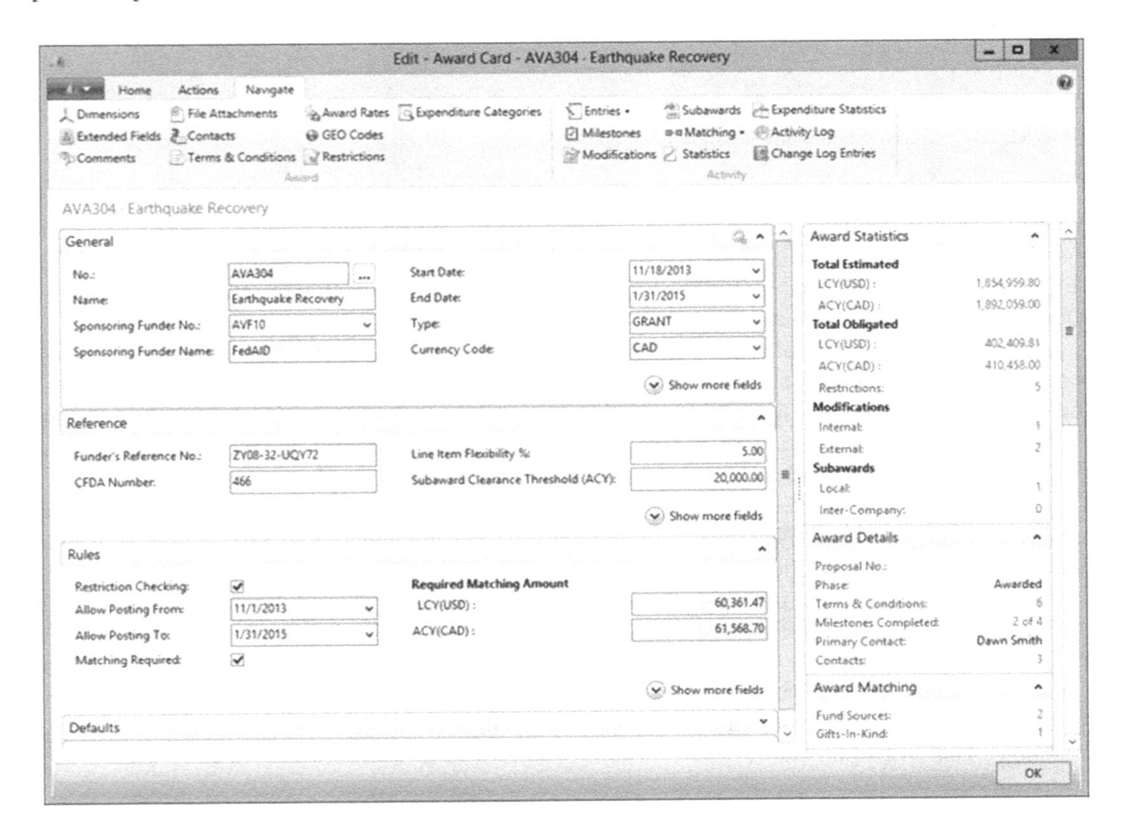

FIGURE 7.16 Serenic grant management screen shot.

* As in earlier chapters, discussing specific vendors in a case is not an endorsement.

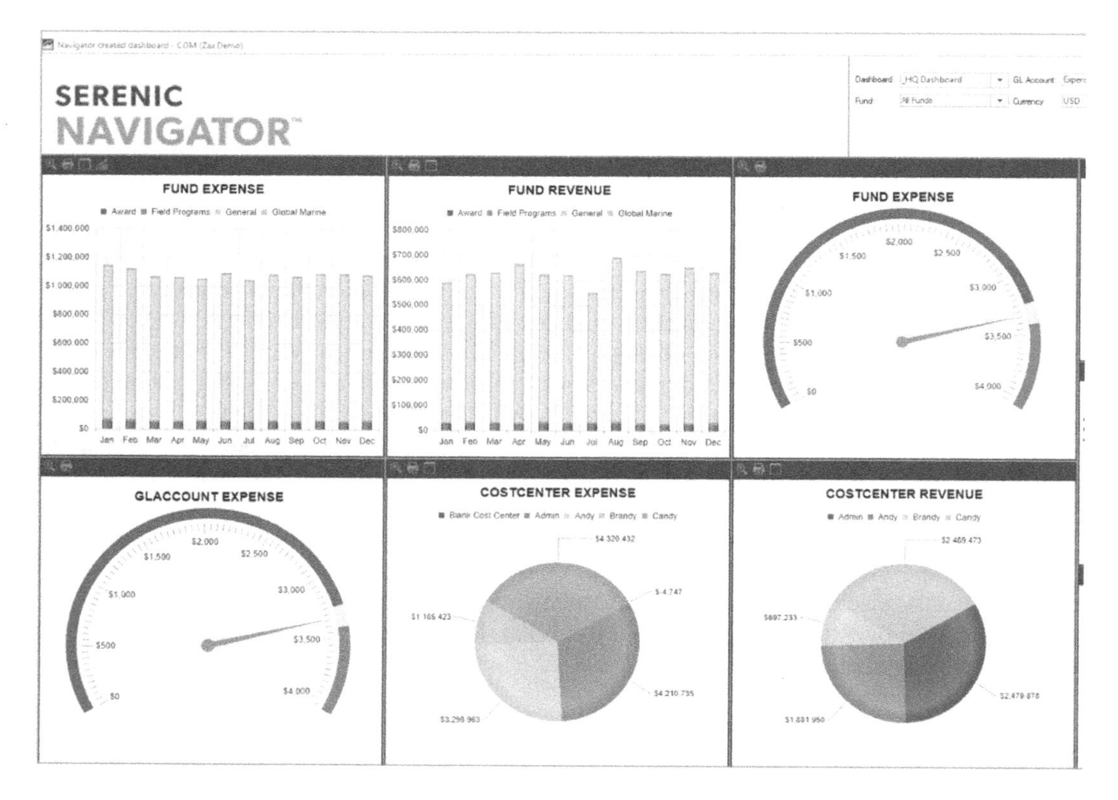

FIGURE 7.17 Serenic Navigator dashboard.

On the award side, there are differences between managing grants and projects. For example, the critical success indicator for a project is completing it on time with the required level of quality. Project reports focus on activities and schedules. For a grant, it is using funds for their stated purpose, so grant reports focus on flow of funds. Such differences mean that software designed for commercial organizations may not suit nonprofits and charities.

Serenic's products are built on a foundation of Microsoft Dynamics. MD includes a database server, an application server that controls its operation, and a client application that runs on the user's computer. Serenic designs database tables and application software that fit this framework. Using an existing, proven framework cuts development effort, and ensures that functions such as keeping users out of each other's way in updating the database are handled properly.

Morgridge Institute for Research at the University of Wisconsin-Madison exists "to address and answer fundamental biomedical questions." It received a startup donation of $50 million from John and Tashia Morgridge in 2006, and is now also supported by external funding such as research grants. The Morgridge Institute uses Serenic to keep track of its funds and their uses.

James Lester, financial analyst at the Morgridge Institute, says "Sponsors require us to track expenditures in great detail." Serenic software, he says, "helps us better manage our stewardship of the money that has been entrusted to us." Lester goes on: "an integrated system helps us do our jobs more effectively." Dashboards, such as the one shown (in part) in Figure 7.17, are vital to this type of management oversight.

Questions

1. Find two other application packages that handle accounting for charities. Using information from their three web sites (*www.serenic.com* and those of the two packages you

found), compare them. For each package, describe either a type of charity for which is it especially well suited or a type of charity for which it would be a poor fit.

2. You work at a small college. Few faculty members get external research grants, so they are managed manually with the help of a few spreadsheets. One of those faculty members has pioneered advances in educating children on the autism spectrum. A donor, who like John Morgridge became wealthy leading a successful high-tech company, knows of her because she helped his son. He has offered your college $65 million (worth about the same as $50 million was in 2006) to fund a research institute in her field. Discuss the pros and cons of three approaches to its information systems:
   a. Developing grant management modules for your existing academic ERP system.
   b. Using a program such as Serenic's and interfacing it with the existing ERP system.
   c. Separating its information systems from those of the college, exchanging data as needed.
3. The Morgridge Institute for Research has a few large donors but awards many small grants (such as research grants for students). Some charities have many small donors but award a few large grants. Discuss how this difference could affect their information systems' needs.

BIBLIOGRAPHY

Belden, "Implementing cybersecurity in offshore oil and gas platforms," 2016, www.iiconsortium.org/pdf/CaseStudy_Belden_Offshore_Oil-Gas_Platforms_v3.pdf, accessed July 8, 2019.

BT, "Long term partnership enables Scotland's capital to transform service delivery," 2006, www.btplc.com/Thegroup/BTUKandWorldwide/BTRegions/Scotland/BTScotlandstory/CECcasestudy-final.pdf, accessed September 20, 2019.

Castellina, N., "SaaS and cloud ERP trends, observations and performance, 2011," *Analyst Insight*, Aberdeen Group, 2011, www.meritsolutions.com/resources/whitepapers/Aberdeen-Research-SaaS-Cloud-ERP-Trands-2011.pdf, accessed September 20, 2019.

City of Edinburgh Council web site, www.edinburgh.gov.uk, accessed September 20, 2019.

City of Edinburgh, "City Plan 2030," www.edinburgh.gov.uk/info/20069/local_development_plan_and_guidance/1821/city_plan_2030, accessed September 20, 2019.

Conexus, "NetSuite for food and beverage distributors," 2017, conexussg.com/wp-content/uploads/2017/06/NetSuite-For-Food-And-Beverage-Distributors.pdf, accessed August 4, 2019.

Dreamgrow, "Top 15 most popular social networking sites and apps [August 2018]," www.dreamgrow.com/top-15-most-popular-social-networking-sites, accessed May 28, 2019.

Facebook Inc., "Facebook reports second quarter 2019 results," 2019, https://investor.fb.com/investor-news/press-release-details/2019/Facebook-Reports-Second-Quarter-2019-Results/default.aspx, accessed De September 20, 2019.

Hammer, M. and J. Champy *Reengineering the Corporation: A Manifesto for Business Revolution* (updated edition), Collins, 2006.

Machfu, "The impact of IoT on smart farming and water usage efficiency," 2018, www.iiconsortium.org/case-studies/IIC-Machfu-Case-Study-Final.pdf, accessed July 8, 2019.

Morgridge Institute for Research web site, morgridge.org, accessed September 20, 2019.

Oracle NetSuite, "#1 Cloud ERP for food distributors," www.netsuite.com/portal/industries/food-beverage-wd.shtml, accessed August 4, 2019.

Planning Edinburgh blog, planningedinburgh.com/category/city-plan-2030, accessed September 20, 2019.

Poindexter, K., "ERP, MES, or both?" *The Fabricator*, November 8, 2017, www.thefabricator.com/article/shopmanagement/erp-mes-or-both-, accessed September 20, 2019.

Quirk, E., "The best distribution ERP platforms of 2018," *ERP Solutions Review*, August 1, 2018, solutionsreview.com/enterprise-resource-planning/best-distribution-erp-platforms, accessed August 4, 2019.

Rao, R., "Intranet of things," *IoT Central*, September 11, 2017, www.iotcentral.io/blog/intranet-of-things, accessed July 4, 2019.

Roberti, M., "Carlo Nizam to lead ICT digital transformation at Airbus Group," *RFID Journal*, June 9, 2015, www.rfidjournal.com/articles/view?13128, accessed July 8, 2019.

RTI, "Siemens wind power," 2015, www.iiconsortium.org/case-studies/RTI_Siemens_Wind_Power_case_study.pdf, accessed July 11, 2019.

SAP, "IoT edge computing: SAP® Edge services," 2017, www.sap.com/products/edge-services.html#pdf-asset=068181cb-ae7c-0010-82c7-eda71af511fa, accessed June 25, 2019.

SelectHub, "Best distribution ERP software comparison," https://selecthub.com/distribution-software, accessed August 4, 2019.

Serenic Software, "Spotlight on higher education foundation operations—there is a better way to manage," 2017, www.serenic.com/wp-content/uploads/2018/07/Position-Paper-Spotlight-on-Higher-Education-Foundation-Operations.pdf, accessed September 20, 2019.

Serenic Software web site, www.serenic.com, accessed September 20, 2019.

Statista, "Number of monthly active Facebook users worldwide as of 1st quarter 2019 (in millions)," 2019, www.statista.com/statistics/264810/number-of-monthly-active-facebook-users-worldwide, accessed May 28, 2019.

Top 10 ERP, "Independent review of ERP software for wholesale distribution," www.top10erp.org/erp-software-comparison-wholesale-distribution-mfgmode-135, accessed August 4, 2019.

Whitehouse, G., "Gantt versus grant," 2018, https://www.serenic.com/wp-content/uploads/2018/07/final_Gantt-versus-Grant.pdf, accessed September 20, 2019.

WorkWise Software, "What is MES (Manufacturing Execution Systems)?" www.workwisellc.com/erp-software/what-is-mes, accessed September 20, 2019.

Ziff-Davis, "Beginner's guide to ERP," 2013, b2b-hosteddocs.s3.amazonaws.com/Custom%20Content/zd_wp_beginners_guide_to_erp_101113_V2_2.pdf, accessed September 20, 2019.

8 Connecting with Customers and Suppliers

CHAPTER OUTLINE

8.1 E-Business and E-Commerce
8.2 Customer Relationship Management (CRM)
8.3 Connecting through Social Networks
8.4 Supply Chain Management (SCM)
8.5 Extranets

WHY THIS CHAPTER MATTERS

In Chapter 2, you read that a firm's balance of power relative to customers and suppliers is one key to business success. A firm whose raw materials cost less than they cost its competitors, or one whose customers will pay a premium price for its products, is in a strong position. But how can a firm make that happen? It can't tell its suppliers "sell for less," or its customers "pay more." It can, however, achieve nearly the same thing through intelligent use of information systems. Employees who can make that happen are valuable. This chapter will show you how that works.

CHAPTER TAKE-AWAYS

As you read this chapter, focus on these key concepts to use on the job:

1. Information systems remove time and distance barriers to many types of sales.
2. Information systems enable a company to make its sales process more effective.
3. Information systems enable a company to treat its customers personally—without personal contact.
4. Information systems enable a company to manage its supply chain, making it more efficient and effective.

8.1 E-BUSINESS AND E-COMMERCE

You've purchased something online: a book from Amazon, music from the iTunes Store, an app from the Android Marketplace, bargain electronics from *woot.com*, hobby supplies on eBay, plane tickets from Expedia or *aa.com* … the list is endless.

You are not alone. In 2018, U.S. consumers bought over half a trillion dollars worth of goods from online sources. (Worldwide, the figure is over twice that.) That's 14.3% of what they bought from physical stores in that year, up from less than 10% in 2014—and the fraction is expected to continue to grow. The growth of *e-commerce*, online buying and selling of goods and services, is shown in Figure 8.1. It is rapid by any standard. Over half of the total growth in U.S. retail sales from 2017 to 2018 was in e-commerce.

Online business activity is more than buying and selling. E-commerce is part of the broader concept of *e-business*. E-business is carrying out any type of business activity online. Filing an automobile insurance claim, registering for courses online, or checking how much your dental insurance will pay toward a filling are e-business, though they are not e-commerce

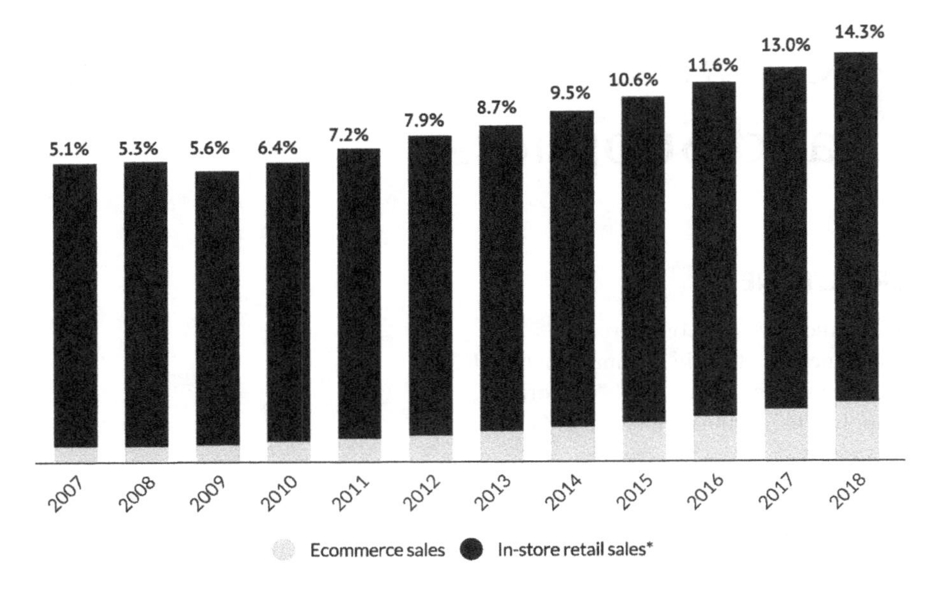

FIGURE 8.1 U.S. e-commerce growth.

since nothing is bought or sold. Figure 8.2 shows the relationship between e-commerce and e-business.

Categories of E-Commerce and E-Business

There are several different types of e-commerce. It's important to understand them because the way an organization approaches e-commerce depends on the type of e-commerce it is involved in. If it's involved in more than one type, it may need to use more than one approach. Besides that, you can expect to hear these terms used in all sorts of business conversations.

Business-to-Consumer

You're probably most familiar with *business-to-consumer (B2C) e-commerce:* A business (M&M) sells something (candies with a photo of your dog) to a consumer (you) via a web site (Figure 8.3).

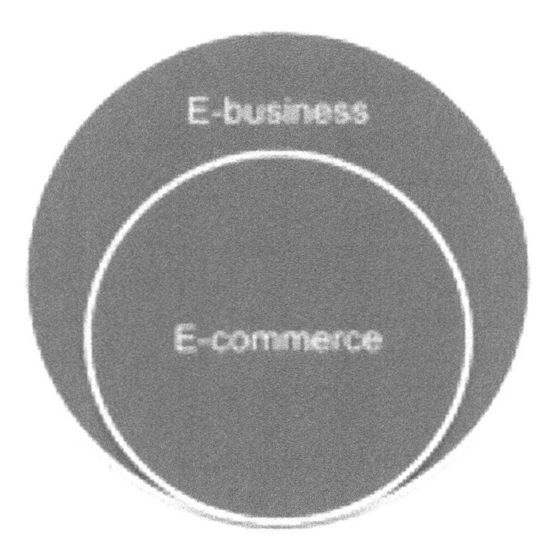

FIGURE 8.2 Relationship between e-commerce and e-business.

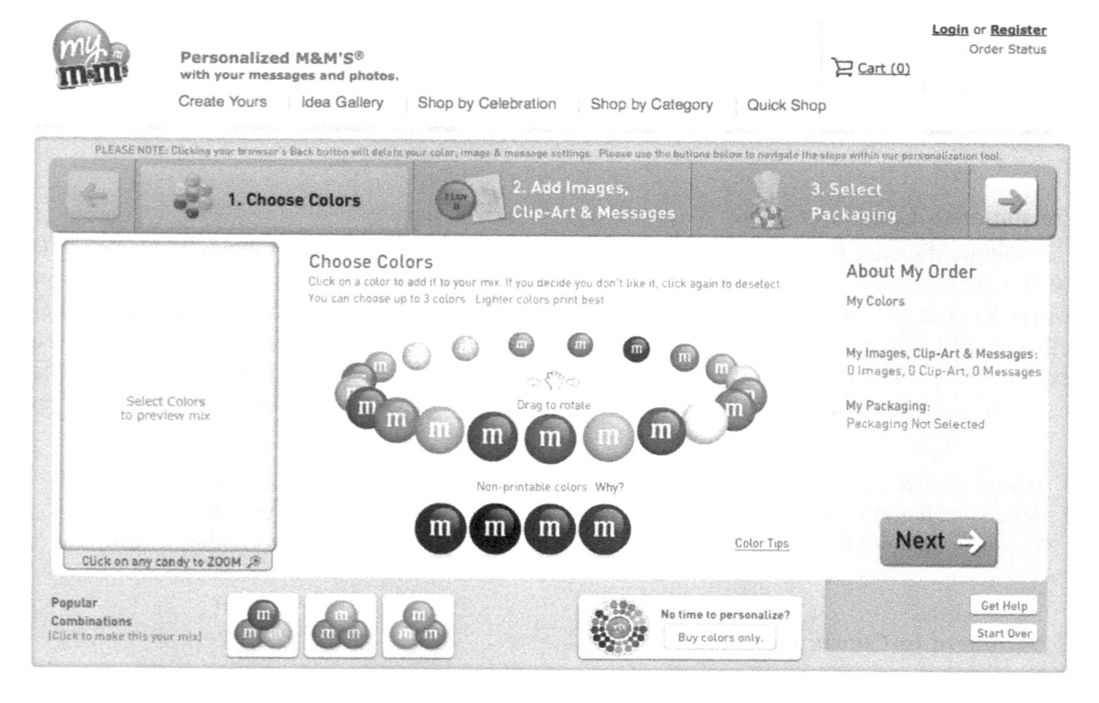

FIGURE 8.3 Screen shot of M&M web site.

B2C is the most common e-commerce type by transaction count, though small transaction sizes keep it from dominating financially. Participants range from large "e-tailers" such as Staples to shops run by individuals on a part-time basis. When an e-commerce company also has physical stores, as Staples does, the combination is known as *bricks and clicks* or *clicks and mortar:* plays on the phrase "bricks and mortar" for a building. E-tailers that don't also have physical outlets are *pure play* e-commerce firms. The iTunes store is a pure play firm. If you want something from it, you must buy it online. You can't walk into a physical iTunes store. There aren't any.

Where you fit in: If your employer sells anything to consumers, it must think about B2C e-commerce as a sales channel.

Business-to-Business

In *business-to-business* (B2B) *e-commerce*, both buyer and seller are organizations. It overlaps B2C e-commerce because many small businesses buy similar products to individuals, in similar ways: paper towels, desk chairs, inkjet printers. Some differences:

1. Some products, such as raw materials for manufacturing, are of little consumer interest.
2. Many organizations have procurement processes for sellers to follow. Consumers don't.
3. Many organizations purchase in large quantities. Consumers buy less at a time.
4. Organizations that buy in bulk expect to negotiate prices. Consumers seldom expect this.
5. Large organizations, knowing how important their purchases are to a seller, expect attention and product customization that consumers don't.

Where you fit in: Businesses that buy anything—what business doesn't?—must consider B2B e-commerce as a way to streamline procurement. Businesses that sell to other businesses must also think of it as a way to broaden their sales efforts.

Consumer-to-Consumer

The central difference between C2C and B2C sales is that consumers don't sell things as a business. (If they do, they're businesses.) So, they need help. Help is at hand: *craigslist.com*, which connects buyers with sellers but does not participate in the sale; *ebay.com*, originally an auction site but now for direct selling as well, which also handles payment and other aspects of sales transactions; focused sites, such as *etsy.com* for crafts; and more. C2C sites make most of their money from transaction fees. Some listings may be free, but free listings attract visitors and thus increase what fee-paying sellers are willing to pay.

Where you fit in: As a consumer, you may use such a site to buy or to sell. As a business student, you should be aware of the business opportunities in setting up and operating a C2C site. Your site may not become as big as eBay, but even a more modest success can be worthwhile!

Government-to-Consumer

Government-to-consumer (or to citizen) e-business, abbreviated G2C and sometimes called e-government, means conducting government business online. It involves online systems to carry out previously paper-based processes: filing tax returns and other forms, obtaining information from government agencies, paying parking fines and other assessments, even checking the public library catalog.

Government-to-Business

Government-to-business, abbreviated G2B and also an aspect of e-government, includes online activities that connect government agencies to businesses. Some of these parallel G2C e-business, such as filing tax returns. Some G2B is for regulatory compliance: learning what has to be done, filing forms, obtaining determinations, and perhaps appealing them. Government agencies also buy from businesses. When they do this online, G2B e-business has an e-commerce element.

Finding an E-Commerce Site

Many e-commerce sites are well known. If you want to buy a book you may go directly to *amazon.com* or *barnesandnoble.com*, though they are far from the only booksellers on the web. (Buyers with specialized needs may do better elsewhere.) If you know who makes what you want, their site may sell it. If it doesn't, it will probably point you to someone who does.

If these don't work, web searches will almost certainly find sites that sell what you want. Search engines have shopping tools. They know your approximate location and will recommend sources near you, especially for location-sensitive items such as movie tickets.

Portals

A *portal* is a focused entry point to the web that provides access to pages in a field of interest. At school you may log into a portal that provides access to registration information, course web sites, announcements from clubs you belong to, and more. Some of this information is on the public university site, but some isn't, and that site is not as focused on your interests.

Other portals are industry portals, such as the one in Figure 8.4 for the trucking industry. It contains news of interest to truckers, several sections (via the tabs at the top) with specialized information, and links to other sites of interest.

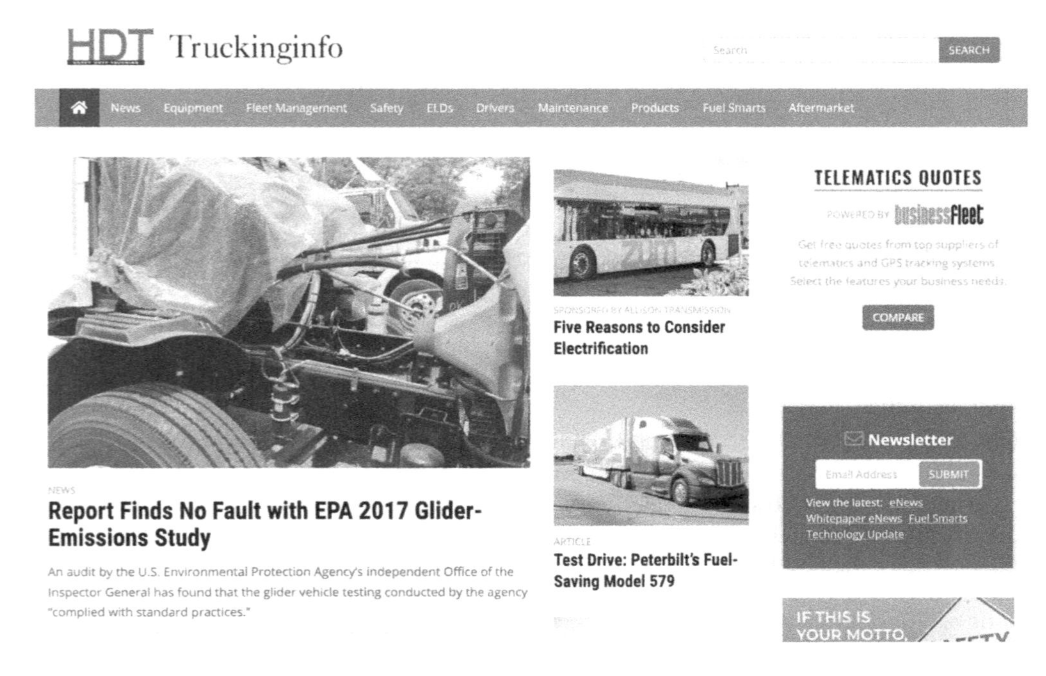

FIGURE 8.4 Screen shot of *truckinginfo.com* portal (accessed August 2, 2019).

This portal, affiliated with *Heavy Duty Trucking* magazine, is supported by trucking-related ads. It has links to the online edition of the magazine, which carries additional ads. Other portals, such as the U.S. Federal Motor Carrier Safety Administration portal in Figure 8.5, are provided by organizations as part of their online presence with no other funding.

Industry portals such as *truckinginfo.com* can direct readers to e-commerce sites. Some portals have product directories, some have articles about various kinds of products, and nearly all have relevant advertising.

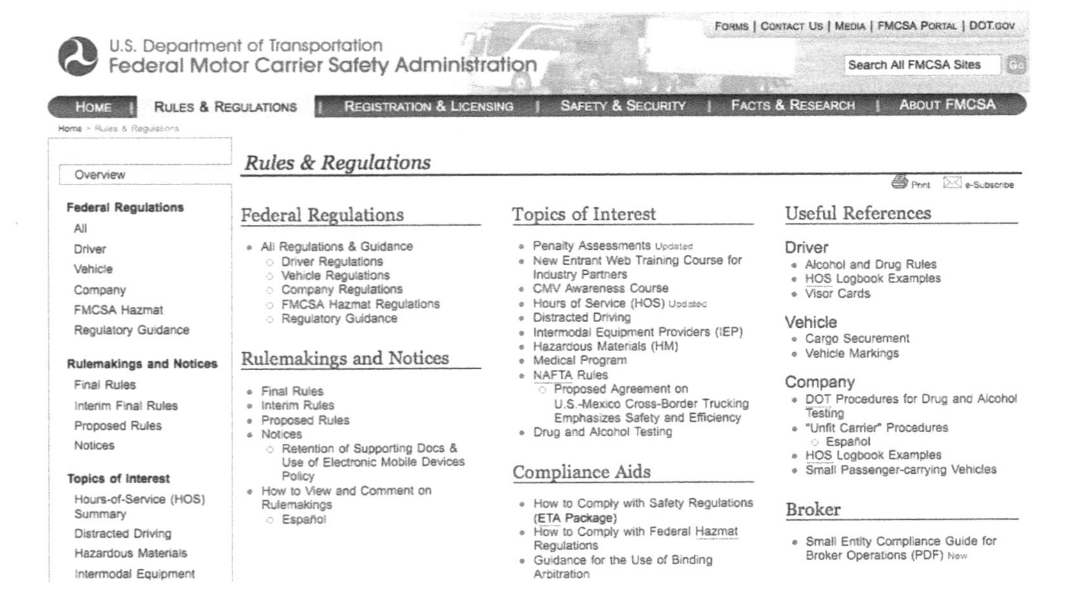

FIGURE 8.5 Screen shot of FMCSA regulations portal.

Where you fit in: Any industry you work in will almost certainly have industry portals. Find one or more. They can help you reach work-related sites quickly and easily.

Auctions

The traditional auction, which goes back at least as far as 500 B.C.E., starts when someone has something to sell. Others offer to buy it, each prospective buyer offering more than the previous bid. The auction ends when no buyer tops the existing high bid or at a predetermined time. A simple program using a shared database can automate this process. Real auction programs are more complicated because of other details, security needs, and more.

Reverse Auctions

A standard auction is run by or for a seller, and seeks the highest price. A *reverse auction* is run by or for a buyer, and seeks the lowest price. Suppose a manufacturer needs a lot of steel tubing. It posts a reverse auction on its web site. Suppliers bid, giving prices for the tubing per the buyer's schedule and specifications. Potential sellers see the lowest bid, but not who made it, and can bid lower if they wish. If one does, the first bidder can bid still lower than that, and so on down.

Exchanges

Exchanges, which typically serve one industry, are clearinghouses for buyers and sellers. They update the concept of Amsterdam's 17th-century tulip exchange, or the even more ancient concept of having all clay pot sellers in the same street, to today's technology.

Potential buyers post their requirements on an exchange so sellers can offer the needed items. Requirements can take the form of a request for bids, a reverse auction, or anything else the buyer chooses. Sellers can post information about what they have, so buyers can contact them. Their offers can be in the form of an auction, an offer to make a bid, or again anything else.

Exchanges are usually private. Firms register to participate. Because its members must trust other members, an exchange must check out prospective registrants to prevent fraud. Even if a firm is well known, such as General Motors in vehicles, the exchange must still confirm that a person claiming to be the GM contact really is, and that his or her contact information is valid.

Getting into E-Commerce

E-commerce requires a web site. It's not hard to set up a basic site. However, if you're serious about e-commerce, you have to do more.

Start by defining what you want your site to achieve. Do you want to sell goods? Perhaps, but that is not a site's only possible purpose. Sites can reduce support costs by automating online help, direct potential customers to stores or dealers, improve a firm's reputation, boost the price of its stock, and more. A site can do any of these, but they won't all be the same site. You can't define the right site for your objectives until those objectives are clear to you and your colleagues.

Having defined your objectives, you can start to design your site. There are millions of sites you can use for inspiration, in terms of what to do and in terms of what not to do. The basic elements of e-commerce sites are well understood in 2019, but each business must decide how to combine those elements, the "story board" of how users will navigate its site, to meet its goals. Working with experienced site developers can help. There are also books and courses on e-commerce site design, though nothing can replace experience completely.

Even a small firm can have an e-commerce site:

- At the lowest level, *online shopping malls* such as eBay—some general-purpose, some focused on specific areas—house e-commerce activity. These malls charge fees for listing

and/or selling items. Some may also have a small monthly fixed cost. There may be little or no ability to customize them.
- Many web hosting firms let customers set up e-commerce sites. A seller can choose from professionally designed templates and customize them further. The cost is covered by a monthly hosting fee. (You'll also need a domain name for a few dollars per year.) Handling credit card payments or other services may carry additional fees.
- E-commerce packages are available from several sources, for all major platforms, for firms to add e-commerce capabilities to their own sites. The web is full of videos on how to use them.
- A company large enough to need features that no existing e-commerce package has is also large enough to employ professional web developers.

Where you fit in: The major issues here are business, not technical. You'll work with technical people to design and develop your firm's e-commerce site, but businesspeople must be in charge of the overall process. As a manager, you'll have to make sure they are.

Mobile Commerce (M-Commerce)

Mobile commerce can be defined as e-commerce using a mobile device such as a laptop, tablet, or smartphone. However, that definition is too broad. When a coffee shop patron orders a book from Amazon via her laptop, to be shipped to her home in the normal way, the e-commerce is exactly as if she had been at home or in her office. The fact that she used a mobile device is irrelevant.

Mobile e-commerce is e-commerce that utilizes a mobile device in a way that reflects its mobility. This involves one or both of:

- Using location information to customize a transaction. Servers obtain location information in several ways. Some are more accurate than others, but all are better than not knowing the location at all. An m-commerce planner must decide if the available accuracy is sufficient: that required for directions to the nearest emergency room is higher than that required to show movie times at area theaters.
- Performing an activity in a way that it could not be performed with a non-portable device. One example is paying with a smartphone instead of cash or a credit/debit card. Device screens can also display discount coupons, concert tickets, airline boarding passes, and other content that would otherwise need to be printed.

The market for m-commerce is growing more slowly in the U.S. than in Europe or Japan. The North American mobile phone market is fragmented among incompatible suppliers. North Americans adopted credit and debit cards sooner than European and Japanese consumers, so they already had a form of mobile payment. And Japanese consumers, if one can generalize about them, tend to be eager to try the latest innovations.

Mobile commerce offers many opportunities to customize the shopping experience. While GPS is not useful in stores—signals may not be strong enough, and it doesn't pinpoint location closely enough to tell one aisle from the next—alternative approaches have been developed. Apple's iBeacon technology, which uses Bluetooth communication, is one such. The latest versions can tell the distance from a sensor to a device to within a few inches/cm. With multiple sensors, a store can not only tell that a customer is in the TV section, but what brand of TV that person is looking at. If a competing brand has a promotion, a coupon for it can be sent directly to the customer's phone. This technology has many possible applications: custom tours of museums focusing on an area of interest, homes that heat only the room you're in, finding people who share your interests at a convention, or an alert that your bicycle just moved.

Where you fit in: People who can come up with innovative applications for new technologies such as this will be well rewarded by their employers!

Location-sensitive informational sites are an aspect of mobile business, if not commerce. Most store chain sites have a feature to find the nearest store to a given location. You can enter a location near which you want it to search, but it defaults to where it thinks you are.

Mobile Payments

Payment is an e-commerce necessity. Payment need not involve passing coins and bills across a counter. Most payments today transfer information about who has money.

Some types of mobile commerce don't require special payment methods. Devices from vending machines to parking meters can be set up to accept credit/debit cards. In many parts of the world, though, these cards are not as common as they are in the U.S. As an alternative, a mobile "wallet" in your phone permits cashless purchases. Examples include Google Wallet, Apple Pay, Square Wallet, Chirpify, Venmo, and more. Most are linked to a bank account or credit card. Digital wallets require a merchant to have a device that can communicate with your phone, and some only work at certain merchants. Some have added features, such as transferring money to another user of the same system. Mobile payments will evolve rapidly during your career.

E-Commerce Issues

Getting the Word Out

It used to be easy to tell prospective customers about you. They'd pass your store and see your ads in the local paper. If they weren't close enough to your store to do that, they didn't matter.

National brands changed this picture. Airlines and car companies need a nationwide clientele. Fortunately, their emergence paralleled the rise of national magazines and broadcast networks.

Things are more complex online. There is no easy way to tell all your potential customers about you. There are things a firm can do, but it is impossible—and, fortunately, unnecessary—to do all of them. A businessperson must be aware of the range of possibilities to make an informed choice based on a company's size, target markets, competitive position, and needs. Some options:

Portals. Portals such as Figure 8.4 exist in many markets. Find those that apply to your business and consider advertising in them.

Links. If you know of sites that prospective customers are likely to visit, you can ask the owners of those sites to post links to yours. Such links are often exchanged by small businesses that do not compete directly but can help each other. In addition to their value in directing visitors to your site, links from other sites will also improve your site's search engine ranking.

Search engines. E-commerce sites are often found via web searches. Make your site attractive to search engines by including key words that people search for and having links to it. There are firms devoted to *search engine optimization* (SEO): designing sites for high search rankings. Some SEO is useful, but some is based on outdated information about how search engines work, and some is outright fraud. Search engine operators know about these services and watch them carefully. When they find a site that has a ranking it doesn't deserve, they adjust their ranking methods to disregard whatever it did, and penalize that site by lowering its ranking.

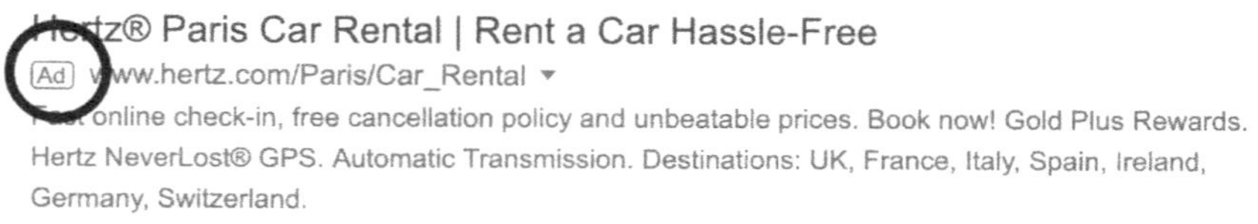

Hertz® Paris Car Rental | Rent a Car Hassle-Free
Ad www.hertz.com/Paris/Car_Rental ▾
Fast online check-in, free cancellation policy and unbeatable prices. Book now! Gold Plus Rewards. Hertz NeverLost® GPS. Automatic Transmission. Destinations: UK, France, Italy, Spain, Ireland, Germany, Switzerland.

FIGURE 8.6 Hertz targeted ad on Google.

Advertising. Many sites that potential customers visit accept ads. A common arrangement is *pay per click*: when a visitor to their site clicks on your ad and goes to your site, you owe the referring site a few cents. Other arrangements are possible, such as a monthly fee or payment per purchase.

Targeted search engine advertising. Search engines can show your site, as an ad or a sponsored link that resembles search results, when users search for terms you specify. A Google search for "car rental paris" displayed (in mid-2019) six ads from firms such as Autoeurope and Hertz. The only thing identifying them as ads is the small icon circled in Figure 8.6. Payment for such ads is due when they are shown. The cost varies with search term popularity, but is known in advance and can be controlled by setting an upper limit. If many firms buy the same search terms, ads are shown in rotation.

Category-specific channels. Do your customers respond to social media influencers? It may be worth sponsoring them or involving them in some other way in your strategy. Are there new channels, such as e-sports (live competition in video games) that your customers are attracted to? Do they listen to podcasts, subscribe to video channels? These approaches aren't just if your targets are under 30 years old. Model railroading, generally considered a hobby for older males, has its share of YouTube channels as well. While not all model railroaders are into YouTube, nearly everyone who sees an ad there will be in its target audience.

The physical world. Don't forget non-electronic media! Print and broadcast ads, conference give-aways, trucks, the products themselves, should carry your site's URL. (Joe Boxer takes advantage of the trend to wear trousers low on the hips to display its URL on underwear waistbands.)

No list such as this can ever be complete. Next week someone will think of a new way to promote an e-commerce site. You must keep your eyes open to what other firms do, even if they're not in your line of business, and be ready to use good ideas from any source.

Where you fit in: E-commerce has a technical component, but choices such as these are business decisions that must be made by businesspeople. That's *you.*

Disintermediation

An *intermediary* is an organization that performs a function between two other organizations. In commerce, a distributor (like Khoury Candy Distributors) is an intermediary: It comes between the manufacturer, who doesn't want to handle small orders, and individual stores, which don't order enough at one time to deal directly with manufacturers.

E-commerce can make it practical for a manufacturer to handle small stores directly. E-commerce can even make it practical for manufacturers to deal directly with customers without any stores at all. This elimination of intermediaries is called *disintermediation.* Figure 8.7 shows a bicycle manufacturer selling directly to small stores, eliminating the distributor, but the term applies to the elimination of any intermediary or intermediaries.

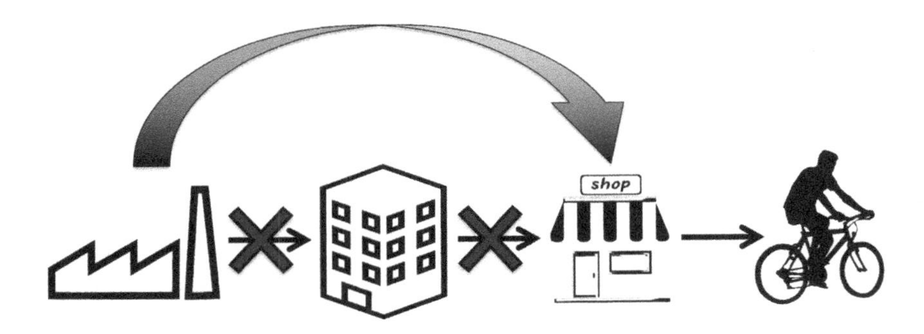

FIGURE 8.7 Diagram of disintermediation.

Manufacturers should not rush into disintermediation. Intermediaries perform useful functions. A vacuum cleaner distributor might replace units that fail under warranty, accepting them from stores and returning them to manufacturers when it has a full shipment. Manufacturers who don't use distributors will need to handle store returns. Such costs can offset much of what they hope to gain by eliminating distributors. Intermediaries can often demonstrate that they add value.

Channel Conflict

Channel conflict is one result of disintermediation. Channel conflict arises when a consumer can buy an item through more than one channel: typically, online or through a physical store.

Companies that sell through independent stores must be careful with channel conflict. If they sell online too aggressively, their retail outlets will be upset. Those stores will promote other firms' products and discourage their sales staff from selling that company's. Few manufacturers have enough brand strength to get away with alienating an important sales channel.

The concept of *omnichannel* helps deal with channel conflict. It means an integrated approach to all the channels through which a business reaches its customers, going to where the customers are by whatever means is best to reach them there.

For example, when you buy a sweatshirt through a company's site and take it back to a local mall, you don't want to hear "I'm sorry, we don't take online returns. You'll have to ship it back." That store may save a dollar in shipping costs, but will lose its best opportunity to turn you into a loyal customer via customer service. With omni-channel marketing, this wouldn't happen.

Fulfillment

It is not difficult to order a pizza online. It is not possible (in 2019) to receive that pizza online. Until we have Star Trek technology, products must be delivered in the physical world. That's *order fulfillment.* Small vendors must maintain their professionalism throughout the fulfillment process, through shipping and customer service.

Many companies specialize in this sort of shipping, from household names such as FedEx and United Parcel Service to small local firms. Since consumers are used to seeing parcels from even the largest shippers delivered by such firms, using them does not label a store as "small fry."

Taxation

In the U.S., taxation of Internet purchases has been contentious since B2C e-commerce became economically significant. Stores in states that collect sales taxes (45 out of 50, plus the District of Columbia) feel that not taxing Internet purchases gives online vendors a price advantage over local stores. Online vendors point out that Internet sales are interstate commerce, which the U.S. Constitution puts under the control of the Federal government, so states cannot tax it.

In 1992, the U.S. Supreme Court ruled that sellers do not have to charge sales tax in states where they do not have a physical presence (called a *nexus*) such as a store, office, or warehouse. Its decision was in the context of mail order sales, but applies to online sales as well. When Amazon decided to open a software development facility in Massachusetts, they knew that would require them to collect sales tax on sales to that state. They began to do so on November 1, 2013. They had previously not collected Massachusetts sales tax. If the Boston area didn't have a valuable pool of technical employees, they might have chosen differently.

In June 2018, the U.S. Supreme Court reversed that policy in *South Dakota v. Wayfair*. By late 2018, 31 states had passed laws taxing Internet sales. Most specify a minimum business volume, such as revenue of $100,000 or 200 transactions, for collecting sales tax. Many such laws took effect in 2019. Several software firms now sell applications to track taxable online sales.

Taxation varies widely from country to country. Countries that use a VAT (Value-Added Tax) system don't have this problem. Among those that don't, laws vary from one country to another.

8.2 CUSTOMER RELATIONSHIP MANAGEMENT

You may have heard "nothing happens in business until someone sells something."

That being the case, a business's customer relationships are important. Businesses that nurture these relationships benefit from repeat purchases. Businesses that don't must find new customers. It's harder and more expensive to find a new customer than to keep an existing one.

Information systems are important to maintaining customer relationships. Systems that have this as their main purpose are called *customer relationship management* (CRM) systems. Other systems may be useful in maintaining these relationships, but it is the *raison d'être* of CRM.

There are two main types of CRM: *operational* and *analytical.*

Operational CRM

Operational CRM is used in working with customers: contacting them, selling to them, helping them use a product or service. These may be transaction processing systems, or they may be used for information retrieval as in reviewing customer history before placing a call. When operational CRM is used solely or primarily by the sales force, it is also called *sales force automation.*

Operational systems can be viewed in three levels, with increasing capability as one moves up:

Level 1: Contact Management

A contact management system is a step up from a personal contact list. It should also handle information such as call and purchase history, call-back dates, conversation notes, and more. Single-user contact managers can use a personal database, but a shared database makes sales contact information available to a team. Figure 8.8 shows a contact manager.

Big-ticket sales such as shopping mall construction or aircraft involve teams on both sides of the table. Everyone on the team must know what everyone else is doing, who they talk to, what commitments they make, what the next steps are. Given the time from initial inquiry to signed contract, turnover is likely; a historical record can bring new team members up to speed. Contact management for such sales is used by a sales team as a group, and needs a shared database.

Level 2: Sales Management

Sales management CRM systems coordinate the selling process, reminding salespeople to go through the right steps in the right order and providing a variety of aids—standard presentations, "boilerplate" content for proposals, competitive analyses, and sales arguments—in doing so.

Level 3: Opportunity Management

Opportunity management extends sales management earlier in the selling process by identifying new or potential customers. Opportunity management software can track the type of sale, the person or group responsible for it, its expected value, the probability of closing it (subjectively, or based

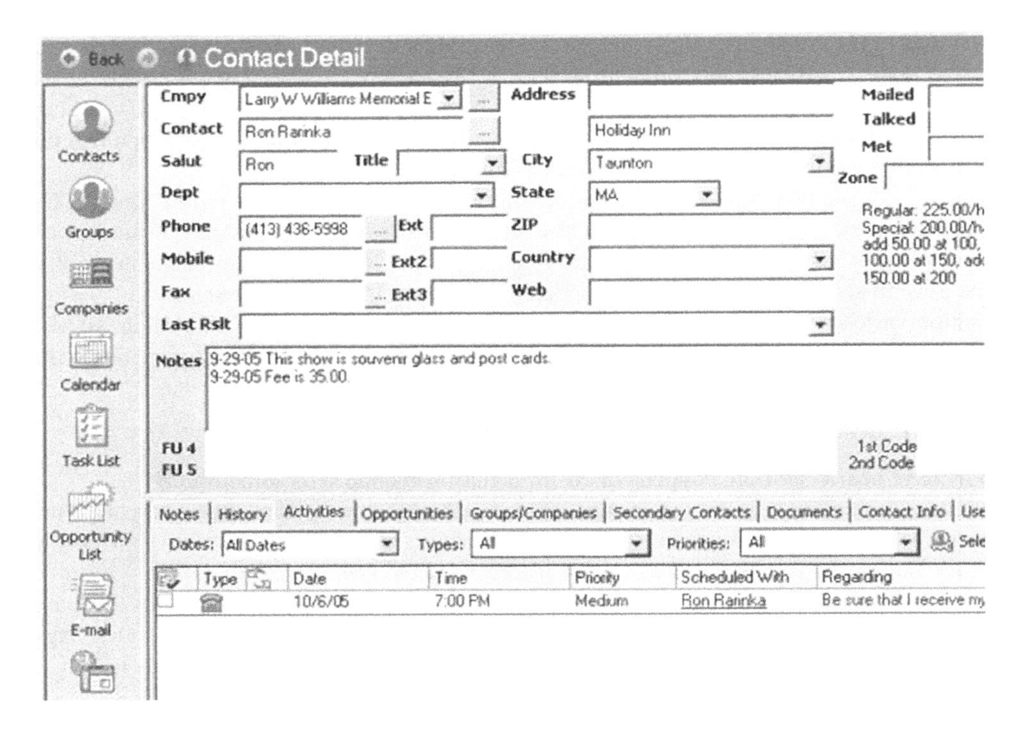

FIGURE 8.8 Screen shot of contact management software.

on milestones such as being on a prospect's short list), and expected contract date. It can be a tool for salespeople to optimize their use of time and for sales management to monitor their team's activities. It can also be used to forecast demand: If a firm has 1,000 opportunities to sell fences, and typically turns a quarter of its opportunities into sales, it should prepare to install about 250 fences. Knowing this allows it to plan staffing, material, and equipment needs.

Regaining the Personal Touch via Operational CRM

When your great-grandparents needed a door hinge, they went to their local hardware store. The proprietor greeted them by name, asked how their basement-finishing project was coming, and suggested that they insulate the walls before paneling them. Your folks left with the 50¢ hinge they came in for, plus $50 worth of basement insulation in the back seat of their Ford Model A.

In the U.S., growth of the suburbs and increased use of private automobiles in the last half of the 20th century led to shopping malls and large, impersonal "megastores." Home Depot or Lowe's staff don't know their customers. They can direct your parents to door hinges, but have no idea that they're finishing a basement. Your parents leave with a hinge ($5 today, not 50¢) and nothing else, unless a register display reminds them that they need AA batteries or breath mints.

Information systems can bring back some of the personal touch. This requires two things:

1. The business must be able to *identify the shopper.*
2. The business must *know enough about the shopper* to provide meaningful customization.

Both of these are an aspect of customer relationship management.

Identifying the shopper is easiest in e-commerce. It can happen in several ways.

- Web site cookies (see Chapter 6) can provide a preliminary identification on the basis of the computer and browser from which the site is accessed. There is often a button labeled "Not Efrem Mallach? Click here" to inform the site that it guessed wrong.

- E-commerce sites can ask or require visitors to log in. This is usually required to access one's account even when a preliminary identification is made from cookies or other information.
- Sellers can use *loyalty programs* to connect purchases to a person. In the physical world, customers often don't identify themselves early enough to customize their experience. TGI Friday's restaurants solve this problem by offering a free salsa-and-chips appetizer to "Give Me More Stripes" frequent diner program members who show their ID card early on.

Knowing about the shopper means knowing his or her previous purchases. Amazon pioneered in using purchase patterns for recommendations. If you buy something, the next time you sign in, they will recommend items that other purchasers of the same item also bought. If you buy a children's book, they will recommend books for older children as he or she grows.

Stores collect personal information when people sign up for loyalty programs. While providing it is voluntary, many consumers want the seller to know about them. A photo store that knows what camera one owns won't promote Canon-compatible lenses to a Nikon owner. A hotel chain that knows what vacations one likes won't offer ski lift discounts to a "beach bum." Target gives an across-the-board 5% discount to its REDcard holders. The information they gain by tracking customer purchases, and increased sales to cardholders who feel connected to the store and know they're saving money, more than make up for lost revenue on items that cardholders would otherwise have purchased at full price.

Analytical CRM

Analytical CRM is about customers as a group rather than any one customer, focusing on overall information and patterns. It overlaps the decision support systems we'll cover in the next chapter.

Uses of analytical CRM include:

- **Segmentation,** grouping customers based on purchase patterns and other factors. This helps target sales campaigns. If most members of a group buy certain products, it can propose those products to customers in that group who haven't. Segmentation can include profitability analysis, enabling a company to drop unprofitable customers or charge them more.
- **Personalization.** Information about customers enables a store to personalize its offerings to each visitor. Analytical CRM helps figure out how personalization correlates with profits.
- **Response analysis** enables a company to determine the effectiveness of different marketing approaches, selecting those that work or targeting them to specific customer segments. The sock coupon example, Scenario 3 early in Chapter 1, is an example of response analysis.
- **Attrition analysis** helps understand why customers leave, to reduce future customer losses.
- **Aligning supply** (purchases or production) with expected demand.

Where you fit in: The benefits that a firm hopes to obtain from analytical CRM, and therefore what it should do, depend on the firm: what it sells, how it sells, the types of customers it has, its sales channels, and what it wants to improve. On the job, be awake to what customer data your employer has or could get, and how that data could make its sales processes more effective.

Two approaches used in analytical CRM are the RFM method and the Customer Data Strategy approach. There are many more. For example, analytical CRM can track sales to various types of customers, leading to better understanding of which customers like a firm's products—and perhaps improving its products or its marketing to appeal more to those who don't.

Both online and physical stores can also take advantage of *market basket analysis*: data about items that are purchased together. For example, airlines know that people who buy air travel may need hotel rooms. Some don't, but many do. After you purchase an airline ticket, the site may ask if you are interested in a hotel. If you click "yes" to go to the hotel site, that site pays the airline a few cents. If you book a room, it pays more. Each payment is small, but they add up. (Discussion Question 3 at the end of this chapter is about this scenario.)

Market basket analysis can reveal more about shoppers than they realize. Target inferred from a customer's purchases that she was pregnant and sent her offers for baby products. This caused considerable embarrassment when a family member opened a mailing that contained offerings for new parents, since she had not yet announced her condition. As a result, Target modified its program. It still uses purchase patterns to identify customers who are probably pregnant, but adds items such as motor oil to their offers to make it less obvious that they are for expectant mothers.

Where you fit in: If you understand the examples in this section, you'll have a good idea of what analytical CRM can do and should be able to use any analytical CRM systems your employer has.

The RFM Method

Sellers want to maximize the profit obtained by selling to every customer. It isn't practical for a business with many customers to develop a strategy for each one. It must assign customers to a workable number of groups and develop a strategy for each group. That strategy may then be customized for individuals, reflecting prior purchase patterns and other customer information, but its overall pattern will not change.

One way to group customers is the *RFM method*: Recency, Frequency, and Money (Figure 8.9). This method groups customers on the basis of:

- **Recency:** How recently did they buy something from us?
- **Frequency:** How often did they buy? This applies to a specific time frame, such as one year.
- **Money:** How much did they spend, in total or average per visit, during that period?

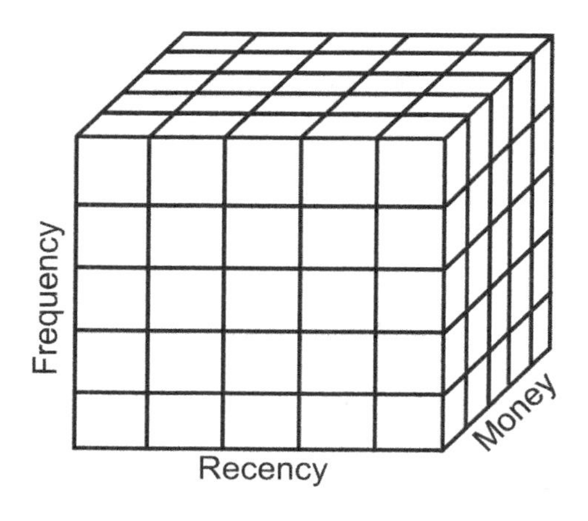

FIGURE 8.9 RFM cube.

Groups can be assigned by finding breaks in the data or by assigning X percent of customers to one group, Y percent to another, and so on. There are usually three to five groups per parameter, 27 to 125 groups in all.

Example: An office supply chain applies RFM analysis to its customers. Its planners use five segments per parameter, placing 20% of customers in each segment. A customer in the top recency group shopped there more recently than at least 80% of its customers, probably in the past few weeks. A customer in the top frequency group was there more often in the past year than at least 80% of customers, probably at least six or eight times. One in the top monetary value group spent more than at least 80% of customers, probably several hundred dollars.

Suppose someone shops there often, because there's a store en route from his home to his office. He stopped there yesterday to get a yellow ink cartridge for his printer. However, he never buys much: an ink cartridge or two, a memory card for his camera, a set of markers. He scores 1–1–5 on the RFM scale. What should the store do?

They don't have to do anything to get him into the store. He's there often, he was there recently. Their challenge is to get him to spend more. They might send him a coupon for a discount on a purchase over a certain amount, say $75: more than he usually spends, but not unrealistic. Some customers who use such coupons will get into the habit of spending more at that store.

Is this the best strategy, how much should the coupon be worth, what size purchase should the coupon require, and what about the other groups? Those are marketing questions. Information systems can't answer all of them, though they can help evaluate the results of testing various options. They can give marketing professionals the information they need to design strategies that will improve customer profitability.

> ***Where you fit in:*** Marketing professionals must be aware of this and other ways to use information systems to improve their effectiveness. Information systems can give them useful information, but using it still requires a business professional's (your!) insight.

Customer Data Strategies

Much of what you should do with the data you can collect from customer interactions depends on two characteristics of your product or service: How frequently the same customer buys it, and how much it can be customized to each customer:

- **Purchased frequently:** A morning cup of coffee. **Purchased infrequently:** A home. This axis tells us if we should focus on retaining existing customers or finding new ones.
- **Highly customizable:** Weddings. **Not customizable:** Bus tickets. This axis tells us if we should try to understand customer needs so we can adapt our offering to those needs.

Piccoli and others who have studied this topic found that the best use of customer information in the four resulting quadrants (see Figure 8.10) is as follows:

High Repurchase Frequency, High Customizability (Upper Right Quadrant)

This quadrant is the ideal place to acquire and use customer data. Returning customers provide a rich source of data, and customization provides a natural way to use it.

Customers in this quadrant have a choice of suppliers. The challenge to their existing supplier is to use what it knows about their preferences to customize its offering in ways that new suppliers can't, or can't as easily. This is a *personalization* strategy.

Consider a banquet facility that hosts business holiday parties. It can collect information about its customers' preferences: size of party, time of day, day of week, entertainment, food and beverage needs, scheduling, and so on. It can use this information to create a preliminary schedule that

Theoretical repurchase frequency	High	***Rewards strategy*** • Loyalty rewards • Reporting, analysis	***Personalization*** • Operations • Differentiation
	Low	**Low potential**	**Acquisition strategy** • Analytics • New prospects
		Low	High
		Customizability	

FIGURE 8.10 Customer data strategy grid.

will meet all its existing customers' needs and customized proposals to send to each customer well in advance of the actual event. This will increase customer retention, compared to the usual approaches of waiting for customers to call or phoning to ask "Gonna have a party again?" By the time they place this call, they'll probably hear "Yes, but we booked the Crystal Room last week."

High Repurchase Frequency, Low Customizability (Upper Left Quadrant)

In this quadrant there is typically little to differentiate suppliers in terms of product or service, yet high repurchase frequency makes keeping customers important. Firms try to reduce customer power, which you read about in Chapter 2. That is often done via a *loyalty program.* Customers who participate in a firm's loyalty program will choose it over competitors if the offers are close.

Airlines are masters at this. Frequent flyers get a variety of rewards: free flights, upgrades to first class, use of airport lounges, and more. A prized benefit is *elite status:* recognition as one of an airline's best customers. That entitles travelers to shorter check-in lines, shorter holds for phone agents, free checked bags, early boarding, and more. After they reach elite status, loyal flyers' target becomes the next status level with even more benefits.

Two potential drawbacks of loyalty programs are:

1. They reward past purchases, but a business wants to motivate future purchases. Expecting rewards only motivates customers indirectly. The purpose of a program must be kept in mind when designing it. It's important to offer rewards that will lead to the desired behavior.
2. People who are motivated by rewards may be bargain-hunters, not a firm's most profitable customers. Loyalty programs may not be effective with their most important targets.

Loyalty programs must be designed and managed carefully to avoid such problems.

Where you fit in: It's easy to dream up a loyalty program. It's harder to plan one that meets its objectives at minimum cost and deals with these two concerns. That requires business savvy.

Low Repurchase Frequency, High Customizability (Lower Right Quadrant)

Big-ticket, once-in-a-lifetime purchases are typically found in this quadrant: a college education, a custom home. These customers choose a supplier on the basis of how well its offering can be customized for their specific needs. Since repurchase frequency is low, however, sellers can't use data to meet the needs of existing customers. They can use analytical tools to identify the needs of typical customers and meet those. A firm's ability to meet the needs of potential new customers becomes its basis for attracting them.

Information systems can be used to analyze customer needs for this just as they can be used to analyze the needs of existing customers. The purpose of the analysis is different, but the data and methods are similar.

Information systems can also be an important part of customizing a product or service. Architects use computer-aided design tools to design any building larger than a doghouse.

Low Repurchase Frequency, Low Customizability (Lower Left Quadrant)

One can't make much use of customer data in this quadrant. There usually isn't much customer data, either. These firms may be better off finding a different basis on which to compete.

However, before dismissing the idea of a customer data strategy, make sure a product or service really is in this quadrant. Consider sports tickets. It seems these can't be customized. Pick a seat and that's it, right? Not necessarily! A team can offer packages including a seat, coupons for food and beverages (at lower cost than if purchased during the game), discounts on team jerseys, and more. The result: a customizable package that moves tickets from left to right in the diagram. Information systems make it practical to offer such packages online.

The same applies to repurchase frequency: People don't have many weddings, but they may have other events that could use the same providers. They may just stay at a hotel in Hawai'i once, but at hotels in general often. There are more opportunities to use customer data than it first appears!

Where you fit in: Technical factors determine what you *can* do. The businessperson must ask: What *should* you do? What customer data to collect, what to do with it, how to use it to improve profitability—that's where businesspeople such as yourself come in.

8.3 CONNECTING THROUGH SOCIAL NETWORKS

As you read in Chapter 7, and as you've known for years, social networks connect members of society. It's not surprising that businesses use social networks to connect with their customers.

TWITTER GETS THROUGH

During its weather-related shutdown on January 6 and 7, 2014, JetBlue Airlines encouraged its stranded flyers to communicate with it via Twitter. That didn't get planes back in the air sooner, but it enabled JetBlue to provide personalized assistance in a way that its customers appreciated.

Visit the web site of a consumer-facing business. You'll probably find a "Find us on Facebook" link or the familiar Like button with their "f" logo. Go to the Starbucks Facebook page; you'll learn that this coffee chain has (in December 2019) almost 37 million Likes. You'll also find content that makes visitors feel part of the Starbucks community. The act of committing by a Like and involvement with content increase customer loyalty, making people more likely to get coffee at Starbucks than a competitor down the block—even if the other coffee is as good and cheaper. This force (reduction in customer power, one of Porter's five forces) is so strong that companies give discounts for Facebook Likes. They ask for Likes, but they want loyalty. And they get it.

Where you fit in: When you need to do social media marketing at work, do a web search for "twitter marketing tips" and so on for other social networks. Some of what you find will be junk, but you'll also get lots of good ideas and food for thought!

THINK BEFORE YOU POST!

In planning your Facebook, LinkedIn, or other presence, focus on what your customers care about. Many such pages focus on making their owner look good. Not the same, and not as effective.

For example, most salespeople's LinkedIn pages say how successful they are, how much they've sold. That might have made sense when LinkedIn was used to find jobs, but not today. Prospects check salespeople's LinkedIn profiles before returning phone calls or replying to emails. What prospect wants to learn that you're a great salesperson, that selling is your top skill? That says you'll just try to sell them something. Prospects want to know how you can solve their problem.

If you're in charge of a company's social networking efforts, you'll be bombarded with offers to improve its standing by adding Facebook likes, LinkedIn connections, or Twitter followers. These can increase the apparent attractiveness of a person as a potential employee, of a record or video as worth playing, of a restaurant as a fun place, or of a business as having customers who support it, so some people will try anything to raise their social network standing—including *click farms.*

A *click farm* is in the business of improving clients' social network standing for a price. Consider *ibuyfans.com*, which focuses on Google+. (This firm existed in August 2019; there is turnover in this business.) You can buy 5,000 Google Plus followers for $15, or 500 in the U.S. for $25. Rather have Instagram followers? 500 will cost you $7; for English speakers, $10. You can buy SoundCloud plays, YouTube views, Twitter followers, and more.

Social network firms know about click farms and try to stop them. Computer programs that generate fake clicks can be blocked. However, people in low-wage areas can create real accounts to use on behalf of clients. Today's click farms are often based in Bangladesh, Indonesia, etc. Ways will be found to detect them. Click farmers will find another method. The race goes on. …

Where you fit in: Using click farms is tempting, but can backfire. It's seldom a good idea.

USING TWITTER

A few ideas from social network marketing experts on using Twitter:

- Keep your content fresh and engaging.
- If you have articles, tweet links to them. Retweets will create links to the article page. Links to a page improve its search rank.
- Make sure your personality comes through. Don't be all business all the time.
- Encourage customers to tweet back, such as by asking questions. Then respond to those who do.
- Use targeted Promoted Accounts and Promoted Tweets (both paid) to extend your reach. Tie them into trending tags, but only if they're truly relevant to those tags.
- Don't stop tweeting when the person who usually tweets is on vacation.
- Don't stop learning about how to use Twitter and how it changes over time.

Tony Maws, chef and owner of the restaurant Craigie on Main in Cambridge, Mass., responded to a blizzard by tweeting "We're open tonight. Walk, snowshoe, sled or ski—come over and brave the storm with us!" Instead of missing out on business because people didn't want to go out, his restaurant was packed.

8.4 SUPPLY CHAIN MANAGEMENT

CRM works downstream, from supplier to customers. *Supply Chain Management* (SCM) works upstream, from customer to suppliers. These are the companies to the left in Figure 8.11 and in the value chain of Figure 2.8. A company's *supply chain* consists of its suppliers, their suppliers, those suppliers' suppliers, and so on back. In the figure, the bicycle manufacturer's supply chain starts with rubber plantations and iron mines. It works through rubber manufacturers and steel mills to the tire and tubing producers who supply the bicycle producer directly.

Customers have choices. They may not care much who they buy from. You need to motivate them to buy from you, not a competitor. Suppliers, on the other hand, want all the customers they can get, not just one. You don't need to make them want to sell to you. They already do.

Suppliers have choices too, though. They can pick the terms under which they will do business. As long as they don't discriminate illegally, prices need not be the same for all customers. When their product or service is in short supply, they can decide who gets it. Since businesses depend on having the materials they need when they need them, and profit depends on getting them at the lowest possible cost, the way suppliers make these choices is important.

SCM helps a business become a desirable customer, which suppliers will treat better than other customers. This reduces *supplier bargaining power*, one of the forces you studied in Chapter 2.

The purpose of SCM, conceptually, is to match supply and demand. When these are not matched:

- If supply exceeds demand, the excess must be sold at a loss, written down (documenting loss of value) or written off entirely.
- If demand exceeds supply, some customers will be unsatisfied. The supplier loses the profit on a sale. Customers may go elsewhere for what they need and may stay there.

Left to themselves, firms in a supply chain will optimize their own operations. Without sharing information, that's all they can do. If they share information, they can optimize the entire chain. That might require one partner to accept increased costs, perhaps by carrying more inventory than it otherwise would. Since all members of the supply chain understand this, the net cost reductions can be shared.

Supply chain members can collaborate in two dimensions:

- *Horizontal collaboration* is between two elements at the same level of a supply chain. *Vertical collaboration* is between suppliers and consumers along a supply chain.
- *Internal collaboration* is between parts of the same enterprise. *External collaboration* is across enterprises.

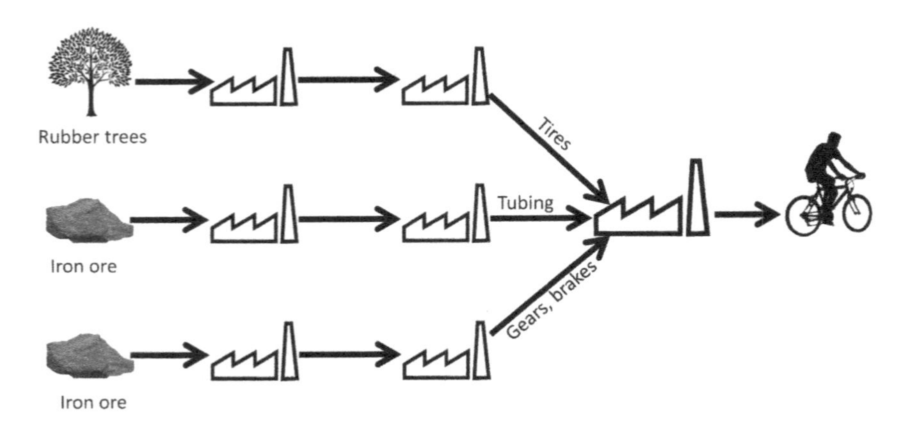

FIGURE 8.11 Diagram of supply chain.

	Internal	External
Vertical	Automobile firm makes own tires, plans demand based on vehicle orders	Company shares production plans with external suppliers
Horizontal	Divisions of one firm coordinate orders to obtain quantity pricing	Firms in an industry collaborate to ensure continued supplies of essential materials

FIGURE 8.12 2 × 2 collaboration grid.

Figure 8.12 shows examples of all four types of collaboration.

Absent collaboration, a company can try to manage its supply chain by itself. The results won't be as good as they would be with collaboration, but they'll be better than nothing.

Electronic Ordering

A desirable customer is easy to sell to. Customers can achieve this is by streamlining the ordering process. If a supplier works with customer A via an automated ordering process, but with B via more cumbersome methods, which will it want to keep? If keeping a customer means offering discounts, is A or B more likely to get those discounts?

The first standard for electronic ordering was called EDI, *electronic document interchange.* It defines messages to be understood by a computer, not a person. EDI standards specify several hundred message types. Each type corresponds to a stage in commerce, from asking for bids to sending an invoice. Figure 8.13 shows an EDI request for payment from a healthcare provider to an

ST*820*0001*005010X306~

Beginning of an 820 transaction set, Control number = 0001

BPR*I*220*C*ACH*CCP*****01*199999999*DA*98765*20140604~**

Total Payment Amount is $220. The payee's bank transit routing number is 199999999 and their bank account is 98765. The Check Issue or EFT effective date is June 4, 2014.

TRN*3*78905~

Reassociation Key provides a sender unique Check or EFT trace number "78905".

REF*38*123456~

The Exchanged assigned QHP Identifier is 123456.

REF*TV*12565496~

The Issuer assigned QHP identifier is 12565496.

N1*PE*BATA INSURANCE CO.*FI*012222222~

The payee's name (BATA INSURANCE CO.) and Federal Tax ID number (012222222).

N1*RM*GOVERNMENT AGENCY*58*123ABC~

The payer's name is Government Agency with an identifier of 123ABC.

FIGURE 8.13 Screen shot of EDI transaction (Source: Data Interchange Standards Association, examples.x12.org/005010X306/different-types-of-payment-made-by-the-hix).

insurance company. It would be followed by other messages that break down the total payment of $220 into individual payments on behalf of specific patients for specific conditions or care.

There are several international standards for EDI messages. The main U.S. standard for EDI is ANSI (American National Standards Institute) ASC X12. Figure 8.13 is in that standard. The UN/EDIFACT standard is used in most of the rest of the world. Industries in specific areas, such as the German automotive industry, have local standards. Some companies are moving to newer e-commerce technologies, such as Extended Markup Language (XML) which is based on the HTML language used in web pages, but EDI is still popular.

EDI message formats are unlikely to match the internal data formats of a company's ordering or financial systems. *EDI translation* software can convert the data between formats.

Older EDI systems used expensive private networks. Today, most EDI messages use the Internet.

Providing Insight

You're a supplier. Would you rather have a customer who keeps you informed of its plans and what it expects to need in the future, or one who just orders when it needs something? One way SCM can make customers easier to deal with, hence more desirable, is by helping suppliers anticipate their needs.

This is done by sharing information along the supply chain. There are several ways to do this. All are, or can be, facilitated by information systems. This section offers examples.

SCM Dashboard

Figure 8.14 shows a supply chain dashboard, annotated by a Johnson Controls executive to show what each "instrument" means. Almost any company can use such a dashboard. The items on it will vary from company to company, but the concept of giving managers insight will be the same.

You'll read more about dashboards in general in the next chapter.

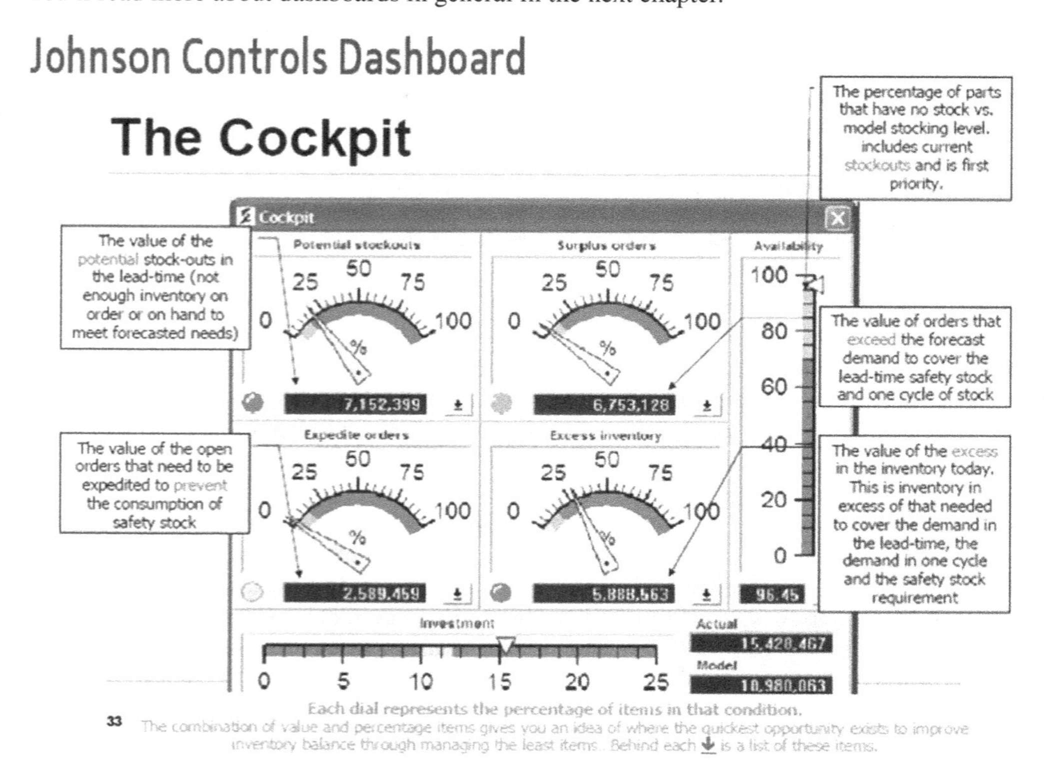

FIGURE 8.14 Supply chain dashboard.

Automatic Replenishment

Once people roamed stores with clipboards, noting what was running low so it could be ordered. Reorders may have been guided by intuition, or purchasing agents may have used reorder points and reorder quantities, but it was a human process.

In modern stores, point-of-sale systems at checkout update the store's inventory. Its inventory management system informs a buyer when something needs to be reordered. The reorder is usually approved, unless the buyer knows that it shouldn't be. (Perhaps the product will soon be replaced by an improved version. Perhaps demand is seasonal.)

Even that approach introduces delay into what can often be automated. What if a supplier could check store inventory and schedule deliveries? Walmart and Procter & Gamble (P&G) were among the first firms to collaborate in this way, moving from electronic ordering to external, vertical collaboration. Walmart gives P&G controlled access to store inventory data. When a store runs low on liquid Tide with bleach, P&G adds it to the next delivery. After the detergent has been delivered, P&G sends Walmart an electronic invoice and Walmart's computers request its bank to transfer money to P&G's account. The system is audited regularly to make sure it's working properly, but most of the time it runs itself. Result: faster inventory replenishment, fewer errors, lower cost.

This cooperation affects the competitive forces you read about in Chapter 2. The link between the two firms makes it harder for Walmart to replace Procter & Gamble as a supplier. This increase in supplier power may be something to strive for (if one is a supplier) or to avoid (if one is a buyer). If a burger chain receives automatic shipments of potatoes from one supplier, it is harder to switch to another supplier whose potatoes are less expensive.

Manufacturing firms that use just-in-time production often have such arrangements with their suppliers. An automobile manufacturer can let a tire company access its production schedule to learn what types of tires it will need and in what quantities. The tire company can then deliver the right tires at the right time. In Japan, where the just-in-time approach is popular, assembly plants are ringed by supplier factories to facilitate frequent small deliveries.

These are examples of *collaborative supply chain management*. Collaboration means working toward mutual objectives through sharing information, risks, and rewards. It's not collaboration unless everyone benefits in some way. Collaborative SCM differs from EDI in that it requires advance planning to benefit everyone, while an EDI transaction is sent out when a customer wants to place an order or a seller has something to say about one.

Supply Chain Control Tower

Air traffic controllers in a control tower see all that goes on at an airport. They manage air and ground traffic to achieve the best results. The term *control tower* was adopted to describe an information system that gives managers visibility into the supply chain: what's on order, from whom, where it is, when it's expected, what it's needed for. That enables managers to fine-tune delivery plans, keep up with changed needs while reducing the amount of inventory kept on hand at each location, and produce goods faster in response to rush orders. (Rapid response to customer requirements helps leverage the competitive force of rivalry with other firms.)

A *supply chain control tower* (Figure 8.15) can support horizontal and vertical, internal and external, supply chain collaboration. It can show the benefits of such collaboration over doing without it.

The idea of a control tower overlaps the idea of a dashboard. It provides visibility into critical factors that affect a certain area. The main difference is that a control tower provides tools to change things, whereas a dashboard provides information that must be acted on separately.

Once a firm can see what's in its supply chain, control can be effected through a *supply chain operating network*. Such a network, according to Patrick Lemoine of E2open:

- Connects all companies and allows them to operate as one virtual supply chain.
- Allows all parties to work off the same set of information, always up to date.

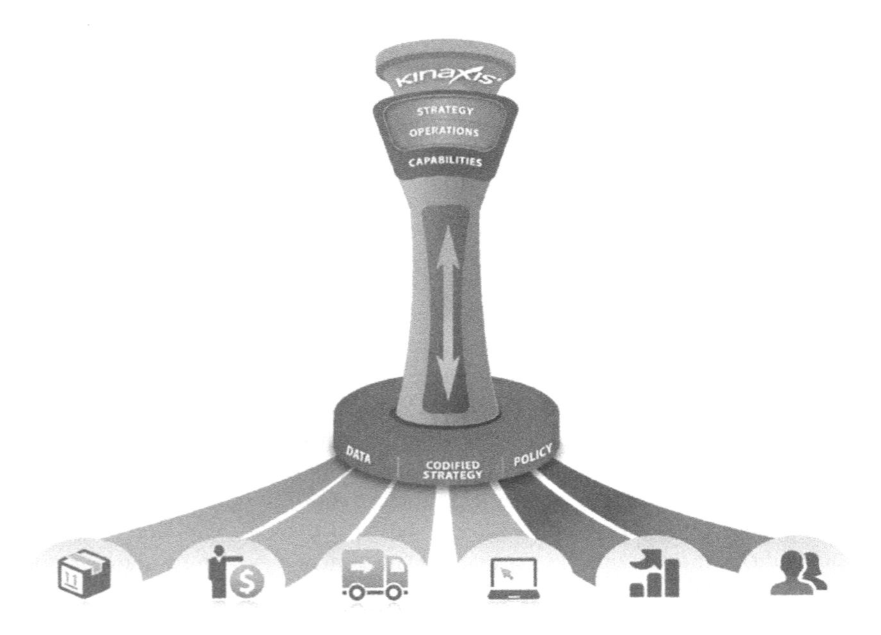

FIGURE 8.15 Supply chain control tower (Source: blog.kinaxis.com/2012/09/the-smart-phone-of-scm-no-synching-sessions-required).

- Supports the different lines of business (procurement, supply chain, demand planning, order management, etc.) with solutions to make better and faster decisions using this up-to-date information from all the supply chain partners involved.
- Extends what ERP has done for the company … provides a secure network for companies to coordinate supply chain activities in real time.

The result, Lemoine feels, is faster and more accurate supply chain decisions.

The Bullwhip Effect

The *bullwhip effect* refers to how small demand fluctuations can, as a result random interactions, look much larger when from further up the supply chain. The name comes from the behavior of a whip, where a small flick of the wrist is amplified into a much larger movement of the whip's tip.

To visualize the bullwhip effect, suppose a store runs low on TV sets. It orders ten to restock. As a result, its distributor is low on 55″ sets and places its standard order for 100. The manufacturer sees this and schedules its standard production run: 1,000. Its ERP system calculates that this run requires more 1,000 μF capacitors than it has in stock. Since several models use this capacitor, the manufacturer orders 100,000. The capacitor manufacturer sees a spike in sales and increases production to handle the anticipated demand. Most of those capacitors gather dust for years.

The reason for this effect, first described by Jay Forrester in 1961, is that companies carry a *safety stock* to avoid stock-out situations due to random fluctuations in demand. Larger fluctuations lead companies to carry a larger safety stock. If companies know what causes a fluctuation, they won't do this. Changes with known causes do not lead firms to expect large future fluctuations.

Here, if the retailer had passed information about its order up its supply chain, excess ordering and production wouldn't have happened. A small TV store is unlikely to manage its supply chain, but a distributor can. The TV manufacturer might still order 100,000 capacitors—it might expect to use them somewhere, and capacitors don't take up much space—but the manufacturer would know there was no demand shift and wouldn't make more than necessary. SCM can reduce or eliminate the bullwhip effect.

Where you fit in: It's tempting, and often simplest, to order what one needs when one finds that one needs it. Many businesses work that way and survive. However, most of them could be more successful if they paid more attention to their supply chains. You can make that happen.

8.5 EXTRANETS

In the last chapter you read about *intranets:* networks that use Internet protocols to give internal users access to organizational resources. Supplier-customer connections can be facilitated by similar access. For example, automatic replenishment requires access to a customer's inventory database.

When an intranet is extended to include an organization's trading partners, the resulting network is called an *extranet*. Figure 8.16 extends Figure 7.14 to show this.

Accessing an extranet resembles accessing an intranet from outside the organization: log on with a user ID and a password. (This information need not be entered by a person. In applications such as automatic replenishment, it is sent by a computer.) As in an intranet, the user ID determines what resources the user may access and what he, she, or it may do. Suppliers can read inventory information and edit their own contact information. Customers may also be allowed to update their contact information, saving the supplier time and effort, and increasing accuracy. Customers may have access to inventory information, such as airline seat availability—but so would the general public, since suppliers want to make it easy for people to buy what they sell.

Where you fit in: Depending on your industry, there may be opportunities to ask suppliers and customers if they have an extranet, or consider ways to extend your intranet to an extranet and make it available to your customers and suppliers. Your management will appreciate these ideas.

KEY POINT RECAP

Connecting with customers and suppliers is one of the three key ways that information systems provide major strategic benefits.

This is one of the three areas in which you, as a businessperson, must look for creative ways to use information technology.

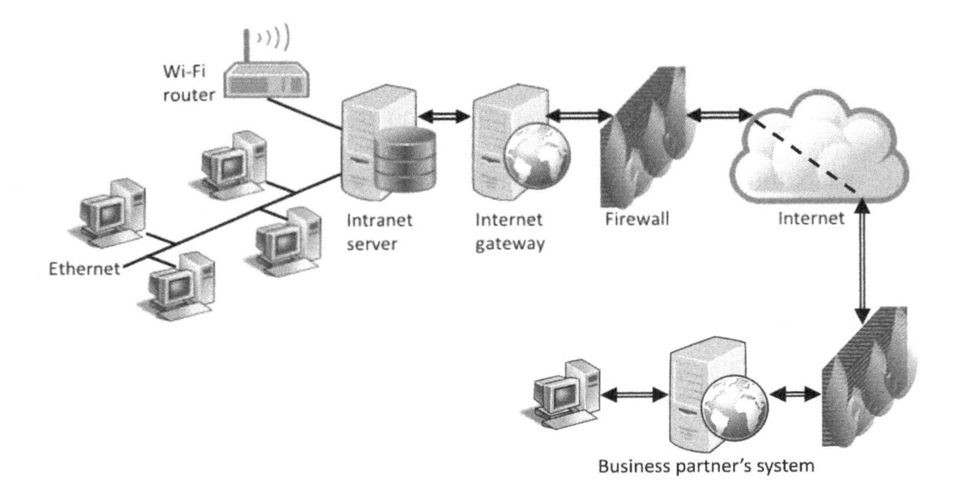

FIGURE 8.16 Diagram of intranet with extranet added.

CRM systems help a company work with its customers effectively. *They are of two types: operational CRM (information about a specific customer, used in sales activities) and analytical CRM (analyzing many customers, used to plan marketing and sales activities).*

As a businessperson, your employer's customer relations affect you even if you're not in a sales or sales management position. The more you know about its CRM systems, the better the position you'll be in when you're called in for sales support or another sales-related task.

Sales information can be used in specific ways to benefit a company. *This chapter discussed the RFM and Customer Data Strategy approaches.*

As a businessperson, your awareness of frameworks such as these, and others you can find, will enable you to contribute to the effective use of information systems in your company.

Every organization has a supply chain, but many do not try to manage it. *SCM systems enable a company to steal a jump on its competitors by being more effective in this area.*

Most positions you might be in as a businessperson have some connection to the supply chain. Your understanding of SCM concepts will enable you to perform more effectively in them.

KEY TERMS

Analytical CRM: Category of *customer relationship management* systems that focuses on using information about customers to plan how to sell to that customer or similar groups of customers.

Automatic replenishment: The process through which a supplier monitors a customer's needs and ships products or materials as needed, without a specific request.

Bricks and clicks (also *clicks and mortar*): E-commerce seller who also sells through physical outlets.

Bullwhip effect: The magnification of small changes in demand as they move back along a *supply chain.*

Business-to-Business (B2B): *E-commerce* in which both buyer and seller are businesses.

Business-to-Consumer (B2C): *E-commerce* in which the seller is a business and the buyer is an individual consumer.

Click farm: A business that provides, for a fee, clicks on online ads or affinity indications in social networks.

Collaborative supply chain management: *Supply chain management* in which companies work together to reduce total costs to all of them together.

Consumer-to-Consumer (C2C): *E-commerce* in which both buyer and seller are individual consumers.

Customer data strategy: Approach to planning the use of customer information based on how frequently a product or service is purchased and how much it can be customized to each buyer.

Customer Relationship Management (CRM): (a) Information system through which a company manages information about its customers; (b) the process of using such an information system to sell more effectively.

Disintermediation: The process of removing intermediaries between producer and consumer.

E-business: Carrying out any business activity online.

E-commerce: Online buying and selling.

EDI translation: The process of converting information between the format of standard EDI messages and that of a company's internal systems.

Electronic Document Interchange (EDI): An international standard for electronic communication of purchase and sale information among trading partners.

Extranet: An intranet extended to a company's trading partners.

Government-to-Business (G2B): *E-business* in which a government agency provides services to businesses online.

Government-to-Consumer (G2C, also *government-to-citizen*): *E-business* in which a government agency provides services to individuals online.
Intermediary: An organization that performs a function between two other organizations.
Loyalty program: Program through which a company provides benefits to repeat customers.
Market basket analysis: Analyzing the items shoppers purchase together in order to increase sales by facilitating purchase of related items in the future.
Mobile commerce (M-commerce): *E-commerce* which takes into account the buyer's use of a mobile device.
Nexus: Presence of a seller's physical facilities within a taxing jurisdiction.
Omni-channel: Taking an integrated approach to dealing with multiple distribution channels.
Online shopping mall: Web site that houses e-commerce activities for multiple small firms.
Operational CRM: Category of *customer relationship management* systems that provides information to use during sales or other customer contact activities.
Order fulfillment: The process of delivering a product or service to its purchaser.
Pay per click: A method of paying for advertising, where an advertiser pays for each time a user of another site clicks on its ad.
Personalization strategy: Competing by using customer information to personalize an offering.
Portal: Web entry point that provides access to many related pages of potential interest.
Pure play: E-commerce seller who only does business online.
RFM (Recency-Frequency-Money) method: Approach to segmenting customers based on how recently they shopped at a store, how frequently they shop there, and how much they spend.
Safety stock: Additional inventory carried to avoid stock-outs due to fluctuations in demand.
Sales force automation: *Operational CRM* systems used primarily by salespeople.
Supply chain: An organization's suppliers, their suppliers, and so on back to raw materials.
Supply chain control tower: Information system that provides managers with information about their supply chain.
Supply Chain Management (SCM): (a) Information system through which a company manages information about its supply chain; (b) the process of using such an information system to obtain supplies more effectively.

REVIEW QUESTIONS

1. Define e-*business* and *e-commerce*. What is the difference between them?
2. Describe five different types of e-commerce identified by three-character abbreviations.
3. What is a *portal*? Give an example of one.
4. What is the difference between a regular auction and a reverse auction?
5. What is the business function of an *exchange?*
6. What is the first step in setting up an e-commerce site?
7. Must a firm be large in order to sell via e-commerce?
8. What are the two characteristics that can define m-commerce?
9. Describe the e-commerce issues of *disintermediation* and *channel conflict.*
10. Define, briefly, the two major types of CRM. What is the basic difference between them?
11. What is an *opportunity management system*? How does it differ from sales management and contact management systems?
12. Identify five ways a business can use analytical CRM.
13. In the RFM method, what do those three letters stand for?
14. Describe the three customer data strategies and the types of products that each one fits.
15. Define an organization's *supply chain.*
16. What competitive force do SCM systems leverage?
17. What are the benefits of EDI, compared to non-electronic commerce?
18. Explain how automatic replenishment works.

19. How does SCM help companies avoid the *bullwhip effect*?
20. Compare an *extranet* and an *intranet* (Chapter 7).

DISCUSSION QUESTIONS

1. Fonderia Pontificia Marinelli has cast bells the same way in the Italian hills since the year 1040. Visit *campanemarinelli.com*. Then discuss: Does the Marinelli foundry need an e-commerce strategy? Why or why not?
2. Pick a consumer product that starts with the same letter as your given name. If your name is Susan or Steve, you can use stoves (not grills), shoes (not boots). Pick something typical consumers buy, not something mostly for businesses. Visit three sites that sell this type of product online. Pick an item and go through the ordering process, stopping before you place an order. "Buy" the same product, or as close to it as possible, on all three sites. Then answer:
   a. Which site made it easiest for you to find the information you needed to select what you wanted to buy?
   b. Which site made the online shopping experience best overall? Why?
   c. If you were going to buy this product online, which site would you buy it from on the basis of your experience with their sites? Why? (Ignore differences that aren't about their site or the online shopping experience, such as price, selection, reputation, or previous experience with the seller.)
   d. If the other two firms hired you as a consultant to improve their customers' online shopping experience, what would you recommend? (Discuss only the sites and the online shopping experience. Ignore other aspects of their offering, such as price and selection.)
   e. Inspect the privacy policies of all three sites. How do the sites compare in ease of finding this policy? Telling you what their polices are? In the policies themselves? Which site do you feel safest with? Why? Are there any you would prefer not to do business with because you don't like their privacy policy? Which, and why?
3. (This is an exercise in making and using business assumptions.) A hotel site pays an airline for referring ticket buyers. Assume it pays the airline 3¢ for each user who clicks through to its site, 50¢ for each reservation such a user makes. Estimate the airline's annual revenue from this site. You'll have to estimate:
   - How many airline ticket purchases are there per day? (Find out or estimate how many flights the airline has per day and the number of passengers on an average flight; from that estimate how many passengers it has each day; then estimate how many purchases those reflect considering round trips, connecting flights, and groups traveling together.)
   - What fraction of those purchases were made on the airline's site, as opposed to other sites or in other ways (such as telephone)?
   - How likely is someone who buys an airline ticket to click through to the hotel reservation site? (Assume it has a range of hotels in all cities that the airline serves.)
   - How likely is a person who clicks through to the hotel site to make a reservation?
4. Consider a store that finds a customer is looking at cereal brand A. Brand B paid the store to offer coupons to customers who look at Brand A for more than a few seconds. The store sends a coupon for Brand B to the customer's phone. The customer moves down the aisle and buys Brand B. Should we have any ethical concerns about this scenario? Why or why not? Would you feel differently if it involved a $300 television set rather than a $3 box of cereal? Is this different from a manufacturer giving store salespeople a bonus for selling its TV sets?
5. A chain of coffee shops is thinking of developing a smartphone app where customers will receive special offers in a text when they are near one of its shops. These offers will

provide discounts on specific items, good for a half hour after receiving the text. Discuss the benefits and drawbacks of this app to the chain. All things considered, is it a good idea?

6. You want a cup of coffee in a new city. You use your smartphone to search for coffee shops. The search engine finds four within walking distance, checks to see which pays the most for referrals, and moves it to the top of the list. In 2019 no search engines are known to do this. However, there are no technical barriers, and it is legal (in the U.S.) if disclosed. Discuss:
   a. The potential business value of this process to coffee shops.
   b. The potential business value of this process to search engine owners.
   c. Any ethical concerns you have with this process. If you have none, explain why.
7. Members of the loyalty program at a chain of casual restaurants earn one point for each dollar they spend. They can use points for awards, from 50 for soup up to 200 for steak. A customer, after waiting 90 minutes for a table, suggested that it offer a "skip the line" award: a member using this award will be seated next, behind only others using that award. Is this a good idea? How could they test it? How could it decide how many points this new award should require? Is it practical to have the number of points depend on how long the wait is at that time? How about having people in line bid points, via a smartphone app, for the next available table?
8. You work for Finch Plumbing Supply. Your customers tend to be of two types: plumbers and plumbing companies who buy in large quantities, know exactly what they want, and insist on competitive prices or they'll go elsewhere; and individuals, who buy less at a time and need help, but are less price-sensitive and tend to be loyal if they can get what they need. About 80% of your $25 million annual revenue comes from plumbers, but about 80% of your customers by count are individuals.
   a. Describe an operational CRM system that this business could use. State its benefits.
   b. Describe an analytical CRM system that this business could use. State its benefits.
9. Think about a manufactured product that you bought at a physical store (not online):
   a. Where did you purchase that product?
   b. Where was it made?
   c. What manufacturing processes do you think were used to produce that product?
   d. What raw materials/components do those processes use? Where were they made?
   e. In how many different locations do you think it was stored before it reached you?
   f. What transportation mode(s) do you think were used to deliver the product to you?
   g. How many elements does this suggest are in the supply chain of the company in (a)?
   h. How can the company that sold you this product manage its supply chain, other than simply ordering more from its manufacturer or distributor when it runs low? What would be the benefits of managing its supply chain in this way?
10. How could Finch Plumbing Supply (see Question 8) use an extranet?

KHOURY CANDY DISTRIBUTORS CUSTOMER SEGMENTATION

Jason had arranged for Jake and Isabella to meet Harvey Leonard, KCD's vice president of marketing, the following week. When they met him at the reception area to be escorted into the building, after the usual opening pleasantries he asked, "What do you two know about customer segmentation?"

"Not a whole lot," Jake admitted sheepishly. "We looked at it in our marketing course, but not in detail."

"But you know what it is," Harvey continued. "That's probably enough. I need someone to bounce ideas off of, and my own people think too much like each other, and like me, for that."

"We'll do what we can," said Isabella. "What did you have in mind?"

"Well," Harvey began, "back when I got here as our first full-time sales rep, we had just a handful of customers. I knew all of them and talked to them personally every so often, even if they usually phoned in their orders—or mailed them in, as a few still did then. Then we grew, and grew. You know where we are today, and I don't think we're done growing. I can't possibly know all our customers any more. No one could. Our regional salespeople know the customers they're responsible for, but that doesn't give them the big picture. I get the big picture in summary reports, but I don't have much personal contact any more.

"Anyhow, we got to where we are mostly because of great customer service along with fair pricing, even if we're not always the cheapest. I'm not sure that will take us to the next level. I'd like to roll out focused promotions to specific groups of customers, but we don't have a good idea of what our customer groups are or what promotions would appeal to them. That's where segmentation comes in."

"And you're trying to figure out what segments make sense, and how you can assign stores to those segments?" asked Isabella.

"That's exactly right," Harvey confirmed.

"In that case, given our bias to information systems, we'd probably look for segments you can assign based on information that's already in your database," Jake said.

"Right," said Harvey. "But there's one more consideration, and it's a big one. We don't want to just assign segments as an academic exercise. They have to be of practical use. I mean, we can segment our customers by name into A–M and N–Z. That's easy. But so what? There won't be any difference between them in how they'd respond to incentives or what it would take to get them to do more business with us. What we need are segments that let us target promotions where they will do the most good."

"What sorts of things are good indicators of that?" asked Jake.

"One is how well they responded to past promotions. We've had promotions before, of course," Harvey said. "We just haven't tried to target them to specific customer segments. Still, we know who bought what during a promo, and what they probably would have bought without it. Customers who bought more during a promo, without buying less later to make up for it, are good candidates for another one. Customers who bought what they always did, but paid less because we ran a promo, aren't.

"So, that might be one way to segment them. We might also look at regional differences, especially U.S. versus Canada. We can also tweak promos to see what sorts of differences affect response, which we've done of course, but then look at the kinds of customers and how each kind is affected. That would be another basis for segmentation if we can figure it out. Do you think there's software that can do that?

"There certainly ought to be," said Isabella with assurance. "Your problem, if you generalize it a bit from candy, is a pretty common one. Lots of people would like to solve it. That's what it takes to create a market for a software package. Most of them also ought to be able to plug into pretty much any SQL database, because so many of the people who'd want those packages would have one."

"In that case," Harvey asked, "do you think you could turn up a few and tell me a bit about them? Ideally, we'd like to use this segmentation for more than just planning promotions. We'd want a package that will also give us a handle on customer preferences in general, so we can tailor other aspects of what we offer as well. You know we use an Oracle database, so they'd have to be able to work with that, but with all the Oracle users out there that shouldn't be a problem."

Questions

1. Is customer segmentation operational or analytical CRM? Why?
2. Whichever way you answered Question 1, can KCD use the other type? If so, how? If not, why not?
3. Jake and Isabella followed up on Harvey's request and wrote a report on three customer segmentation applications. Unfortunately, their report wasn't ready in time for this book. Write your version of it.

CASE 1: ALIMED EDI

EDI is old. Its roots go back to shipping manifests of the 1960s. Still, it's not ready for the scrap heap. Extended to cover today's commerce needs and using the Internet for data transmission, it is used to some extent by the majority of large companies and many smaller ones. One such firm is AliMed, a medical supplies company in Dedham, Mass.

Most of what AliMed sells comes from about 20 suppliers. Over the years they developed EDI relationships with most of those, using the Delta data translation and Electronic Commerce Server (ECS) packages from Liaison, an Atlanta, Ga. software firm. When Allmed moved to a new ERP system, Microsoft Dynamics AX, they also replaced the Liaison packages. The new EDI software was supposed to be more closely integrated with Microsoft Dynamics AX. Unfortunately, it didn't work out well. Neither did a second EDI system.

Eventually Fred Fish, AliMed's MIS manager, recognized the problem. "We tried two different EDI vendors," Fish said. "Both approaches failed to meet our expectations. ... We were never able to get EDI working without many hours of manual effort every day. We were never able to get all of our trading partners implemented after 18 months. I wanted a solution that was easy to use and could run the transactions unattended."

Fish's solution was to return to Liaison Delta/ECS. This created a new problem: They weren't designed to work with Microsoft Dynamics AX, which was why AliMed changed EDI software in the first place. Fish contacted Aurora Technologies of Harrisville, R.I., and Green Beacon Solutions of Boston, Mass. Aurora brought Liaison Delta/ECS and general EDI experience to the project, while Green Beacon had expertise in Microsoft Dynamics AX.

Together, these two firms created an automated interface so that Liaison EDI software would work seamlessly with Microsoft Dynamics AX. Aurora defined the *business object models* that specify the objects with which an activity works, so EDI transactions would be transmitted with the correct information. Green Beacon then interfaced these objects with Microsoft Dynamics AX to let EDI transactions share the ERP database. Finally, Aurora trained AliMed personnel to monitor EDI activity and set up new trading partners.

The net effect was to give AliMed the EDI capability they needed, working with the ERP system they needed. Their business is now well supported and can continue its growth. Faith Lamprey of Aurora Technologies, thinking back over this project, says, "We're pleased that AliMed recognized the Return on Investment that full data integration and automation of the process provides."

Questions

1. AliMed used external contractors (Aurora, Green Beacon) instead of hiring staff. This is called *outsourcing*. Discuss why outsourcing was, or was not, a good idea here.
2. AliMed's network uses EDI. That gives EDI inertia. These firms might like to move to a new technology, but they won't all want to move at the same time. Does this lock in EDI forever? In general, how can a group of firms move from an old technology? (EDI is just an example.)
3. Microsoft offers an EDI module for Microsoft Dynamics AX developed by Scalable Data Systems of Spring Hill, Queensland, Australia. Since Microsoft Dynamics AX is popular, a web search will find many more EDI programs that are designed to work with it. Find three, including that one if you wish. Compare their features. For each, try to identify one type of company for which it is especially well suited, and one type of company for which it is not.

CASE 2: THE RISKY SUPPLY CHAIN

Supply chains are vulnerable to many types of risk. SCM systems can both increase and decrease them.

SCM systems introduce a new risk: *digital misinformation*. This could disrupt more than supply chains. For example, digital misinformation can cause a stock market panic. In SCM, it can mean passing off inexpensive fish as a more expensive variety, or diverting a shipment from its purchaser to a thief.

One way to combat digital misinformation is through visibility into the supply chain. End-to-end supply chain visibility, however, is rare. A widely cited Geodis study found that only 6% of companies had such visibility in 2017. Others give higher figures, but all well under half. Andrew Atkinson, Director of Product Marketing at SCM software supplier E2Open, likens lack of supply chain visibility to playing poker without being able to see the cards in your hand. Martijn Lofvers, Founder and Chief Trendwatcher of Supply Chain Media, adds that "Most companies know their direct suppliers personally, but rarely know their suppliers' suppliers."

The Business Continuity Institute's 2018 Supply Chain Resilience Report, written together with Zurich Insurance, found that 56% of businesses had suffered at least one supply chain disruption over the previous year. While this is high, it is down from 65% a year earlier and 70% in 2016. Of these disruptions, just over half were in a respondent's immediate (Tier 1) supplier, while 34% were further up the supply chain. (Some respondents did not know where a disruption occurred.) Losses due to disruption exceeded one million euros (US$1.1 million in 2019) in about 14% of the cases.

The leading cause of supply chain disruptions was unplanned technology outages. Cyberattacks were third. (Weather was second.) Zurich Director of Strategic Research Linda Conrad points out, "As the supply chain gets more global, it starts looking like more of a spider web." It is no wonder that Zurich has named supply chain disruption as the top risk "blind spot." Visibility into the supply chain can reduce the likelihood of such disruptions.

To manage supply chain risk, *Spend Matters* Managing Director of Research Jason Busch says that the ideal SCM system should be *predictive*, *prescriptive*, and *proactive*. Predictive: It can forecast the outcome of a supply chain decision. Prescriptive: It can make recommendations. Proactive: It pushes these recommendations to users, not waiting for users to ask, "What should I do now?" when it may be too late. Today's SCM doesn't have those capabilities, but Busch feels they're coming—well within your first decade in the work force.

John Herr, CEO of supply chain risk management software supplier Avetta, writes "managing a supply chain requires processes and capabilities to gather and analyze a vast range of data." He continues "The breadth and depth of the data required to manage all aspects of a complicated supply chain ... would be essentially unmanageable without strong data and analytics." Deloitte partner Larry Kivett agrees: "The complexity of today's environment has made the use of supply chain data analytics critical for identifying supply chain waste, fraud, billing anomalies, and risk patterns."

Supply chain risk management pays off. For example, in the lead-up to Hurricane Maris in 2017, multinational biotechnology company Biogen used software from Resilinc to identify suppliers likely to be hit by the storm, which raked the Caribbean and skirted the U.S. East Coast. Based on this analysis, the company decided to proactively move materials to an inland assembly plant in Kentucky, ultimately preventing damage and ensuring continued timely delivery.

In the final analysis, some risk is unavoidable. Companies can, however, minimize both risk and the potential disruption due to risk. Effective supply chain management systems help do that.

Questions

1. Discuss how being predictive, prescriptive, and proactive were involved in the Biogen example, or, if they weren't, how they could have been.
2. Your manufacturing firm has agreed to provide a school with 100 cafeteria tables. To make them, it depends on suppliers of molded plastic, metal tubing, hardware, and more. These, in turn, depend on their own suppliers (your Tier 2). How would you minimize supply chain risk? What would you do differently if, instead of this being a one-time job, you planned to produce 100 tables every week?

3. Using overseas suppliers increases supply chain risk compared to using suppliers located in one's home country. Identify three such increased risk factors. State how you could avoid or minimize them.
4. The Business Continuity Institute *(thebci.org)* has issued Supply Chain Resilience Reports annually since 2009. When you read this, its 2019 report should be available and later ones may be. Find the latest available report, compare its findings with information from the 2018 report in the case, and discuss the differences. Try to go beyond the discussion of trends in the report itself.

REFERENCES

24/7 Staff, "The emerging business of supply chain risk management," *Supply Chain* 24/7, February 5, 2019, www.supplychain247.com/article/the_emerging_business_of_supply_chain_risk_management, accessed September 21, 2019.

AliMed web site, www.alimed.com, accessed September 21, 2019.

Arnett, A., "Eats and tweets," *The Boston Globe*, December 26, 2014, page B5.

Arthur, L., "We're ready for the omnichannel revolution—Are you?" *Forbes*, May 8, 2012, www.forbes.com/sites/lisaarthur/2012/05/08/were-ready-for-the-omnichannel-revolution-are-you, accessed September 21, 2019.

Aspentech CRM, "Benefits of customer segmentation part 1: Segmentation defined," February 28, 2018, www.aspen-tech.com/blog/Benefits-of-Customer-Segmentation-Part-1---Segmentation-Defined_AE153.html, accessed September 25, 2019.

Baldwin, H., "Supply chain 2013: Stop playing whack-a-mole with security threats," *Computerworld*, April 30, 2013, www.computerworld.com/s/article/9238686/Supply_chain_2013_Stop_playing_whack_a_mole_with_security_threats, accessed August 3, 2019.

Bunyan, D. and K. Sarkis, "Making connections with the new digital consumer," *Strategy+Business Tech & Innovation section*, May 6, 2019, www.strategy-business.com/feature/Making-connections-with-the-new-digital-consumer, accessed September 25, 2019.

Busch, J., "Supply chain and risk management: '3Ps'—Predictive, proactive, prescriptive," *Spend Matters*, spendmatters.com/2013/08/12/supply-chain-and-risk-management-3ps-predictive-proactive-prescriptive, accessed September 21, 2019.

Business Continuity Institute, "BCI supply chain resilience report 2018," 2018, www.thebci.org/uploads/assets/uploaded/c50072bf-df5c-4c98-a5e1876aafb15bd0.pdf, accessed August 3, 2019.

Buy Plus Followers, www.buyplusfollowers.com, accessed September 21, 2019.

CRMforecast.com, "Healthcare industry CRM software solutions," www.crmforecast.com/healthcare.htm, accessed September 21, 2019.

DeAngelis, S., "Have you heard about supply chain control towers?" *Enterra Insights*, July 27, 2011, enterpriseresilienceblog.typepad.com/enterprise_resilience_man/2011/07/have-you-heard-about-supply-chain-control-towers.html, accessed September 21, 2019.

Deloitte, "Understanding risk management in the supply chain," 2019, www2.deloitte.com/us/en/pages/risk/articles/risk-management-in-supply-chain.html, accessed September 21, 2019.

Erhun, F. and P. Keskinocak, "Collaborative supply chain management," Chapter 11 of *Planning Production and Inventories in the Extended Enterprise*, K.G. Kempf et al., eds, *International Series in Operations Research and Management Science*, Vol. 151, Springer, 2011.

Forrester, J.W., *Industrial Dynamics*, MIT Press, 1961.

Herr, J., "Making a case for supply chain risk management," *Utah Business*, August 6, 2019, www.utahbusiness.com/supply-chain-risk, accessed September 21, 2019.

Hill, K., "How target figured out a teen girl was pregnant before her father did," *Forbes*, February 16, 2012, www.forbes.com/sites/kashmirhill/2012/02/16/how-target-figured-out-a-teen-girl-was-pregnant-before-her-father-did, accessed September 21, 2019.

Internet Retailer, "Ecommerce represented 14.3% of total retail sales in 2018, according to Internet Retailer's analysis," March 13, 2019, www.digitalcommerce360.com/article/us-ecommerce-sales, accessed August 2, 2019.

Ivey, J., "Compare healthcare CRM software," *Software Advice*, December 2, 2014, www.softwareadvice.com/crm/healthcare-crm-comparison, accessed August 2, 2019.

Khalamayzer, A., "Zurich: Supply chain management is 2013's 'blind spot,'" *Property Casualty* 360, December 14, 2012, www.propertycasualty360.com/2012/12/14/zurich-supply-chain-management-is-2013s-blind-spot, accessed September 21, 2019.

Liaison Technologies, "AliMed rediscovers the power of the Liaison Delta & ECS data transformation products," 2012, cdn2.hubspot.net/hubfs/126065/Alimed%20Case%20Study%20GB.pdf, accessed September 21, 2019.

Lemoine, P., "A better approach to managing your extended supply chain," *E2open*, www.e2open.com/a-better-approach-to-managing-your-extended-supply-chain, accessed September 21, 2019.

Lofvers, M., "Framework for end-to-end supply chain visibility," *Supply Chain Movement*, January 18, 2019, www.supplychainmovement.com/framework-for-end-to-end-supply-chain-visibility, accessed August 3, 2019.

Maximizer CRM, "Successful segmentation – how to work your data assets," October 2, 2018, www.maximizer.com/successful-segmentation-how-to-work-your-data-assets, accessed September 25, 2019.

McFarland, M., "How iBeacons could change the world forever," *The Washington Post*, January 7, 2014, www.washingtonpost.com/blogs/innovations/wp/2014/01/07/how-ibeacons-could-change-the-world-forever, accessed September 21, 2019.

McNicholas, K., "Nordstrom disses 'omni-channel' term, but Alexandra Mysoor says it's real," *PandoDaily*, May 3, 2013, pandodaily.com/2013/05/03/nordstrom-disses-omni-channel-but-entrepreneur-contends-its-rea.

Pastoor, I., "A look behind the scenes of click farms," *Diggit*, January 11, 2019, www.diggitmagazine.com/articles/look-behind-scenes-click-farms, accessed August 2, 2019.

Piccoli, G. and F. Pigni, *Information Systems for Managers*, Prospect Press, 2018.

Piccoli, G. and R. Watson, "Profit from customer data by identifying strategic opportunities and adopting the 'born digital' approach," *MIS Quarterly Executive*, vol. 7, no. 3 (August 2008), www.researchgate.net/publication/220500603_Profit_from_Customer_Data_by_Identifying_Strategic_Opportunities_and_Adopting_the_%27Born_Digital%27_Approach, accessed September 21, 2019.

Practical Ecommerce, "Cisco exec on growth of worldwide ecommerce, cultural differences," June 10, 2011, www.practicalecommerce.com/articles/2841-Cisco-Exec-on-Growth-of-Worldwide-Ecommerce-Cultural-Differences, accessed August 3, 2019.

Pruitt, J., "How to derive more value from customer data," *Inc.*, December 7, 2017, www.inc.com/jeff-pruitt/how-to-derive-more-value-from-customer-data.html, accessed September 21, 2019.

Resilinc web site, www.resilinc.com, accessed August 3, 2019.

Rosenberg, J., "Internet sales tax collection laws gradually taking effect," *USA Today*, December 23, 2018, www.usatoday.com/story/money/business/2018/12/23/sales-tax-online-retailers-begin-collect-internet-customers/2387450002, accessed August 2, 2019.

Salesforce.com, "Getting more out of your business with customer segmentation," undated, www.salesforce.com/products/marketing-cloud/best-practices/customer-segmentation, accessed September 25, 2019.

Salesforce.com healthcare page, www.salesforce.com/industries/healthcare, accessed August 3, 2019.

Spend Matters, "Enabling agile supply chain management to combat natural disasters: A case study of Biogen and Resilinc,", November 9, 2018, spendmatters.com/2018/11/09/enabling-agile-supply-chain-management-to-combat-natural-disasters-a-case-study-of-biogen-and-resilinc, accessed August 3, 2019.

Surana, S., "5 ways to automate customer segmentation," *Agile CRM*, November 25, 2016, www.agilecrm.com/blog/5-ways-automate-customer-segmentation, accessed September 25, 2019.

Tatum, S., "Two things technology sales & business development pros can do right now to generate more leads on LinkedIn," *The Conversion Company*, February 13, 2014, www.theconversioncompany.com/two-things-sales-business-development-pros-can-do-right-now-to-generate-more-leads-on-linkedin, accessed September 21, 2019.

Voo, B., "Digital wallets—10 mobile payment systems to take you there," *Hongkiat.com*, January 9, 2019; www.hongkiat.com/blog/digital-wallets, accessed September 21, 2019.

Wildfire, "Social advertising part 1: Facebook, Twitter, and LinkedIn,", August 19, 2014, www.slideshare.net/trustful88/social-advertising-best-practices-38143897, accessed September 21, 2019.

World Economic Forum, "Global risks report 2019," reports.weforum.org/global-risks-2019, accessed September 21, 2019.

9 Making Better Decisions

CHAPTER OUTLINE

WHY THIS CHAPTER MATTERS

The success of any business depends on decisions that its managers make. Good decisions about who to hire, how to raise capital, where to locate, what to produce, and how to market it lead to success. Bad decisions lead to failure.

Intelligent use of information systems helps companies make better decisions. Being able to use information systems for this purpose is important to successful companies, and to those that want to be. Businesspeople who can use information systems to make better decisions will make their companies more successful and will have more successful careers as well.

CHAPTER TAKE-AWAYS

As you read this chapter, focus on these key concepts to use on the job:

1. Decisions can be categorized in useful ways.
2. When decisions have structure, *models* can help make good decisions.
3. When decisions have little or no structure, proper access to, analysis of, and presentation of data can help make good decisions.
4. Data warehouses, which can be used in several ways, provide a good way to access data.

9.1 DECISION-MAKING CONCEPTS

Life is full of choices. That is as true in business as it is in your personal life.

Choices have consequences. That is even more true in business than it is in your personal life.

Therefore, it is important to make business choices carefully. Since decisions should be based on information, it stands to reason that information systems can help improve decisions.

A *decision* is a choice among two or more options. Before you make any business decision, you need a clear *decision statement*:* What do you want to decide? A decision statement should define a decision as narrowly as possible, but no more narrowly than that. "Where should we eat lunch?" is narrower than "What should we do for lunch?," but should be used only if you want to rule out cooking, ordering take-out food, or skipping the meal in favor of a run.

All parties to a decision must agree on the decision statement. If they don't, disagreements that seem to be about the decision might really be over what is being decided.

Before we can understand how information systems can help people make decisions, we should see how decisions are made.

---

* This term has a different meaning in computer programming. Context will tell you which is meant.

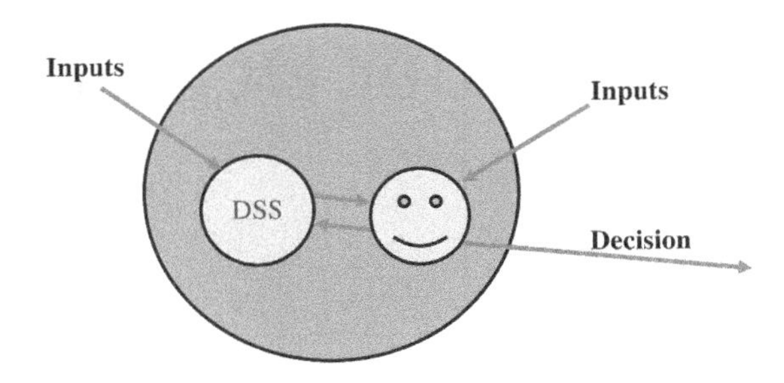

FIGURE 9.1 Diagram of human + machine decision-making system.

The Decision-Making System

Few decisions can be made by computers alone. Even when one usually can be, people may want to review it if it is affected by factors the computer wasn't programmed to consider. However, computers can improve both the quality (effectiveness) of decisions and the efficiency with which they are reached. The entire decision-making system consists of both technology and people, as shown in Figure 9.1. The outer circle shows the system boundary. Information crosses this boundary in both directions, into it and out of the system.

This diagram shows inputs to the electronic part of the decision-making system. They come from internal and external databases, unstructured sources such as Twitter feeds, responses to external queries such as requests for bids on a part, and more. Other inputs come to the person or people who will make the decision. Those inputs are usually unstructured.

Both parts of the system also have internal information. The electronic side might have a database of sales history by product and region. People might have opinions about the importance of cost versus capacity or speed.

The decision is eventually made by a person. It will leave the system, by crossing its boundary to the rest of the world, to be put into effect.

Decision Phases

The process of problem solving is generally broken down into the five phases shown in Figure 9.2. (This concept was first put forth by Herbert Simon in 1960.) The first three phases constitute decision making. The last two take place after a decision has been made

The first phase is *intelligence*. This word is used in the sense of military intelligence: collecting information without necessarily knowing how it will be used. This stage consists of whatever leads to the realization that a decision must be made, and a clear decision statement.

Example: It's 5 p.m. You're hungry. You realize that you must decide what to do about dinner.

Most experts recommend choosing evaluation criteria here, before you know the options. That reduces the likelihood of biasing the criteria, perhaps subconsciously, to favor a pet option.

Example: You decide that the important factors are cost and time. You then decide that cost will account for 70% of the decision; time, 30%.

In the second phase, *design*, alternative choices are formulated.

Example: You can cook dinner, bring in cooked food, drop in on a friend whom you know will invite you to stay for dinner, or eat at a restaurant.

FIGURE 9.2 Diagram of problem-solving phases.

The third phase is *choice*. In this phase, the decision maker or makers pick one of the alternatives that were developed in the previous phase.

Example: You score the four alternatives on the basis of the criteria that you established earlier, and choose cooking dinner as being both fast and inexpensive. Visiting a friend would be a little cheaper but a lot more time-consuming. Since you chose the criteria earlier, the attractions of seeing a friend or of having someone else do the cooking don't influence your decision.

The three decision phases are not as separate as this description implies. As Simon writes:

> *The cycle of phases is ... far more complex than the sequence suggests. Each phase in making a particular decision is itself a complex decision making process. The decision phase, for example, may call for new intelligence activities; problems at any given level generate subproblems that in turn have their intelligence, design and choice phases, and so on. There are wheels within wheels.*

Decision making is an iterative process. We make a decision and try to put it into effect. We may encounter an unanticipated difficulty: a shortage of parts for our product design, a price tag of $60,000 for that "cute little sports car," or no discount fares on Flight 392 to Chicago. We return to an earlier phase and consider more alternatives in the light of new information. We may even revise our criteria if we realize, having looked at choices in more detail, that we left some out.

The last two phases of problem solving, *implementation* and *monitoring*, are not part of decision making. Implementation must follow any decision if that decision is to matter. Monitoring will, sooner or later, lead to realizing that a new decision is necessary: sooner if the decision didn't work out well, later if it worked out well but is no longer suitable.

Many decisions must be made regularly even if they work out well. Cafeteria managers plan menus every week. Clothing store buyers plan purchases every season. Managers plan salary changes annually. History may guide decisions—if few diners chose baked haddock, maybe we shouldn't serve it again for a while; if most of the jeans were still on the shelf after two months, maybe our customers don't buy jeans—but doesn't affect the need to make them.

> ***Where you fit in:*** Knowing the decision-making phase helps choose tools to help at that phase. During the intelligence phase, for example, information systems that give a broad view of what's going on in your business may help recognize the need for a decision. In the choice phase, you may need information systems that help meld group members' views into an acceptable solution.

Decision Structure

Think over decisions you've made. Some were easier than others. The decision to take FIN 301 at 11 a.m. Tuesday–Thursday was easy: It's a required course, and that was the only open section that fit your schedule. Others, such as where to go to college or whether to go at all, are not as simple.

This difference is *decision structure*. Structure refers to whether a computer could, conceptually, be programmed to carry it out. This concept is applied to each decision phase separately. If all three are structured, a decision is said to be *structured*. If none is structured, the decision is *unstructured*. If one or two phases are structured, it is *semistructured*.

Many fully structured decisions are routinely carried out by computers. Consider, for example, the decision: "How much should we withhold from this paycheck for income taxes?"

- The intelligence phase is structured. This decision is required for every paycheck.
- The design phase is structured. Alternatives include any amount of money from zero up to the amount of the check. The answer cannot be "November 6, "BMW 340i," or "Dallas."

- The choice phase is structured. In the U.S., the Internal Revenue Service publishes rules for tax withholding. Other countries have similar procedures.

Structured decisions are often reviewed. Suppose a furniture factory has an order for 50 tables. At four legs per table, they will use 200 braces. Inventory will drop below the reorder point, so the computer prepares a purchase order. However, the firm will soon stop making these tables, so it won't need such braces in the future. The purchasing agent knows that, so he cancels it.

Less structured decisions require human judgment. Information systems can augment judgment by providing information to help make a decision, assessing likely consequences of each choice, and coordinating the activities of people involved in making it.

Decision Scope

Decisions can also be categorized by their *scope*: how much of an organization they affect, for how long they affect it, and how much they constrain other decisions. From the top down:

Strategic Decisions

An organization's *strategy* is its overall approach to its business. Choosing and implementing a strategy require decisions. Strategic decisions commit an entire organization, or a large part of it, for a long time into the future. Such decisions are made by an organization's top management.

Examples: A Korean automobile firm decides to build a factory in the U.S. A university decides to change from a commuter school to a 50:50 mix of commuters and residential students.

Tactical (Managerial Control) Decisions

Once an organization's top management chooses a strategy, others make decisions that move the strategy forward within their areas of responsibility. Such decisions commit part of the organization for a while, but can be changed occasionally without changing the overall strategy. These decisions are usually made by managers.

Examples: The Korean firm's production department figures out how large the factory must be and how it should be equipped, its real estate department looks for suitable locations, its finance department plans how to raise capital for its construction, and its public relations department prepares a press release about the plan. The university decides how many dorms to build, where to build them, how to pay for their construction, and how to attract residential students.

Operational Decisions

Operational decisions are made on a regular basis during normal operations. They affect a small part of the company, do not constrain any other decisions, and have little long-term impact. They are made by lower-level managers or by non-management employees.

Examples: The car firm orders 10,000 tires from Bridgestone in the right sizes for its production plan. The university decides to equip its first dorm with water-saving toilets and keycard access.

Relationship among Decisions of Different Scope

Each type of decision is made within the framework of higher-level decisions. For example, suppose a company decides (1) that its human resource strategy will include creating the best possible work environment. That is a *strategic decision*. Lower-level decisions follow:

2. We will have an employee cafeteria.
3. There will be three entrées every day: one meat, one fish or poultry, one vegetarian.
4. Next Tuesday the entrées will be sirloin tips, baked haddock, and eggplant lasagna.

Decisions 2 and 3 are in the managerial control range, with Decision 2 nearer the strategic level. (People typically talk about three categories of decision scope, but some situations involve more or fewer levels.) Decision 4 is operational. It has no impact beyond next Tuesday, unless the cafeteria manager thinks leftover sirloin tips would make a good beef stew on Wednesday, and does not constrain any other decisions.

Putting the Concepts Together

With decision structure and decision scope having three levels each, there are nine possible combinations. They are shown in Figure 9.3.

Most decisions lie near the diagonal from the upper left of the figure to its lower right, as the oval indicates. Strategic decisions tend to have less structure. Operational decisions tend to have more. However, there are many exceptions. That is a general tendency, not a rule.

> ***Where you fit in:*** If you know something about a decision, you're in a better position to know what types of information and information systems will be helpful in making it. That can save you and your employer time and money.

9.2 MODEL-DRIVEN DECISIONS

When a decision has structure, an information system can use that structure to predict the outcome of various choices. You may have done this with a spreadsheet program such as Excel in an accounting or finance course: What will be the effect, year by year, of different depreciation schedules for an asset? For a given cost of capital, which schedule has the best net present value?

When you did this you created a *model*. The numbers on your spreadsheet behave, under certain assumptions, as the real world will. Part of this model is accurate: Tax laws specify depreciation rules. However, in comparing the worth of a dollar saved today with the worth of more dollars in the future, it uses a discount rate (as you probably learned in finance courses). A model developer must assume discount rates for several years into the future. These assumptions may be wrong.

In other cases, we can't be certain how a system behaves. Market research data may indicate that a 10% price cut will increase sales by 20%, but market research can be wrong. Sales will probably go up, but perhaps by more or less, and a competitive announcement could affect the outcome. This model is based on our best understanding.

It is often a good idea to try out several values of critical variables. In the spreadsheet model of depreciation, we could try discount rates higher and lower than our best estimates. In the market research example, we might try several possible values for the sales increase, perhaps ranging from 10% to 30% by steps of 2%, to estimate the effect of our assumptions on profits.

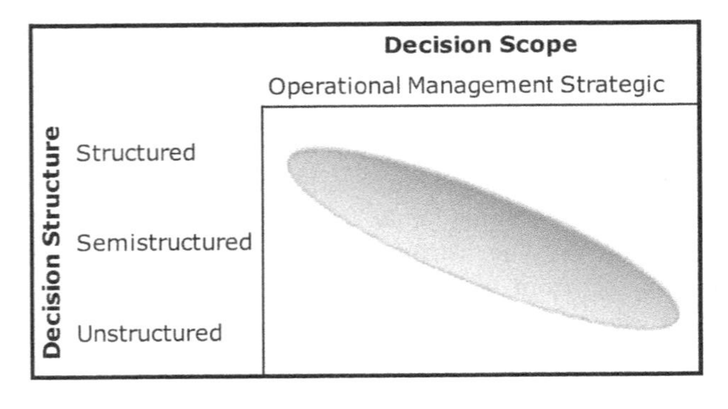

FIGURE 9.3 Decision-making grid.

We can make a model-driven decision when a decision has enough structure that we can infer something about the likely outcome of each choice. Consider the design of a road intersection: number of lanes in each direction, their use (left, right, straight, multiple uses), and the signal cycles. We can get information about traffic at that intersection by counting cars. Driver behavior is well understood statistically: for example, how soon the first driver enters an intersection after the light turns green. Knowing this one can model different designs and optimize any desired measure, such as minimum land use for an acceptable waiting time.

Where you fit in: You can create a simple forecasting model in a spreadsheet. Look for chances to use this knowledge on the job. It will improve the quality of your work and bring you to the attention of your management as someone who understands using technology for business results.

If a modeling situation is too complex for you, being able to explain the business side of the problem to technical professionals will improve the odds of getting useful results.

Simulation Studies

The behavior of a complex system cannot be predicted by a mathematical formula that one might put into a spreadsheet, even allowing for statistical variability in the input data.

Consider, for example, the problem of designing a production line within constraints of available floor space and funds. This involves many decisions, such as:

- Should we have two stations for production step A and one for step B, or one for step A and two for step B? There isn't room for two stations for each step.
- Should we get faster conveyor belts or faster welding equipment? We can't afford both.
- How much room should we leave for partially completed products between steps? If there's too little, work will stall until space opens up. If there's too much, it will waste space that could be put to more productive uses.
- Should we get enough milling machines to leave one set up for each product, or save cost and space by having fewer machines and taking time to change the setup between runs?

And hundreds more. The only practical way to answer questions like this is *simulation.*

Simulation involves following a system through time to observe what happens to it. We do this when we need to evaluate the impact of decisions we must make about a system, but we can't build all possible versions of the system in real life to see how they work. Instead, we "build" them in a computer and observe the computer's output to see what the system would do.

Simulation Example

A barber shop owner is thinking of hiring a third barber. Customers arrive, on the average, 15 minutes apart with a known statistical distribution. Each haircut takes an average of 30 minutes, with a range of 15 to 45 minutes. In the long run, two barbers keep up with demand, but variations in arrival and haircut times force some customers to wait. If a potential customer arrives and sees others waiting, he may get a haircut somewhere else (Figure 9.4).

The simulation study begins by waiting for the first customer. A computer does this by choosing a random number from the arrival time distribution and noting "Customer in seat A at 9:05 a.m." It then chooses a random number from the haircut time distribution and calculates that his will take 27 minutes, so it will be over at 9:32 a.m. It also calculates another arrival time from the arrival time distribution, noting that the second customer—who will be in seat B—arrives at 9:21. From 9:21 to 9:32, both seats are occupied. If a random choice from the arrival distribution brings a third customer before 9:32, he will have to wait.

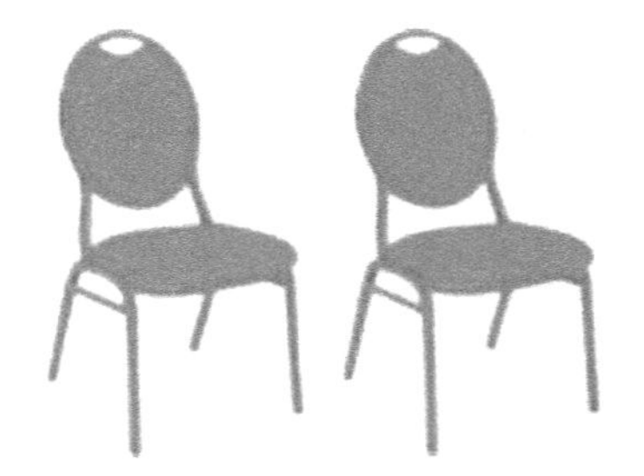
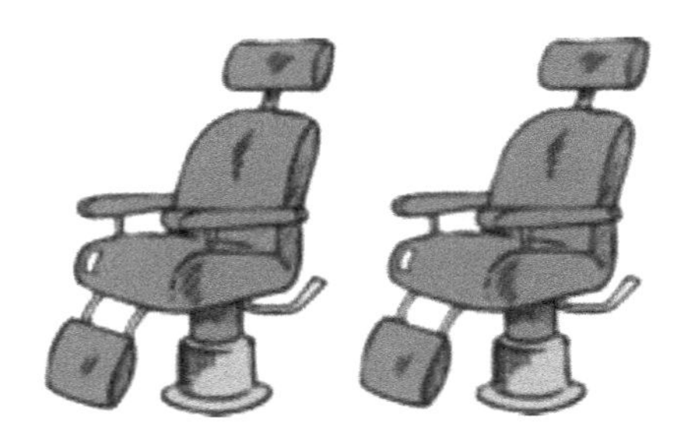

FIGURE 9.4 Illustration of barber shop for simulation example.

This process continues through the day. As it models each step, the computer tracks how many customers are served, how long customers have to wait, how many customers leave because they don't want to wait (based on the barber's opinions about customer behavior or observing a few days' traffic), and how much money the barbershop takes in. After running this simulation several times with different random number sequences, the shop owner has a picture of his current situation. If the output doesn't match reality, he can calibrate the model—perhaps adjusting his assumptions about potential customers leaving if they see others waiting—until it does.

The next step is to add a third barber to the model and watch what happens. There will be less waiting, hence fewer lost customers, hence more income. But will this increased income justify the cost of a third barber? That's what the owner hopes to learn. Using simulation is cheaper and faster than putting in a third chair, hiring that barber, and letting him or her go if things don't work out.

The Simulation Process

Developing a simulation starts by determining the *state variables* that specify what the system is doing at a given moment. Here, that's defined by how many customers are in the shop. If there are two or fewer, nobody's waiting. If there are three, one is waiting. The owner may observe that there are never more than three because customers don't stay if one is waiting when they arrive, even though there are two seats in the waiting area.

More complicated systems have many state variables. Collectively, they define the *system state.* The system state changes as the model runs, reflecting what would happen in the real world.

The modeler next figures out what events can change the system state. There are two such events in our barbershop: a customer arrives, a haircut ends. Arrivals increase the number of customers in the shop by 1. (If it's already 3, it doesn't change.) Finishing a haircut reduces it by 1.

The modeler then defines the timing statistics of those events. Many activities have been studied: Supermarket checkout takes, on the average, 41 seconds plus 3 seconds per item. The first driver at a traffic light takes, on the average, 2.2 seconds to move forward after the light turns green. In other cases, one can often measure them.

Putting this information into a model requires some skill with computers, but less than you might expect. Many simulation packages are usable by businesspeople, though complex studies require professional help. Simulation should be part of a business analyst's "bag of tricks."

> ***Where you fit in:*** If you encounter a business situation that you'd like to model, simulation is probably not beyond your ability. Even if your preferences or other responsibilities mean you won't develop this model yourself, knowing that it is possible can help you suggest it to your management. And understanding how simulation works will help you describe the business situation to a simulation professional who will model it.

Artificial Intelligence

Artificial Intelligence (AI) was defined informally in Chapter 4 as "making computers do things that, if we saw a person do them, we'd say the person was using intelligence." Non-trivial business decisions fall under that umbrella.

Artificial intelligence applications can be placed on a three-point scale: assisted, augmented, and autonomous intelligence.

- *Assisted intelligence* refers to carrying out tasks such as robotic spray-painting of vehicles in an assembly line. Its objective is usually to improve operational efficiency. These tasks could be programmed without AI, but using AI methods is often the best way to develop them.
- *Augmented intelligence* involves give-and-take with people. Decision makers benefit from the insight of AI. The AI system, in turn, learns from its interactions with people. Its output is recommendations to decision makers.
- *Autonomous intelligence* allows a system to make decisions on its own, as with self-driving cars. People are often wary of the implications of autonomous systems "going rogue" (in the movie *2001: A Space Odyssey*, HAL refuses to open a door for Dave) or making a bad choice (following a map despite hand-painted signs telling drivers that a bridge was washed out).

Here, we are primarily concerned with augmented intelligence. It fits Figure 9.1 nicely.

Several different approaches to augmented intelligence have been used over the years. Early AI relied primarily on the *expert system* approach, in which rules derived from decision making of human experts are coded into an AI system by *knowledge engineers*. Non-experts can learn from what such systems do, since expert systems can explain a conclusion by showing which rules led to it. Going the other way, inability to solve a problem or an incorrect conclusion can help expand the rule set so the system will deal with that situation correctly later.

One success story with expert systems is the Authorizer's Assistant of American Express. It improved both the speed and quality of decisions to grant one-time credit increases for major purchases. (Quality is a combination of fraction of requests approved, which should be high, and fraction of approved requests that are not paid as agreed, which should be low.)

More recently, the trend is to use *machine learning* rather than the knowledge of human experts to create artificially intelligent systems. This is possible because of the vastly increased power and storage capacity of today's computers compared with those of, say, the year 2000. That power and storage capacity can be used to handle the vast number of training examples that practical AI applications require. For example, a self-driving car must be able to not only recognize a clear Stop sign, but also to recognize one that is partially obscured by snow.

Machine learning is based on the concept of a *neural network:* a series of simulated "neurons" that behave similarly to the neurons of a biological system. In such a system, the output of each neuron goes to one or more neurons in the next layer. (The first layer gets input from sensors or other data sources.) Those neurons have formulas that define how their output should respond to different combinations of inputs. Simple neural networks, such as the one in Figure 9.5, have just a few layers and are limited in what they can do. Deep learning systems, though, may have twenty layers or more. Training can adjust the connections among neurons and the requirements for a neuron to 'fire." After training they can understand a great deal about a business situation, or behave as if they do. (We leave the question "what is true understanding?" to philosophers.) They can then produce the appropriate output, which might be "Slam on the brakes," "Sales will likely increase 5% with that promotion," or "Buy back stock if its price drops below $70 a share."

The biggest limitation of deep learning networks in practice is having enough training data. For example, they are used in language translation, where they can be trained on existing translations of

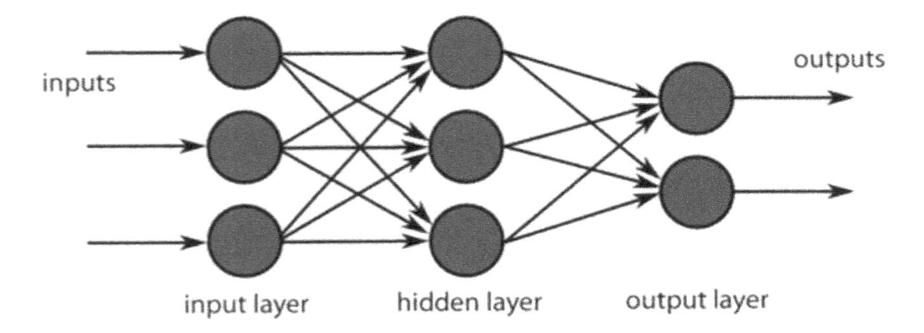

FIGURE 9.5 Simple neural network.

books and other documents. (Translation system errors keep humor bloggers busy. However, so do the errors of word processors' autocorrect features, which deal with a far easier problem.) Google Translate has an "Improve This Translation" button for users to provide corrections, and displays a shield with a check mark next to translations that have been confirmed as correct.

9.3 DATA-DRIVEN DECISIONS

We often have data even when a decision doesn't have enough structure to build a model and there doesn't seem to be a useful way to analyze it. In that case, an information system can help by organizing and presenting the data even if it can't predict what a particular choice will lead to.

You may hear the term *business intelligence* (BI) for interpreting business data in order to make decisions. That's another term for what we're talking about here. BI can be used for other purposes, but researchers such as Howard Dresner have found that its top objective is making better decisions. (Revenue growth and improved operational efficiency came next.)

Data-driven decisions are often made with the help of *data warehouses*.* You read in Chapter 5 that a data warehouse is "a historical database used for decision making." Information for a data warehouse comes from the organization's transaction processing databases, often augmented by other internal and external sources. It goes through the ETL process shown in Figure 9.6.

Since data warehouses store an organization's historical data over time, they get large. Several terabytes is small for a data warehouse. One of eBay's data warehouses is 40 petabytes in size: 40,000 terabytes, about ten thousand times larger than the biggest disk drives in personal computers today. That's big enough to store 10 billion photos. In 2016, its total data warehouse size was reported to be "over 200 PB"; it is surely larger today. One of its tables has over a trillion rows. And eBay's is far from the largest data warehouse around: Facebook's was over 300 PB a few years ago and is also surely larger today. Organizations such as the National Security Agency in the U.S. probably have larger ones, but they don't talk about them.

Business intelligence isn't beyond the reach of small organizations, though. One study found that organizations with 1–100 employees reported more success with BI than larger ones. Its authors attribute this to greater management "ownership" of the BI initiative, and the ease and speed of deploying BI in a smaller company. They found that large companies use data warehouses more than small ones do, but BI can be used without one. Small firms may not have data warehouses, but they can still interpret and use business data!

Data warehouses are often organized by the dimensional model of Chapter 5 because it facilitates the "slicing and dicing" analyses for which they are often used. This is not mandatory. The data warehouse concept is independent of how content is organized.

* Not to be confused with warehouses as places to store goods and materials.

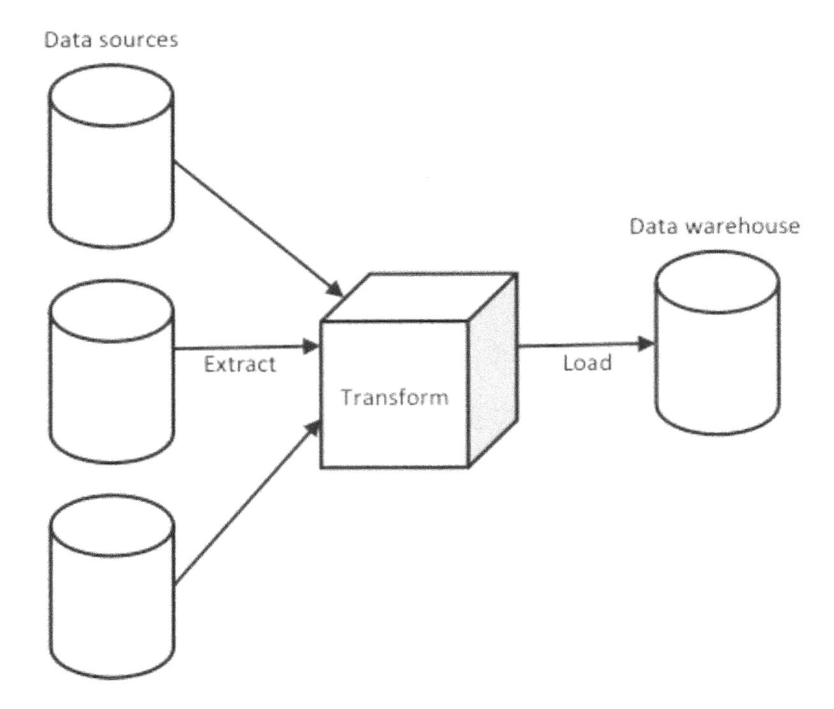

FIGURE 9.6 Extract-transform-load diagram.

The trend in 2019 is to put data warehouses in the cloud. All advantages of cloud storage apply to data warehouses. Its biggest disadvantage, delay in moving lots of data between cloud and user, is eliminated if cloud applications perform queries and return only results, since an added second or two to transmit results isn't noticed if analysis takes several seconds or more. Results may even reach users faster if the cloud supplier has high-performance computers that are optimized for data analysis, and which it can justify by spreading their cost over many customers.

A *data mart* is a small data warehouse focused on a specific part of the organization. A data mart can be easier to set up and easier to use than a full corporate data warehouse. Ease of setup and ease of use correspond to the two ways data marts come into being:

- ***Easier to set up:*** An organization with no experience in data warehousing creates a data mart for one department or division. If all goes well, they grow it into a full data warehouse.
- ***Easier to use:*** An organization extracts part of its data warehouse for one department.

Another term you may encounter is *data lake*. While a data warehouse is structured, just not structured the way an operational database is, a data lake is not structured at all. It is a repository for whatever data comes in, in whatever format it has. When data in a data lake is needed for a particular purpose, it is transformed to the structure that the purpose calls for. That purpose might be loading data into a data warehouse for the kinds of analysis that a data warehouse supports, or it might be something else. Conversely, a data warehouse—or a smaller data mart—might be thought of as part of a data lake. Thus, the concepts are complementary.

The concept of *big data*, which you read about in Chapter 5, is related to data warehouses and data lakes. Data warehouses are more structured than typical big data. After all, one of the three Vs that make for big data is *variety*. Data warehouses are usually used for one or more specific types of analysis, while big data tends to be used for information gathering (the intelligence stage of decision making) and getting the "big picture."

A *data warehouse system* includes tools through which users access the data it contains. These tools include online analytical processing, data mining, and predictive analytics.

OLAP

Online Analytical Processing (OLAP) is a data analysis method that relies on a human user to ask increasingly informed questions until analysis is complete.

Consider a business planner deciding how many vacuum cleaners her firm should produce in the next year and how it should allocate that production among models and colors. She has a data warehouse with historical information on vacuum cleaner sales, breaking them down by model and color as well as other factors such as sales region. She might proceed in these steps:

1. She realizes that her first decision is how many to make in each major category: upright and canister cleaners. She asks for the trend in sales of these types over the past three years. Her computer displays Figure 9.7.

   The figures for the past two years and for this year through October summarize thousands of transactions. Each transaction is stored in the data warehouse, so they can be analyzed in other ways as well. This year's November and December figures include firm orders and forecasted sales, with a higher fraction of firm orders for November than for December.
2. She extrapolates the curves, keeping in mind that the approximate nature of any planning process does not justify several significant digits. She chooses 1,500,000 canister cleaners and 800,000 uprights as a starting point.
3. Her second decision is the fraction of canister cleaners that will be equipped with power brushes for better rug and carpet cleaning. (All but the two least expensive canister models have power brushes.) She asks for a breakdown of canister sales with and without power brushes, and gets Figure 9.8.

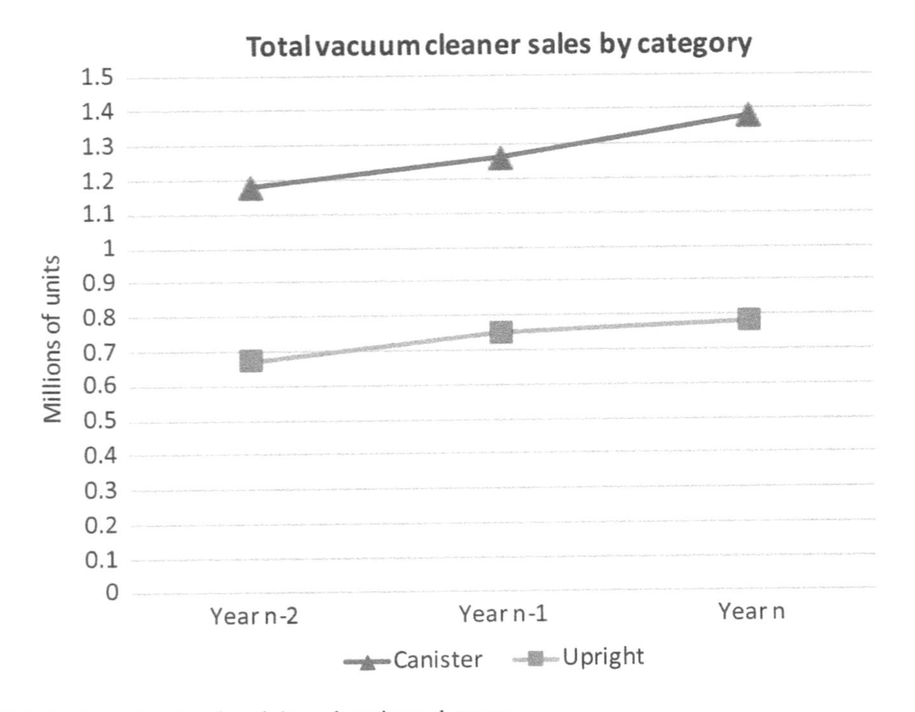

FIGURE 9.7 Sales trends of upright and canister cleaners.

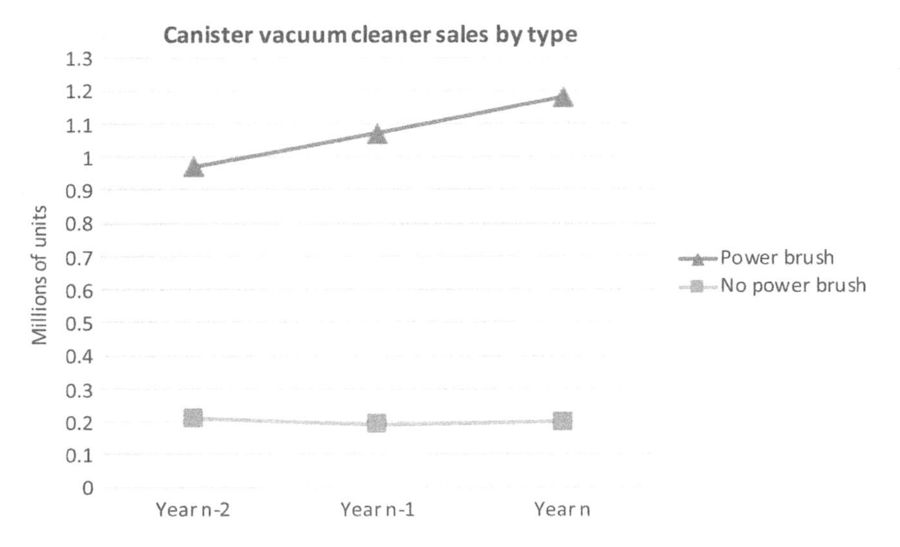

FIGURE 9.8 Sales trends of canister cleaners with and without power brushes.

4 She realizes that sales of canister cleaners without power brushes are steady while those with power brushes are increasing—perhaps related to the flattening in upright cleaner sales. (People buy upright cleaners primarily to clean carpets. Canisters with power brushes also clean carpets well.) She'll hold next year's production of canisters without power brushes at this year's level, 200,000 units. The balance, 1.3 million, will have power brushes.

5 She now switches to uprights to plan the breakdown by model. The firm's upright cleaner models haven't changed much in three years, so a trend analysis should be meaningful. It is shown in Figure 9.9.

6 There is a slight trend toward more expensive models, which she will put into her plan for next year. That creates a breakdown of 200,000, 250,000, and 350,000 from bottom to top.

However, the planner knows the products will change. The most powerful motor, only in the Model 300 today, will go into the Model 200 to respond to competitive moves.

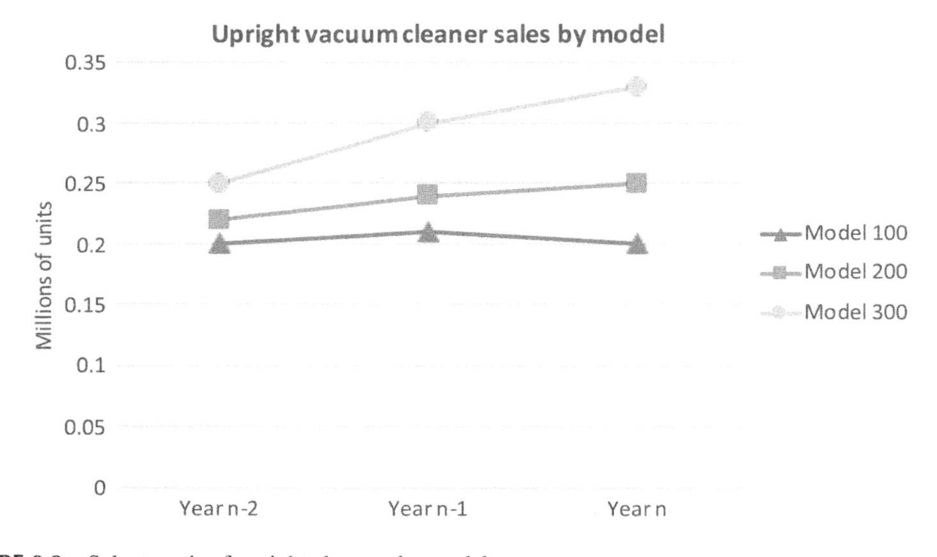

FIGURE 9.9 Sales trends of upright cleaners by model.

Model 300 will remain the only self-propelled cleaner. Some purchasers who chose it for its large motor, but don't care about self-propulsion, will now buy Model 200. Lacking relevant historical data, she decides to move 10% of the Model 300 forecast to the 200. Her final figures are 200,000, 285,000, and 315,000.

7 Since she is working with the upright cleaners, she decides to select their color breakdown next. She obtains sales data for upright cleaners in the three available colors, maroon, gray and dark green, as shown in Figure 9.10.

8 On an educated hunch, she now asks the computer to show color preferences by model for last year. This is where the capability of OLAP tools to group and correlate data shines.

This format is called a *pivot table*. A spreadsheet program created the pivot table of Figure 9.11 from the data in Figure 9.12. It's easier with full-fledged data analysis software, and spreadsheet programs can't handle millions of rows of data, but the concept is the same.

9 She sees a relationship between color trends and models. Sales are largest, every year, along the diagonals from lower left to upper right. Model 300 buyers prefer green. Model 200 sells best in gray, and Model 100 in maroon. To help her see if this is a fluke or a real pattern, she asks for a graph of the data in the pivot table. Figure 9.13 shows this graph for last year.

Similar graphs for the other two years, as well as comparable graphs for canister vacuums, confirm this aspect of customer preferences for color by price range. While the charts don't look exactly the same, the pattern is similar. More expensive models sell better in green, less expensive cleaners in maroon. Gray is preferred in the middle. She therefore decides to plan upright vacuum cleaner production as follows:

Model	Dark Green	Gray	Maroon	Total
Model 100	50,000	50,000	100,000	200,000
Model 200	75,000	140,000	70,000	285,000
Model 300	155,000	85,000	75,000	315,000
Total	280,000	275,000	245,000	800,000

10 She continues the process for the five models of canister cleaners: three with power brushes, two without. While the plan will be fine-tuned as the year progresses, a plan that is close to

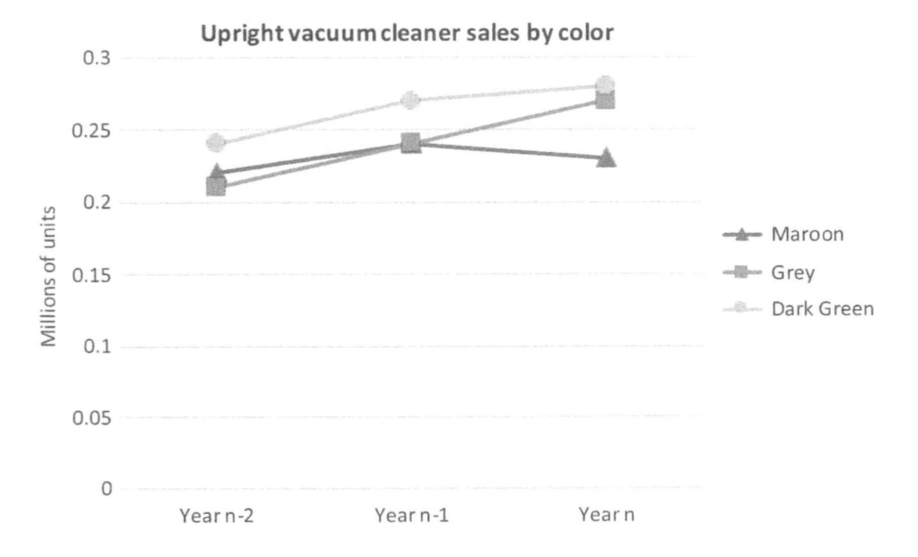

FIGURE 9.10 Sales trends of upright cleaners by color.

Sum of Sales (Mil.)	Dark Green	Grey	Maroon	Grand Total
Year n	**0.28**	**0.27**	**0.23**	**0.78**
Model 100	0.05	0.06	0.09	0.2
Model 200	0.07	0.12	0.06	0.25
Model 300	0.16	0.09	0.08	0.33
Year n-1	**0.27**	**0.24**	**0.24**	**0.75**
Model 100	0.05	0.06	0.1	0.21
Model 200	0.07	0.1	0.07	0.24
Model 300	0.15	0.08	0.07	0.3
Year n-2	**0.24**	**0.21**	**0.22**	**0.67**
Model 100	0.05	0.05	0.1	0.2
Model 200	0.06	0.1	0.06	0.22
Model 300	0.13	0.06	0.06	0.25
Grand Total	**0.79**	**0.72**	**0.69**	**2.2**

FIGURE 9.11 Sales trend pivot table of upright cleaners by model and color.

what is eventually sold can save a great deal of money. A process that can create such plans enables a firm to save via volume purchasing and longer, more efficient production runs.

Later, distribution planners will use the data warehouse to analyze shipments from each regional distribution center. This will enable the firm to match each center's inventory to demand in its region. If customers in the western part of the country like maroon vacuum cleaners, based on orders coming into the western warehouse from retailers, it will receive more than a proportional share of that color. This, in turn, will reduce the frequency of stock-outs and back orders, thus reducing costs and increasing customer satisfaction.

An important aspect of OLAP is the ability to *drill down* in data for increasing detail. Suppose the sales vice president of a U.S. computer store chain notices that sales in the Midwest lag the other regions. Absent drill-down capability, this VP can only ask the regional director "How come?" With drill down, the VP can look at districts in the Midwest region and learn that the problem is centered on Illinois. Another step pinpoints the Chicago store. A look at Chicago sales by product category narrows the issue to systems, where a deeper dive shows that only high-end systems didn't sell well. Armed with this knowledge, the VP can place an informed call to Chicago and learn that a competitor is test-marketing a new high-end system prior to national rollout. The chain can now develop a response to the competitive product before it is public knowledge.

OLAP systems can set limits to prevent managers from drilling too many levels down, to prevent executives from contacting people deep in the organization directly. Instead, an executive must contact someone at an intermediate level. That person, a manager in his or her own right, doesn't find an email from the president frightening. The person at the end of the line doesn't find an email from the manager frightening, either, though he or she would have been intimidated by receiving an email directly from the company president. Such restrictions can help people accept a new system.

OLAP systems also present an opportunity for the use of augmented intelligence. The vacuum cleaner planner just above went through this planning process on her own. Each result led her to the next question. But what if the database could have said: "There seems to be a relationship between region and color preferences?" Granted, she figured this out herself, but someone less experienced might have missed it. Most of us could benefit from a bit of help once in a while.

Year	Model	Color	Sales (Mil.)
Year n-2	Model 100	Maroon	0.10
Year n-2	Model 100	Grey	0.05
Year n-2	Model 100	Dark Green	0.05
Year n-2	Model 200	Maroon	0.06
Year n-2	Model 200	Grey	0.10
Year n-2	Model 200	Dark Green	0.06
Year n-2	Model 300	Maroon	0.06
Year n-2	Model 300	Grey	0.06
Year n-2	Model 300	Dark Green	0.13
Year n-1	Model 100	Maroon	0.10
Year n-1	Model 100	Grey	0.06
Year n-1	Model 100	Dark Green	0.05
Year n-1	Model 200	Maroon	0.07
Year n-1	Model 200	Grey	0.10
Year n-1	Model 200	Dark Green	0.07
Year n-1	Model 300	Maroon	0.07
Year n-1	Model 300	Grey	0.08
Year n-1	Model 300	Dark Green	0.15
Year n	Model 100	Maroon	0.09
Year n	Model 100	Grey	0.06
Year n	Model 100	Dark Green	0.05
Year n	Model 200	Maroon	0.06
Year n	Model 200	Grey	0.12
Year n	Model 200	Dark Green	0.07
Year n	Model 300	Maroon	0.08
Year n	Model 300	Grey	0.09
Year n	Model 300	Dark Green	0.16

FIGURE 9.12 Raw data for Figure 9.11.

Where you fit in: There's a good chance that the company you work for will have something like a data warehouse. You'll probably be able to access it via OLAP tools. By being aware of that, you won't miss opportunities to use OLAP. That will benefit your work directly. It will also show your managers that you appreciate the value of technology, boosting your career.

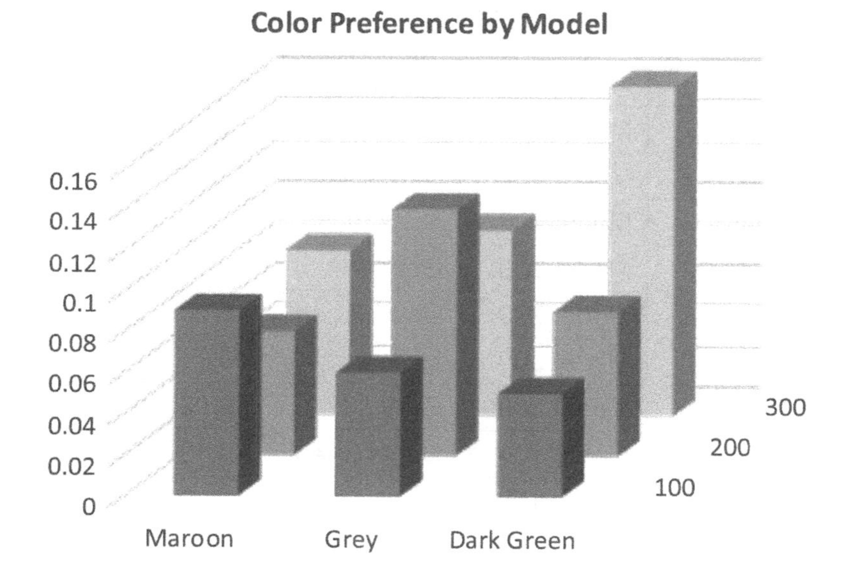

FIGURE 9.13 Sales graph of upright cleaners by model and color.

Data Mining

Data mining, in effect, points an application program at a database and says "Tell me what you find there." A request is usually more specific, such as "Find factors associated with released convicts who are arrested again." The software mines the data to find the hidden nuggets. Here, if it can identify characteristics of those who tend to be re-arrested, public safety agencies can focus on them (within legal and ethical bounds) rather than trying to watch every released convict. Such characteristics may not be obvious or intuitive.

In one application of data mining, investment firm Morgan Stanley is trying to lighten the reading load of its money managers by using computers to help read prospectuses for municipal bonds backed by specific development projects. (Municipalities hardly ever default on bonds backed by general revenues, but a development project can fail.) These prospectuses, typically 120,000 words long, must be studied to decide if an investment firm should buy one bond or another. Morgan Stanley strategists used textual analysis principles to find patterns that predicted a higher likelihood of eventual default. They found, for example, that projects with longer prospectuses are less likely to fail, while projects that use the title "Mr." more often in executive biographies are more likely to.

Figure 9.14 shows the results of a data mining classification study that separated the subjects into three groups based on a variety of characteristics. When the groups were plotted for having a fourth characteristic, which could not have been inferred from the data originally used, they came out in three distinct—though not perfectly separated—groups.

Target used data mining to identify customers likely to be pregnant based on purchase history, to send them focused offers—and take advantage of people's greater readiness to change shopping patterns when their lives change. (This was mentioned in Chapter 8.) They knew new parents are bombarded with offers and wanted to get a jump on competitors by reaching out before the birth. They identified a group of products, such as unscented lotion and calcium supplements, that as a group tend to be purchased by pregnant women though they show no such tendency individually. Target could even estimate a baby's due date, enabling them to time promotions to the week.

Credit card firms use data mining to identify usage patterns that often precede fraud. Universities use it to identify people who are most likely to complete a program. One university found that the

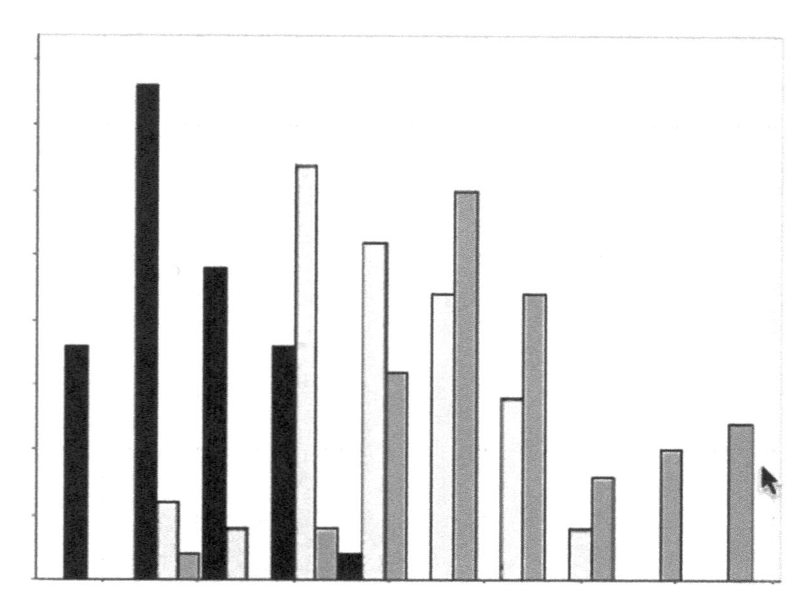

FIGURE 9.14 Sample data mining screen shot.

best predictor of finishing its part-time MBA program was not undergraduate GPA, work experience, GMAT score, or any other academic or professional factor. It was distance from the applicant's home to campus. Those with shorter trips were more likely to stay in the program.

An example will illustrate the difference between OLAP and data mining. In a story which has been associated with both Walmart and 7-11 and may not be true, a data mining market basket analysis showed high correlation between purchases of diapers and beer. The store concluded that fathers went to the store to buy diapers and, while there, picked up a six-pack. It moved the diaper display next to the beer (easier than moving a refrigerated beer case) and beer sales spiked.

Assuming the story is true, what if they had used OLAP? An analyst can list items purchased together with a specific product, sorted by frequency of appearance in the same market basket. Suppose the analyst started with hot dogs. Most purchasers of hot dogs also buy hot dog rolls; many buy mustard; quite a few buy relish, sauerkraut, and baked beans. No surprises there. The analyst would then move on to sugar or suntan lotion. Again, no insights. Finding the beer-diaper connection would require a lot of patience or a lot of luck.

Where you fit in: What patterns, if found, would be helpful in your job? You don't know yet, but once you're in a specific job, you will. Your next question should be "Can we figure out these patterns by analyzing historical data?" If the answer is "yes," you may be able to use data mining.

Predictive Analytics

Predictive analytics moves past OLAP and data mining to forecast what will happen in the future if certain actions are taken. It combines the use of historical data with business modeling. For example, historical data might show a relationship between price and sales volume, with its trend over time and how it depends on external factors. Knowing this relationship, a planner can forecast sales volume from a given price under given assumptions, and the resulting profit.

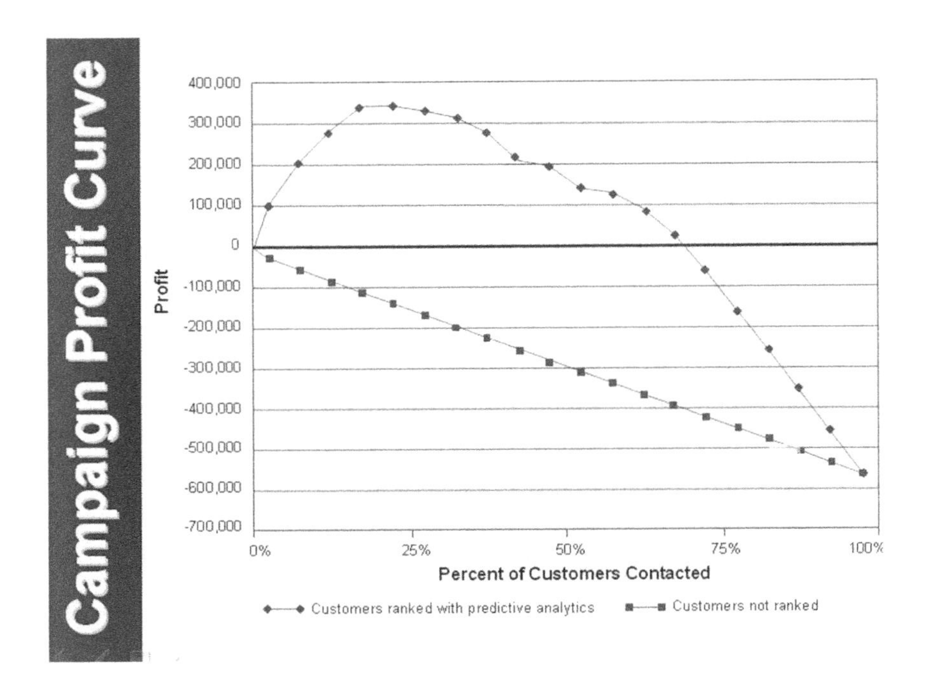

FIGURE 9.15 Predictive analytics screen shot (Source: Prediction Impact).

Figure 9.15 shows the impact of predictive analytics. Predictive analytics techniques were used to predict the probability that a customer would respond to a sales campaign. If all customers were contacted with this resource-intensive sales campaign, the company would lose $565,000. Contacting a random subset would just reduce this loss. Ranking customers by their likelihood of responding changes the picture. This campaign is profitable when up to 70% of customers are contacted. Profit is maximized by contacting the top 20–25%.

This analysis consists of several stages:

1 A connection is found between the likelihood of responding to this sales campaign (the dependent variable) and one or more known factors (the independent variables).
2 The customers are ranked by the combination of independent variables that is found to provide the best prediction of responding.
3 The probability of each group's responding to the sales campaign is used to calculate the profitability of contacting that group of customers with this campaign.
   The ratio of this probability to the overall probability of responding is called the *lift* of that group. If a group is twice as likely to respond as a randomly selected group, it has a lift of 2.
4 The company then contacts groups whose lift is above the profitability threshold.

The PODS moving and self-storage company uses predictive analytics to decide when to promote their service in each geographic area, rather than relying on the overall U.S. tendency to move in spring or fall. A case at the end of this chapter describes Dannon's use of predictive analytics.

Data mining and predictive analytics complement each other. With diapers and beer, there is unlikely to be enough data to predict the effect of moving the diapers. In other cases, once data mining has found a relationship, predictive analytics can be used. Target used predictive analytics to estimate the long-term benefit of giving discounts on products that some customers would have bought anyhow. Customer priorities in Figure 9.15 were probably determined by data mining.

PREDICTIVE ANALYTICS IN ACTION: 1

Baseball may generate more statistics than any other sport. Fans thrive on charts such as the one below. It shows standings in the American League's East division on August 28, 2018.

American League								
EAST	**W**	**L**	**PCT**	**GB**	**RS**	**RA**	**DIFF**	**POFF**
Boston	91	42	.684	–	714	507	207	99.9
NY Yankees	84	48	.636	6.5	686	531	155	98.1
Tampa Bay	70	62	.530	20.5	556	523	33	9.3
Toronto	60	72	.455	30.5	589	681	−92	0.2
Baltimore	39	94	.293	52	519	725	−206	0.0

This chart uses historical data up to its last column (POFF). POFF is the estimated probability of each team being in the American League playoffs. (Five teams reach the playoffs: three division winners and the two remaining teams with the best records.) Where does POFF come from?

To create POFF, an analyst simulates the remaining games of the season 1,000 times. In 999 of those 1,000 simulations, the Boston Red Sox won their division or were one of the two "wild card" teams. On that basis, the analyst gives them a 99.9% probability of being in the playoffs.

How does this work? The next Red Sox game was against the Miami Marlins in Boston. Since Miami had won 39.6% of their games at that time, the analyst might give Boston a 75% chance of winning on their own home field. A computer then generates a *pseudo-random number* x where $0 \le x < 1$, as a spreadsheet program can do with its RAND function. If $x < 0.75$, Boston would "win" that game in the simulation. If $x \geq 0.75$, Miami would "win." It does this for all remaining games of all teams, using odds that apply to each. At the end, it sees which teams are in the playoffs. It "plays" the season 999 more times with different pseudo-random numbers, gets 1,000 sets of playoff teams, and counts how many sets each team is in. The Red Sox were in 999 of them.

Different analysts make different estimates of Team A's chances of defeating Team B. (That's the predictive part.) Different software uses different sequences of pseudo-random numbers. If the analysts are competent and their programs work properly, their POFFs will be similar.

What is the decision here? The chart doesn't say. Those who use it may have decisions in mind. In business, analysis is usually focused on a known decision.

(The record shows that Boston won its August 29 game against Miami, 14–6, and went on to defeat the National League's Los Angeles Dodgers in the World Series.)

PREDICTIVE ANALYTICS IN ACTION: 2

United Orthopedic Group, headquartered in Carlsbad, California, designs, manufactures, and sells a variety of orthopedic products such as foot and ankle braces, shoulder, and walking boots. CIO James Clent explains that "our market is a limited one, so customer retention is key. We [wanted] to identify 'silent defectors': customers who don't say they're unhappy, but whose orders trickle down or stop. We hypothesized that we could rescue those customers if we could identify them before they left. We built a tool that, based on order patterns, predicted when a customer's next order ought to occur. If that time came and went, a salesperson would give them a call."

This pilot project alone reduced silent defections by 50%.

Clent recommends keeping it simple, "get the organization used to working with predictive analysis and build enthusiasm over time. The easily overwhelmed will get on board, advanced users will ask for more. Then you can have follow-up projects. … It becomes more than a fad; it's something that lasts."

The new term *data science* refers to the skill set used to plan and carry out predictive analytics studies. The University of California at Berkeley defines it as capturing, maintaining, processing, analyzing, and communicating data in order to uncover useful intelligence for organizations. This definition is broad enough to encompass nearly everything to do with data in decision making. That underscores the fact that "data science" has yet to converge to a single, generally accepted definition. You can expect to hear it in your career, perhaps as an option for further study, but you must find out what a person or organization using this term means by it.

Where you fit in: Opportunities to use predictive analytics arise everywhere that past data are relevant to predicting the future. (The relevance is not always obvious.) Keep your eyes open for ways to use historical data to improve your company's future results.

9.4 GROUP DECISIONS

Many business decisions are made by groups. This can improve decision quality via information sharing, diverse strengths of different group members, and reduced personal bias; increase buy-in from those who have to implement the decision; train new decision makers before they become responsible for their own decisions; and improve consistency from one decision to the next by creating an "organizational memory" of how decisions are reached.

All types of information systems can be utilized by groups, together or separately with the results then shared. However, a group can benefit from additional types of information systems. *Group support software* focuses on the process through which a group arrives at a joint decision. Other types of software are not group support software even if a group uses them.

Group support software can focus on group communication or on the decision-making process.

Group Support Software for the Decision-Making Process

A group is evaluating bids for a new corporate jet. They consider several factors, such as those listed in Figure 9.16. First, they decide how important each factor is. They then evaluate each aircraft on each factor and come up with a weighted total for it. The final evaluation sheet might look like the figure. The fastest aircraft, the Curtiss Hawk, scored low enough on other factors to be ruled out. (All else being equal, faster aircraft use more fuel, need longer runways, and are less roomy.) The Eagle and H-10 scored close enough to be considered tied. The group can then look at other factors, such as supplier reliability, to select one of those two.

Traditionally, this evaluation would be done by assembling the group in a room. First, the group would decide on the importance of each factor. This could lead to a shouting match between a member from Accounting, who thinks cost should count at least 50%, and one from the CEO's office, who thinks speed and range are most important and nothing else matters much.

Group support software can help in this process. It can ask group members to suggest percentages and compute an average. It can support the *Delphi method*, in which those whose suggestions are far from the average in either direction are asked to explain their reasoning, after which group members offer new percentages. Eventually a consensus is reached. Then, the software can ask

Aircraft Proposal Evaluation Summary Sheet							
Item	Weight %	Curtiss Hawk		Wright Eagle		Atlas H-10	
		Score	Wtd. Score	Score	Wtd. Score	Score	Wtd. Score
Cruising speed	18	10	180	8	144	9	162
Max. altitude	2	10	20	10	20	10	20
Shortest runway	10	5	50	8	80	10	100
Range	25	8	200	10	250	7	175
Purchase cost	10	10	100	9	90	9	90
Oper. cost/mile	5	8	40	9	45	10	50
Maint. per 800 hrs.	8	10	80	10	80	10	80
Max. passengers	15	7	105	8	120	10	150
Max. cargo	7	6	42	9	63	10	70
Totals	**100**		**817**		**892**		**897**

FIGURE 9.16 Aircraft proposal evaluation summary sheet.

members to rate the candidates on each of the decision factors. As before, the Delphi method can be used to reach a consensus. A decision can be reached in less time and with less shouting.

Another application of group support software in the decision-making process is in *brainstorming* to come up with ideas. In face-to-face brainstorming sessions, only one person can talk at a time, and ideas can be presented no faster than a person can record them. With group support software, each group member enters ideas into a shared database. After a member enters an idea, the system displays some other members' ideas on his or her screen to accelerate the creative process, much like hearing people's ideas in a face-to-face session. Once a sufficient number of ideas has been generated, or the flood of new ideas has slowed to a trickle, such software can help users group and merge ideas for consideration and analysis.

Group Support Software for Group Communication

Any type of communication software can help groups reach group decisions.

Some companies have installed *decision rooms* in which groups can meet, use computer tools to help them reach a decision, and have technical support present or on call. Such rooms are used less today than they once were, because the availability of remote meeting tools reduces the advantages of the whole group being in the same room or the need for special software. You may still see one in a large company, though, such as the one at Procter & Gamble in Figure 9.17.

Telepresence meeting rooms, which you read about in Chapter 6, are becoming more popular as their cost drops. Telepresence rooms aren't limited to decision making, so it is easier to justify their cost than to justify the cost of a single-purpose facility such as a decision room.

Concerns with Group Decision Making

Group decision making is not without issues. Three of them are:

- Group decisions can be slow because people have to meet, even if online; discuss issues; and resolve possible differences of opinion. Faster implementation because of buy-in may make up for some of the lost time, but doesn't always.
- Groups can fall into the trap of *groupthink*, where disagreement violates the group's social norm. This can lead to a decision that nobody likes, if no member wants to be the first to disagree.
- Groups are subject to possible *group polarization*: a tendency to converge on extreme solutions. For example, a group may make a riskier decision than any member would make individually, because group members may not feel as responsible for the decision as each would if it were his or her individual responsibility.

FIGURE 9.17 Photo of Procter & Gamble decision room.

Where you fit in: Group decision making is common in organizations. You will be in decision-making groups throughout your career. The better you know what sorts of information system support is available to such groups, the better position you will be in to get it (if your employer doesn't already have it) and to use it. This will reflect well on you.

9.5 DASHBOARDS

A *dashboard*, according to a classic definition by Stephen Few, is "a visual display of the most important information needed to achieve one or more objectives, consolidated and arranged on a single screen so the information can be monitored at a glance." This definition does an excellent job of capturing the essential features of a dashboard while avoiding being so general as to mean almost anything or getting bogged down in the specific technologies that someone finds useful.

A dashboard, as Few continues, is an approach to presentation—not a type of information or of technology. Managers in nearly every type of organization can use dashboards.

Dashboards, by presenting the same information to everyone, can align an organization around metrics that it most needs to track. In a sales organization, those might be the rate at which leads are being generated, how they progress through the sales cycle and lead to revenue, win/loss ratios versus specific competitors, and how campaigns are performing. A manufacturing division might measure production volumes, delays from various causes, defects found in final inspection, and scrap rates. Hotel managers care about room occupancy, bookings at various rate levels, restaurant sales per guest, and so on. Figure 9.18 shows a finance dashboard.

Metrics of this type are called *key performance indicators* (KPIs). A good set of KPIs is one where succeeding on them all but ensures success, while failing at them makes success impossible. For a dashboard to be effective, the organization must agree on the metrics it shows. Just creating a dashboard can, by forcing this agreement, have a positive effect. As people use the dashboard and refer to it in business discussions, it will build more momentum.

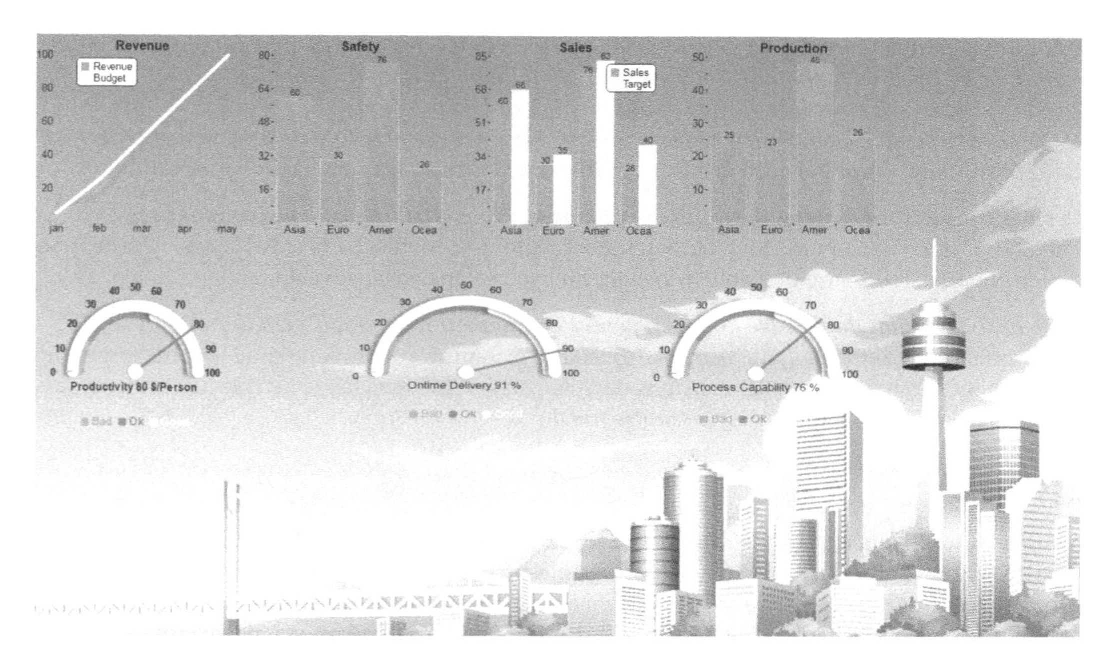

FIGURE 9.18 Screen shot of finance dashboard.

Most enterprise software (ERP, CRM, etc.) packages enable users to create dashboards using their databases. A supply chain control tower is such a dashboard, using SCM data. Where this capability is insufficient, business intelligence software usually has good dashboard capabilities. This stands to reason: BI is about presenting information, and dashboards do that. Since BI software can also read data from a variety of sources, it can provide dashboard capabilities to nearly any organization. Should you ever want a dashboard and find that the capabilities of your employer's existing software aren't sufficient, BI software will almost certainly do the trick—or specialized dashboard software, which will be less expensive if additional BI capabilities aren't needed. Homegrown dashboards are seldom necessary or cost-effective.

> ***Where you fit in:*** Dashboards are cost-effective for nearly every organization that is large enough for its head not to be able to watch all of it personally. If you work for such an organization, and it doesn't have one, look into creating one using inexpensive software to access its existing data.

KEY POINT RECAP

Making better decisions is one of the three key ways that information systems provide major strategic benefits.

This is one of the three areas in which you, as a businessperson, must look for creative ways to use information technology.

Decisions can be categorized on the basis of scope and structure. *They are usually divided into three levels of each.*

As a businessperson, you will frequently be faced with decisions. Understanding where they fit along these dimensions will help you figure out what kind of information system support might help you make good decisions. (People are rewarded more for good decisions than for bad ones.)

It is often possible to model the outcome of decisions that have a lot of structure. *Simple models can often be built with spreadsheet formulas. Simulation is another useful modeling technique. There are more, which you can learn about as you need them.*

A businessperson should know about tools like these—to be able to use them, perhaps after specialized training, when the need arises, or to find others to use them on his or her behalf.

Deep learning is increasing the ability of computers to contribute to decision making. The ability of computers to carry out intelligent-seeming activities will only increase during your career. Be prepared to use that ability to your, and your employer's, advantage.

Even decisions that don't have much structure can benefit from intelligent data analysis. Data warehouses and "big data" analysis tools are becoming common in large organizations and, increasingly, in smaller ones. As a businessperson, you will be expected to be able to use them—and to consider information quality issues when you do.

KEY TERMS

Artificial Intelligence (AI): The ability of computers to exhibit seemingly intelligent behavior.

Assisted intelligence: Using AI to enable machines to execute predetermined complex actions.

Augmented intelligence: Using AI to contribute to a decision-making process.

Autonomous intelligence: Using AI to make decisions and act on them independently.

Brainstorming: Method of generating ideas for the *design* phase of decision making by free association and recording ideas without criticizing them.

Business Intelligence (BI): The process of interpreting data to make business decisions.

Choice: Third phase of decision making, in which one alternative is chosen.

Dashboard: A visual display of key business information, arranged to be monitored at a glance.

Data mart: A small data warehouse focused on a specific business area.

Data mining: Automated process by which a data warehouse system searches for answers to specific questions by going through its content. Contrast with *online analytical processing.*

Data warehouse: A historical database intended to be used for decision making.

Data warehouse system: An information system that includes a data warehouse and tools through which users can access it.

Decision: A choice among two or more options.

Decision room: Room designed and equipped for group decision making.

Decision scope: The breadth and time frame of a decision's impact.

Decision statement: A clear, concise written expression of what must be chosen and the set of options from which it must be chosen.

Decision structure: The degree to which the three decision phases can be carried out by a fixed procedure, which can be programmed into a computer.

Decision: A choice among known options.

Deep learning: A subset of machine learning in which the neural network has many (typically more than ten) internal layers.

Delphi method: Approach to reaching a consensus by having group members state positions, explain them, and repeat until the process converges.

Design: Second phase of decision making, in which alternatives are formulated.

Drill down: In *online analytical processing*, a series of increasingly detailed queries.

Expert system: An approach to artificial intelligence in which an information system follows rules derived from the behavior of human experts.

Group support software: Information system designed to help a group reach a decision.

Implementation: Fourth phase of problem solving, in which the chosen alternative is put into effect.

Intelligence: First phase of decision making, ending in recognizing a decision to be made.

Key Performance Indicator (KPI): A business metric used to evaluate factors that are crucial to the success of an organization.

Knowledge engineer: A professional who encodes the behavior of human experts into rules that an *expert system* will follow.

Lift: The ratio between a measure of interest in a group selected by data mining to the same measure in the entire population.

Machine learning: Training *neural networks* to produce desired outputs in response to complex combinations of inputs.

Managerial control decision: Same as *tactical decision.*

Model: Representation of the real world used to predict how relevant aspects of it will behave if certain decisions are made.

Monitoring: Fifth phase of decision making, in which results of implementation are observed.

Neural network: An information system whose structure is based on the interactive behavior of biological neurons.

Online Analytical Processing (OLAP): Process by which a data warehouse system responds to queries by a human user. Contrast with *data mining.*

Operational decision: Decision that affects one activity for a short time and does not constrain future decisions. Contrast with *tactical decision, strategic decision.*

Pivot table: Data summarization table in which categories are organized along the vertical and horizontal axes, and can be moved (pivoted) from one to the other.

Predictive analytics: Use of historical data and statistical methods to predict what will happen in the future if certain decisions are made or certain conditions exist.

Pseudo-random number: Number generated by a computer from a formula, so that the set of such numbers behaves statistically as if those numbers were generated by a random process.

Semistructured: Said of a decision if one or two of its three phases are structured. Contrast with *structured, unstructured.*

Simulation: A model that follows what a real-world system would do through time, step by step.

Slicing and dicing: Analyzing database content along some, but not all, of its data dimensions.

State variable: A data item that describes what one part of a system being simulated is doing at a given moment.

Strategic decision: Decision that affects an entire organization, or major parts of it, for a long time. Contrast with *tactical decision, operational decision.*

Structured: Said of a decision if all three of its phases are structured. Contrast with *unstructured, semistructured.*

System state: The set of all *state variables.*

Tactical decision: Decision that affects part of an organization for an intermediate time. Contrast with *strategic decision, operational decision.*

Unstructured: Said of a decision if none of its three phases is structured. Contrast with *structured, semistructured.*

Review Questions

1. What are the three phases of decision making? How do they correspond to the phases of problem solving?
2. Define the three levels of decision *structure.*
3. What is decision *scope*? What are its three levels?
4. What is a *model* in decision making?
5. What is *simulation*? When is it useful?
6. Define the term "business intelligence."
7. What is a *data warehouse?* How does it differ from a database that an organization would use in the course of its daily operations?

8. How are data warehouses and data marts related?
9. Describe the basic concepts of online analytical processing (OLAP).
10. What is the role of a person in data mining?
11. What is predictive in predictive analytics? What is analytic about it?
12. What are some benefits of making decisions as a group?
13. What are some risks in making decisions as a group?
14. What sort of information should be on a dashboard?

Discussion Questions

1. A Florida luxury boat builder is thinking of opening a sales office in Dubai. Write its decision statement and describe the three phases of its decision.
2. Consider the two examples in the decision scope section (Korean car firm, commuter university). Explain the relationship between each strategic or tactical decision and the one below it. How does each higher-level decision constrain the lower-level ones?
3. Fill each cell in the grid of Figure 9.3 with a decision you make as a student. Use decisions, not topics: "financial aid" is not a decision, but "should I apply for a student loan?" is.
4. Give examples of one strategic and one operational decision that could benefit from BI.
5. How many cells are in the data cube of the vacuum cleaner planning example in Section 9.3?
6. Soon after Target began sending promotions to women whose purchases suggested they were pregnant, the father of a high-school student stormed into a Target store to complain about a mailing his daughter had received. "What are you trying to do," he roared, "encourage her to get pregnant?" The manager assured him that it would not happen again. When the manager followed up a few days later, it turned out that the daughter was pregnant, but hadn't yet told her parents. Since no algorithm is perfect, though, she might not have been. Think of another data mining application that is similarly sensitive, identify one consequence of misidentifying a person as a member of a specific group, and describe a solution that would work there.
7. Target's REDcard gives holders a 5% discount on purchases, even though most people who get it shop there anyhow. Some of this discount is offset by not paying fees to other card issuers, but those fees are less than 5%, and operating REDcard isn't free. How can this discount be justified? What does that say about the value of information?
8. Pick a local business whose name begins with the same letter as your given name. Explain how it could use (a) data mining, (b) OLAP, and (c) a dashboard. Your answers should reflect what each does, who typically uses it, what information it provides, and why that is useful. Explain any "they can't use it" answers in a way that reflects this same understanding.
9. A university found that 20% of business majors who earn an A in English 101, and C or below in Math 101, graduate in five years or less. However, 70% of English majors with those grades graduate in that time frame. (This scenario is entirely fictional.)
   a. Is this an example of predictive analytics? If so, what are the analysis and the prediction?
   b. If you aced English but did poorly in math, and plan to major in business, would you want to know this? (Consider a possible "self-fulfilling prophecy" effect.)
   c. In that same situation, would you want your advisor to know this? If he or she knows, what would you want him or her to do with the information?
10. Watch the video "Why Watson Matters—A Personal Reflection on the Significance of IBM's Watson" by Adrian Bowles at *www.youtube.com/watch?v=PaBQ0AmlfJw*. Discuss in your own words the significance of Watson for helping companies make better decisions. On the grid of Figure 9.3, what types of decisions do you think Watson would be most helpful with, what types would it be least helpful with? Give examples, not Bowles's, of two real-world decisions with which Watson could help.

11. Sketch dashboards that some or each of the following groups (as your instructor assigns) could use. Show at least six KPIs on each dashboard. Represent each KPI appropriately, using at least three different types of representations on each dashboard.
    a. The president and provost/vice-presidents/deans/etc. of a university.
    b. The owner of a professional sports team. (Bonus question: Which owner KPIs are also KPIs of the team's managers/coaches, which are not?)
    c. The chief of police and other top police officers in a large city.
    d. The group in charge of organizing the next summer Olympic Games.

KHOURY CANDY DISTRIBUTORS CONSIDERS RECOMMENDATION SW

Retailers must forecast demand. A store that doesn't forecast sales well is stuck with unsold merchandise. Unsold candy gets stale, though not as quickly as some foods. A distributor that tries to sell as much as it can to retailers will find, if it is successful in the short run, that retail outlets whose candy gets stale will resent the tactic and, later, buy less than they would have. The need for forecasts applies up the supply chain. In fact, a poor forecast gave KCD its start.

Therefore, KCD systems planner Brian Greenwood reasoned, if they could forecast demand, they'd earn customer loyalty and be able to optimize their own ordering as well.

His problem was that there isn't enough of a market for a software firm to develop an application to forecast candy sales. Still, Brian thought it might make sense for KCD. KCD's cost would be less than a software firm's, because they wouldn't have to package the software for sale, set up a support structure, and do all the other things that go into a commercial software offering.

With that background, Brian met Jake and Isabella in the now-familiar reception area and escorted them to the MIS conference room. Brian looked familiar to them from the meeting to select ERP software (see Chapter 7 episode); now they knew who he was. The three went to the MIS conference room, where the candy bowl on the table didn't have nearly the attraction it once did.

"What we'd like to do," said Brian, "is to predict a store's baseline sales in the absence of any sales or promotions. Once we have the baseline, we can factor those in. In most cases we can take the previous year's sales as a starting point, but then we have to account for changes from the economy down to local things like population and job growth near a store.

"Candy is also seasonal, of course. Over a quarter of a year's sales are for Hallowe'en, Christmas, Valentine's Day, and Easter. Those are far enough apart that, if you overbuy for one, you can't keep the inventory until the next. Some kinds of candy will keep a few months, but others won't. So you have to deal with each of those on its own.

"I was thinking that maybe that artificial intelligence stuff I read about might help here. And I was hoping that the two of you might have some ideas there."

"Actually, that's probably not a bad application for AI," Isabella said. "But one of the things that AI needs is a lot of data to learn from. You'd need to have actual sales from a lot of stores, probably broken down by the kind of candy, and data about all the other parameters that you think might figure into the forecast."

"Hmmm," mused Brian. "I hadn't thought about that part of it. How much data is 'a lot of data'?"

"There's no one-size-fits-all number," said Jake. "A lot depends on how many parameters you want to use as input. If you have a lot of variables, you need more data for the system to figure out how each matters. Some people say you need ten data points for every parameter in the model. Other people say that's not nearly enough. A third group says it may or may not be, depending on how clean your data set is and how different your data points are from each other. A thousand data points that say the same thing may not be as valuable as a dozen that point to an important relationship."

"What do you think I should do, then?" Brian asked.

"Professor Acton* said you should start in on your model with the data that you have and add more data once you see what that does for you," Isabella replied. "He said data needs become more obvious once a project has some results. He also showed us some examples of AI systems with how many data points were used to train each one. Some of the numbers were in the millions. Those systems were trying to do non-trivial things with natural languages, though. That's probably harder than forecasting candy sales."

"I sure hope so!" laughed Brian. "We don't have nearly that many data points. For that matter, I'm going to have to look at our data to see if we have any useful data points at all. We know what retailers bought from us, but we don't necessarily know what they sold. Maybe they guessed wrong."

"I hate to be the bearer of bad news," Jake said sadly, "but today's AI systems have to be trained. We're not yet where we can tell one to go off and learn about something on its own. So, if you don't have data to train one with, you can't build one that will work."

"Don't feel bad," said Brian. "If you kept me from going off on a wild goose chase, you did me a huge favor. And if the project does come off, at least some of that will be because you pointed me in a useful direction. Do you think your professor might be willing to come in for a chat?"

Questions

1. A software package that could forecast candy sales could also forecast the sales of other kinds of snacks, soft drinks, and more. That suggests a much larger market than forecasting candy sales alone. Therefore, do you feel Brian should have concluded from the lack of commercially available packages that do what he wants that it isn't feasible?
2. Brian clearly has little or no experience with AI. He was chatting about it with the students, but not as anything he expected to devote serious resources to in the near future. Under those circumstances, what would he expect to get out of meeting with Prof. Acton?
3. A different approach to artificial intelligence is an *expert system*. An expert system follows rules based on what a human expert does, to reach (hopefully) the same conclusion a human expert would. Would an expert system be useful here? (You'll probably want to read more about them before answering.)

CASE 1: PREDICTIVE ANALYTICS AND YOGURT

From 2008 to 2013, Greek yogurt went from a tiny share of the $7 billion U.S. market to 40%. It kept growing to over 50% in 2018. With change this rapid, intuition developed over decades can't keep up.

Instead of intuition, Dannon uses a predictive analytics system from IBM. Its software originated at a company named M-Factor, which was acquired by DemandTec, which was then acquired by IBM. For example, suppose they are considering a 49¢ reduction in the cost of a four-pack of yogurt. They know it will increase sales, but will it increase profits?

"More and more consumer packaged goods manufacturers use predictive analytics to ensure that the right products are in the right stores for the right shopper," says John Karolefski, executive director of Shopper Technology Institute, a specialist in technologies that engage shoppers and analyze their behavior. "That's the holy grail for manufacturers: to better forecast demand, help retailers prevent out-of-stock items and develop more effective promotions."

For years, Dannon's sales force used "a very manual, Excel spreadsheet-driven process" to forecast sales and manage inventory, says Dannon CIO Timothy Weaver. "It was a very cumbersome process and not particularly accurate." The time-consuming forecasting left sales reps with less time for selling, and the forecasts themselves were only about 70% accurate.

---

* In appreciation of the late Prof. Forman S. Acton, who opened my eyes to computers decades ago.

With IBM's software, after sales reps design a promotion, they enter factors such as pricing conditions, target customers, length of the promotion, and product SKUs into a computational model. The model crunches the figures, along with historical, regional, and market data, and outputs "the expected base sales you will always see from that pack group, the incremental sales you can expect from that particular promotion, and its financial implications such as gross revenue increase and profitability," says Weaver. And its forecasts are 98% accurate.

For this system to work, its models must be carefully calibrated and tested. That's the *analytics* part of predictive analytics. Only after that does the *predictive* part come into play: when the modeling software gets new conditions, such as this promotion, to predict their effects.

For example, sales data analysis revealed that price sensitivity varies by region. "Reducing price by 10 cents has a very different impact in the central part of the country compared to the East Coast and West Coast," says Jeremie Davis, Dannon's director of sales and marketing systems.

Forecast accuracy is more important with yogurt than with non-perishable goods. "Anything in the dairy case has a very short shelf life, so predicting volume is really important," says Alison Chaltas, vice president of shopper and retail strategy at research firm GfK. "The shorter your shelf life, the more important it is to get [forecasts] right."

Business analysis also requires examining how demand for one product can affect demand for others. When Dannon noticed an increase in sales of its Light & Fit yogurt, it explored how that affected the sales of its other Greek yogurt products and those of its competitors. They found that Light & Fit's popularity led to some switching from other Dannon brands, but it also ate into competitors' sales and thus improved overall profitability.

Analysis is not a one-time activity. "We thought that we'd have to test and validate our models twice a year," Davis says. "Because of the dynamic nature of the category of yogurt, we've found that we need to test more often, on a quarterly basis."

To calibrate its models Dannon takes small sets of data, such as sales of strawberry Oikos yogurt in April, and runs those numbers through the model to see if sales and inventory forecasts using data from preceding months match up with what occurred. That enables Dannon to adjust and improve its models. As Gerhard Pilcher, vice president and senior scientist at Elder Research, a predictive analytics consulting company in Charlottesville, Va., says, "A model is not something you just put in place and forget about."

This opening (condensed) was on *jobs.danone.com* (Danone is Dannon's French parent company) in September 2019, along with other analytics openings from intern on up.

Demand Planner: Provide timely, accurate forecasts by brand, pack group, and sales division. Ensure correct use of forecast models. Provide accurate input for forecast alignment process in terms of baseline forecast, marketing and trade events, working with Sales and Marketing.

Responsibilities *(partial list):*

- Use syndicated data and forecasting models to build customer/brand forecasts.
- Analyze and adjust forecast based on category/product trends and marketing/sales plans.
- Lead and participate in forecast process, building 18-month forecast by brand/product group.
- Develop forecasts for new products and line extensions.
- Develop and analyze short-term forecast and promotional volume utilizing POS trends, promotional plans, marketing plans, and customer inventory levels.
- Maintain scorecard data to monitor and optimize forecast performance on an ongoing basis.
- Analyze short term/long term forecasts vs. actuals by promotion, UPC, and major customer through report development, preparation, and review.

- Manage demand planning for new launches; give feedback to Demand Planning and Brand/Sales teams.
- Develop projects to improve Demand Planning organization and efficiency.

Education, experience, knowledge, skills, and abilities *(partial list):*

- Bachelor's degree.
- 4+ years of Demand Planning or other relevant work experience.
- Expertise in Demand Planning software; SAP APO (Advanced Planning and Optimization) application experience preferred
- Experience in CPG (Consumer Packaged Goods) industry; emphasis on Planning or Supply Chain functions preferred
- Expertise using syndicated data sources such as IRI, Nielsen, Retail Link
- Superior analytical skills
- Ability to communicate technical and analytical concepts
- Presentation and communication skills

Questions

1. What characteristics of the market for Greek yogurt make it suited to the use of predictive analytics? Identify one other market that is also well suited to the use of predictive analytics and one that isn't. For both, explain why you feel that way.
2. Your school wants to predict the effect of changes to its financial aid policy on decisions by accepted applicants to attend. It can't look at historical data: It has followed the same policy for 15 years, so it has no other policy to compare it to. It can, however, test new policies on small groups in the next entering class. As a member of your school's strategic planning department, write a memo to the director of admissions suggesting what he or she should do to provide data to the predictive model. Be as specific as possible.
3. Jobs such as the one Dannon posted carry a great deal of responsibility, make a difference to the success of a company, and offer good promotional opportunities. Do you think you would be good at this work? Do you think you would enjoy it?

CASE 2: DATA WAREHOUSING AT VOLVO

Every Volvo car has hundreds of microprocessors and sensors. Their data is used in diagnosis and repair. It is also captured by Volvo, integrated with the company's CRM, dealership and product data stores, and stored for analysis. This enables Volvo to spot design and construction flaws early, enable proactive correction of faults, and see how its cars respond in accidents.

Volvo sells about 600,000 cars per year in 2019. Each generates 100–150 KB of data per year. Volvo has collected this data since 1999. It now collects about 500 GB of car data per year. That figure increases each year, as older Volvos are scrapped and newer ones, with more electronics, replace them. And cars are far from Volvo's only data source.

Volvo stores all this data in a data warehouse from Teradata. It went live in July 2007, with a size of 1.7 terabytes at the time. A daily mileage calculation that previously took two hours could then be run in five minutes. A report of diagnostic codes by model and year went from two weeks to 15 minutes. Where the previous system could barely process one query per hour, the new platform completed one a minute. This enabled Volvo to extend usage from a handful of users to over 300.

What are some business results of this system?

- Volvo analysts can predict failure rates over time. Each month they look at how many cars that have reached a certain service age and how many of those have experienced specific failures. "This … tells us how many cars in a given population have experienced a particular failure, and how many are at risk," explains senior engine diagnostic engineer Mikael Krizmanic. "It … describes the failure rate over time, and that's what we use for predictive modeling. It helps us understand which faults will produce large warranty impacts if not addressed systematically."
- Cars in urban China and rural Sweden see different driving conditions and behaviors: speeds, engine loads, operating temperatures, environmental conditions, time at idle, etc. "Because we have both error codes and operational log data in the warehouse, we can understand the relationships between geography, patterns of use and mechanical failure," Krizmanic points out. "A problem may be a high priority … in one geography but not in another."

"I would say that today we have only scratched the surface. I don't think we understand yet, from a business point of view, this tool's true potential," says Åke Bengtsson, vice president of quality and customer satisfaction. "I believe that we can better use data to provide early indications. … We must be able to act quickly, to reduce the number of steps to an accurate, proactive response. Every car we produce with a fault costs the company money. And every minute, hour, and day by which we can expedite a solution saves money for the company. The earlier we can resolve an issue the better it is for the customer and the company. So I think our direction is clear."

The Teradata system has enthusiastic support from Volvo's IT side. Jonas Rönnkvist, head of enterprise architecture, helped orchestrate its acceptance in the corporate environment. "Every new development project at Volvo Cars now follows a standard process, including reviews by my team. If the requirements include data consolidation and integration, and the design doesn't leverage the Volvo Data Warehouse platform, the project will not be able to proceed without CIO approval."

Volvo's data warehouse has had an impact on decision-making processes. "Our decision making has become more fact-based," says senior business analyst Bertil Angtorp. "Now, whenever a question arises, people ask 'what is the data telling us?' Once we've verified the existence of a problem we use the data to … to prioritize and scale our response. It helps us … [focus] on the things that are most likely to affect the customer experience."

Questions

1. Does Volvo have "big data" here? Why or why not?
2. Volvo Cars has gone through several changes of ownership. It was part of AB Volvo until 1999, was owned by Ford Motor Company from then to 2010, and is now owned by Geely Automotive of China. How do you think this might have affected the development, use, and acceptance of this data warehouse?
3. Volvo Cars uses *dendrograms* to analyze clusters of alarm data, finding relationships among different signals. If you're not familiar with dendrograms (most people aren't), research them to understand what they are and what they show. How could an instructor teaching a course in Microsoft Office use dendrograms to analyze problem areas for students?

BIBLIOGRAPHY

Adunuthula, S., "Role of spark in transforming eBay's enterprise data platform," *Spark Summit*, 2016, February 18, 2016, databricks.com/session/role-of-spark-in-transforming-ebays-enterprise-data-platform, accessed September 15, 2019.

Albright, A., "Morgan Stanley robot learns by reading unreadable muni documents," *Bloomberg Business News*, September 16, 2019, www.bloomberg.com/news/articles/2019-09-16/a-robot-learns-lessons-by-reading-unreadable-muni-bond-documents, accessed September 17, 2019.

Barnett, T., "Group decision making," *Encyclopedia of Business* (2nd ed.), www.referenceforbusiness.com/management/Gr-Int/Group-Decision-Making.html, accessed August 16, 2019.

Big Data Insight Group, "Teradata case study: A car company powered by data," May 16, 2012, www.industolutions.com/caseStudy/volvo-car-company_150598, accessed September 14, 2019.

Brownlee, J., "How much training data is required for machine learning?" *Machine Learning Mastery*, July 24, 2017; machinelearningmastery.com/much-training-data-required-machine-learning, accessed September 11, 2019.

CIO Executive Council, "3 CIOs reveal how they got started with predictive analytics," *CIO*, November 18, 2013, www.cio.com/article/742867/3_CIOs_Reveal_How:They_Got_Started_ With_Predictive_Analytics, accessed August 16, 2019.

Cohen, D., "How Facebook manages A 300-petabyte data warehouse, 600 terabytes per day," *Adweek*, April 11, 2014; www.adweek.com/digital/orcfile, accessed September 15, 2019.

Dannon web site, www.dannon.com, and Danone career site, jobs.danone.com, accessed September 14, 2019.

Dresner Advisory Services, "2019 Wisdom of crowds business intelligence market study report," 2019, dresneradvisory.com/products/2016-wisdom-of-crowds-business-intelligence-market-study-report-buyers-guide-edition, accessed September 15, 2019.

Duhigg, C., "How companies learn your secrets," *The New York Times*, February 16, 2012; www.nytimes.com/2012/02/19/magazine/shopping-habits.html.

Few, S., "Dashboard confusion," *Perceptual Edge*, March 20, 2004, www.perceptualedge.com/articles/ie/dashboard_confusion.pdf.

Github, "My data is bigger than your data!" 2019, lintool.github.io/my-data-is-bigger-than-your-data, accessed September 14, 2019.

Glanz, J., "Is big data an economic big dud?" *The New York Times*, August 17, 2013.

Grimmer, J. and B.M. Stewart, "Text as data: The promise and pitfalls of automatic content analysis methods for political texts," *Political Analysis*, 2013, web.stanford.edu/~jgrimmer/tad2.pdf, accessed September 17, 2019.

Harris, D., "Why Apple, eBay, and Walmart have some of the biggest data warehouses you've ever seen," *GigaOm*, March 27, 2013, gigaom.com/2013/03/27/why-apple-ebay-and-walmart-have-some-of-the-biggest-data-warehouses-youve-ever-seen, accessed September 14, 2019.

Kanani, R., "Target practice: The power of predictive Analytics," *Forbes*, July 29, 2013, www.scoop.it/topic/internet-of-things-quantified-home/p/4005697100/2013/08/05/target-practice-the-power-of-predictive-analytics-forbes, accessed September 14, 2019 (original page at forbes.com no longer available).

Li, Y., "Augmented analytics: The future of OLAP," *Kyligence*, June 12, 2019, kyligence.io/blog/augmented-analytics-the-future-of-olap, accessed September 15, 2019.

Lindell, C., "Hershey offers 8 insights to take into 2019," *Candy Industry*, January 8, 2019, www.candyindustry.com/articles/88481-hershey-offers-8-insights-to-take-into-2019, accessed September 11, 2019.

Marr, B., "What is a data lake? A super-simple explanation for anyone," *Forbes*, August 27, 2018, www.forbes.com/sites/bernardmarr/2018/08/27/what-is-a-data-lake-a-super-simple-explanation-for-anyone, accessed September 15, 2019.

Mitsa, T., "How do you know you have enough training data?" *Towards Data Science*, April 22, 2019, towardsdatascience.com/how-do-you-know-you-have-enough-training-data-ad9b1fd679ee, accessed September 11, 2019.

Shahbandeh, M., "U.S. Greek yogurt market - statistics & facts," *Statista*, March 15, 2018, www.statista.com/topics/2351/greek-yogurt, accessed August 16, 2019.

Simon, H.A., *The New Science of Decision Making*, Harper & Row, 1960.

Sincavage, D., "How artificial intelligence will change decision-making for businesses," *Tenfold*, August 24, 2017, www.tenfold.com/business/artificial-intelligence-business-decisions, accessed September 14, 2019.

Smith, D., "How much AI training data do you need?" *Lionbridge*, June 12, 2019, lionbridge.ai/articles/how-much-ai-training-data-do-you-need, accessed September 11, 2019.

TDWI, "EBay's revamped customer database creates insight," November 17, 2016, tdwi.org/articles/2016/11/17/ebays-revamped-customer-database-creates-insight.aspx, accessed September 14, 2019.

Turner, N., "The bigger they are, the harder they fall," *DataIQ*, January 7, 2013, www.dataiq.co.uk/articles/articles/201301data-quality-bigger-they-are-harder-they-fall, accessed September 14, 2019.

University of California Berkeley, School of Information, "What is data science?" datascience.berkeley.edu/about/what-is-data-science, accessed August 16, 2019.

Vijay, P., "Cognitive marketing: 'Packaged intelligence' in action," *Flytxt*, January 16, 2018, www.flytxt.com/blog/cognitive-marketing-packaged-intelligence-in-action, accessed September 14, 2019.

Volvo, "Volvo Cars – Retail sales by car model – August 2019," www.media.volvocars.com/global/en-us/corporate/sales-volumes, accessed September 14, 2019.

Waxer, C., "Inside the Greek yogurt wars: Dannon taps predictive analytics," *Computerworld*, August 12, 2013, www.computerworld.com/s/article/9241522/Inside_the_Greek_yogurt_wars_Dannon_taps_predictive_analytics, accessed September 14, 2019.

Woodie, A., "What's driving the cloud data warehouse explosion?" *Datanami*, November 8, 2018, www.datanami.com/2018/11/08/whats-driving-the-cloud-data-warehouse-explosion, accessed September 15, 2019.

10 Planning and Selecting Information Systems

CHAPTER OUTLINE

WHY THIS CHAPTER MATTERS

Every information system begins when someone thinks "We need a new system to do ________." (Fill in the blank with your favorite idea.)

Such ideas can arise from:

- Dissatisfaction with a current information system or business process.
- Seeing or hearing about what other organizations, perhaps competitors, have done.
- A merger or acquisition that changes how an organization works.
- Recognizing that new technology makes new things possible.

Sometimes more than one driver is present. Dissatisfaction can arise from what a competitor did. A merger may create the critical mass to justify a new, but expensive, technology. A software salesperson can show what new technology makes possible, creating dissatisfaction.

Once an idea arises, the next step is a business decision. Technology defines what is possible but cannot determine what is desirable. For that reason, organizations cannot leave these decisions to their MIS staff. Businesspeople, yourself before long, must make them. This chapter will show you how they're made and give you some tools to make them yourself.

CHAPTER TAKE-AWAYS

As you read this chapter, focus on these key concepts to use on the job:

1. Users *must* be involved in information systems (IS) planning to get the systems they need.
2. The organization's existing infrastructure both enables and constrains this planning.
3. There are systematic ways to approach information systems planning.

10.1 THE INFORMATION SYSTEMS STEERING COMMITTEE

This section describes larger organizations. Small ones have less formal structure or no structure at all. Still, everything a steering committee does is important. Even the smallest company should think about all the tasks of the steering committee and make sure someone handles each one.

Organizations that recognize the role of information systems in their success generally establish an *IS Steering Committee* to oversee those systems on behalf of top management. This committee is headed by the organization's *Chief Information Officer* (CIO) or a senior IS manager. It includes

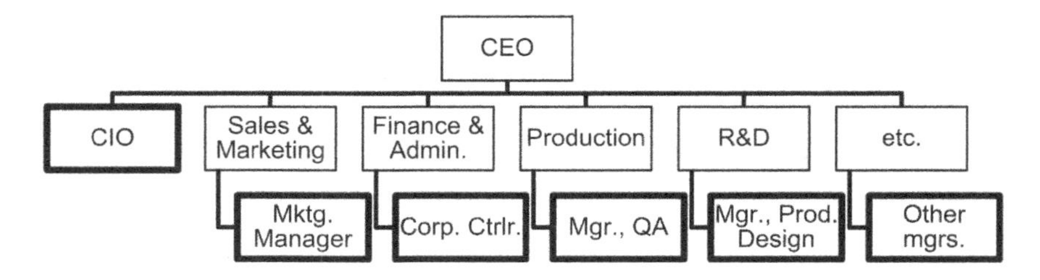

FIGURE 10.1 Organization chart of IS steering committee.

members from all parts of the organization, usually managers or senior professionals who can speak for their colleagues. Figure 10.1 shows a steering committee, with example members in heavy borders.

The purpose of a steering committee is to make sure that information systems investments benefit the organization. It decides what to develop. It reviews plans for major systems, and appoints subcommittees to carry out detailed reviews as needed. It sets policies for information system use, such as security policies, and approves major technology budget requests. By its commitment to effective use of information technology, it sets an example for others. Today steering committees conduct most of their work online, with occasional face-to-face meetings to get acquainted and maintain personal ties.

Employees submit ideas for new information systems or changes to existing ones to the steering committee by filling out a short form, usually online. They describe what the new system could do, why that is better than today's method, and give benefits of the change. Proposals may include rough cost estimates. They may mention factors such as new legislation with which the organization must comply, or a competitor's new system that leaves it at a disadvantage.

The steering committee considers suggestions as they arrive or on a schedule. It starts by dividing them into two groups: those that are inconsistent with the organization's resources, strategies or needs, and those that are worth closer examination.

The committee then assigns systems analysts (see Section 12.3) to conduct preliminary investigations of proposals in the second group. This investigation, described in more detail in the next chapter, yields a report called a *feasibility study*. Proposals that do not get a preliminary investigation when first submitted may be retained for later consideration. Either way, the committee notifies the person who submitted the suggestion of its decision.

When the committee receives the feasibility studies it requested, it evaluates them. Most suggestions that reach the preliminary investigation stage turn out to be feasible.

Where you fit in: As your career advances, you may find yourself on an IS steering committee. This is a sign of confidence in your ability and gives you an opportunity to interact with managers from other parts of the organization at a strategic level.

Before then, you may be on a subcommittee for project review or another purpose. This is an opportunity to contribute outside your direct responsibilities and to show that you are capable of handling bigger assignments.

Whatever your job, know who represents your department on the IS steering committee and learn how it works. When you have an IS idea, this representative can guide you in taking it forward, helping you present it in a way that is likely to lead to acceptance. If you obtain this person's support ahead of time, he or she can also champion your suggestion in committee meetings. It will still have to be evaluated, but this type of support matters when suggestions compete for limited funds.

10.2 SETTING PRIORITIES

There are usually more feasible proposals than there is money to pay for them. The steering committee must set priorities, selecting those that will go forward and those that will not.

Some information systems are mandatory. This includes systems required by new legislation, or that customers insist on as a condition of doing business.

Next come information systems that will provide a measurable benefit. The costs and benefits of some systems can be estimated with acceptable accuracy—that is, as accurately as other future cost and revenue figures are estimated. When this is done, finance methods can determine the expected *internal rate of return* (IRR) of the new system. If this IRR is above the *hurdle rate* that the business uses to justify uses of funds, the business will be willing to fund the new system.

Figure 10.2 shows a sample financial analysis for a new system. This system will cost $400,000 in development expenses during Year 1. In Year 2, when hardware to roll it out to users must also be purchased, it will cost $500,000. It will cost $50,000 per year for support and maintenance as long as it is in use. An additional $50,000 for user training is budgeted in Year 3.

The system's benefits begin in Year 3. They are estimated at $500,000 per year, less in Year 3 because it will not be fully productive for the entire year. Its benefits continue through Year 7, the organization's *planning horizon* for assessing costs and benefits of new information systems.

Adding up the costs and benefits shows total costs of $1,200,000 and total benefits of $2,400,000. The net benefit of $1.2 million suggests that this system would be very profitable. However, as you know if you've taken a finance course, this ignores the time value of money. A dollar (or euro, yen, etc.) today is worth more than the expectation of a dollar in the future for two reasons:

1. A dollar in hand is certain. A dollar in the future is an expectation. The further in the future we expect to receive it, the greater the opportunity for that expectation to change.
2. A dollar in hand can be invested. We'll have more than one dollar in the future if we invest it wisely. Or, it is a dollar that the firm doesn't have to raise by borrowing, issuing stock, etc.

Businesses take these factors into account by applying a *discount rate* to future costs and benefits. Typical rates are 15–25% per year. Applying a 20% discount rate (cell B8) reduces future net cost or benefit by that much for each year that it is in the future. The planned cost of $500,000 in Year 2 is divided by 1.20, yielding a *present value* of $416,667. The hoped-for benefit of $450,000 in Year 7 is divided by 1.20^6, or 2.986, yielding $150,070—about a third of the raw dollar amount. That is what the expectation of receiving $450,000 in Year 7 is worth today.

The row headed "PV" below the raw data shows the present value of each year's net cost or benefit. Summing them shows the project has a positive *net present value* of $200,650. That's about one-sixth of the original difference of $1,200,000. It shows what the project is really worth to the organization.

Experimenting with various discount rates, or using the IRR function of spreadsheet software, shows that this project has an NPV of zero at a discount rate of 28.4%. That is the project's *internal rate of return*. Any discount rate below 28.4% means a positive NPV. Any discount rate above this

Year	1	2	3	4	5	6	7
Cost	$400.00	$500.00	$100.00	$50.00	$50.00	$50.00	$50.00
Benefit			$400.00	$500.00	$500.00	$500.00	$500.00
Net	-$400.00	-$500.00	$300.00	$450.00	$450.00	$450.00	$450.00
All figures in thousands							

PV	-$400.00	-$416.67	$208.33	$260.42	$217.01	$180.84	$150.70
Rate	20.0%						

NPV
$200.65

FIGURE 10.2 Costs and benefits of new information system.

figure means a negative NPV. If the organization changes its hurdle rate—the rate a project must reach to be worth doing from a financial point of view—the new hurdle rate can be compared with 28.4% to see if this project is still worth pursuing.

Where you fit in: Understanding present value concepts is essential to managerial thinking. Showing this sort of thinking in project proposals increases their chance of acceptance, even if a preliminary investigation modifies your numbers. If you find yourself on a committee charged with project evaluations, other committee members may look to you, as a business graduate, to carry out this sort of analysis. Maybe you should keep your finance book!

If there are funds left over after approving mandatory systems and those that provide a positive NPV, management judgment comes into play. Businesses, through their IS steering committees, use several methods to choose among the remaining candidates for development:

Consistency with Corporate Strategy

The overall approach an organization takes to get ahead and stay ahead is its *corporate strategy*. You read about several possible strategies in Chapter 2. An information system that supports an organization's strategy is more likely to be chosen than one which doesn't.

WHAT DOES CONSISTENCY WITH CORPORATE STRATEGY MEAN?

Consider two department store chains: Walmart and Nordstrom. Walmart's strategy is to offer the lowest possible prices. While they do not deliberately provide bad service, they recognize that their customers will tolerate shortcomings. Nordstrom prides itself on providing the best possible service. Its customers come because they want top-notch service. They know they pay for it in higher prices, and they accept that.

If a Walmart employee proposes an information system to improve customer service, the IS steering committee will ask "How much will it cost?" If its cost is significant, the idea will be rejected. Walmart's customer service is good enough. Customers come to get low prices. More costs lead to higher prices. Higher prices go against the corporate strategy.

If a Nordstrom employee proposes an identical system, it will be welcomed. Cost will matter, but is secondary. If this system will improve service, Nordstrom wants it. Its cost will be recouped through customer loyalty.

Balanced Portfolio

The original concept of a balanced portfolio of development projects was to balance project risk and expected benefit. The idea was to have a mix of short-term projects that provide an assured, but limited, benefit, and riskier longer-term ones whose benefit will be immense if they succeed.

The concept of a balanced portfolio can be extended to include balance in other respects. An organization might try to balance the departments that new information systems support. This can counter the fact that some members of the steering committee might be more persuasive in support of their departments' needs than others. For example, the Marketing representative might be trained in persuading people, so she might make an excellent case for a new marketing system. The committee member from Manufacturing might be tongue-tied, but his department's needs are no less important. Balancing funding across departments can counter this effect.

Peter Weill of MIT's Center for Information Systems Research divides systems into the four categories of Figure 10.3 for balance. In order of increasing risk, they are:

1. Transactional systems that reduce costs. These have the lowest risk. Good information systems management can reduce that risk even further, but it is already low.
2. Informational systems used to support decision making. Their risk is that it can be difficult to act on information to create business value. Good IS management can provide a process to capitalize on improved information.
3. Infrastructure enhancements. The risk here is long-term technology uncertainty. Good IS management can reduce this risk by selecting enhancements for maximum flexibility.
4. Strategic information systems to gain competitive advantage. The risk is that such systems may not achieve this objective. Good IS management can reduce this risk by thinking through the business issues and competitive forces involved.

Investments should be balanced among these four categories. While exact balance can never be achieved, organizations should make sure that categories are not ignored or overemphasized.

The percentages in Figure 10.3 are the percentages of investment in the four categories that Weill found in a survey of large companies about a decade ago. There is no reason to think they have changed much since then. He emphasizes that there is no single right number. The right mix for a firm depends on the firm's strategy. For example, a firm with a low-cost strategy that uses IS to reduce costs (Walmart?) could well spend more than the average on transactional systems.

Organizations also try to balance new development with maintenance and enhancement of older systems. This can be difficult, because enhancements to existing systems can often consume the entire budget if they're allowed to. By allocating a minimum budget to new development, an organization can ensure that it doesn't get stuck with lots of old systems and no modern ones.

These are not the only types of balance that are possible. Some experts suggest balancing the project portfolio along four dimensions, as illustrated in Figure 10.4. Saunders suggests balancing these four factor aspects of projects:

- **Financial:** How do we look to those who provide funds, such as investors?
- **Innovation and learning:** How can we continue to improve?
- **Customers:** How can we improve how we look to them, and to our other stakeholders?
- **Internal processes:** What processes must we excel at to support our strategy?

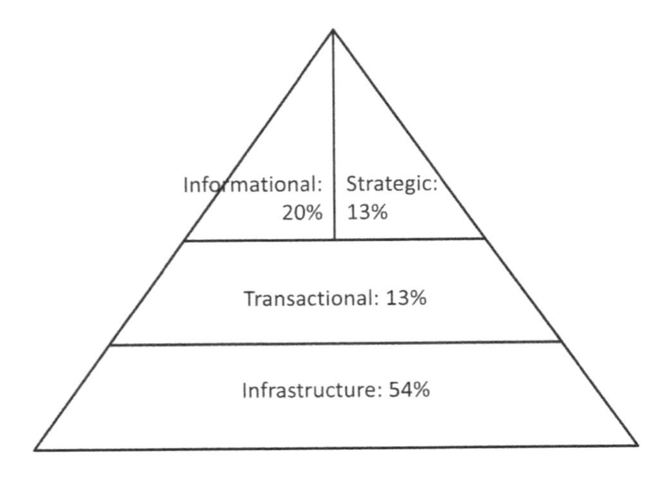

FIGURE 10.3 CISR balanced portfolio concept.

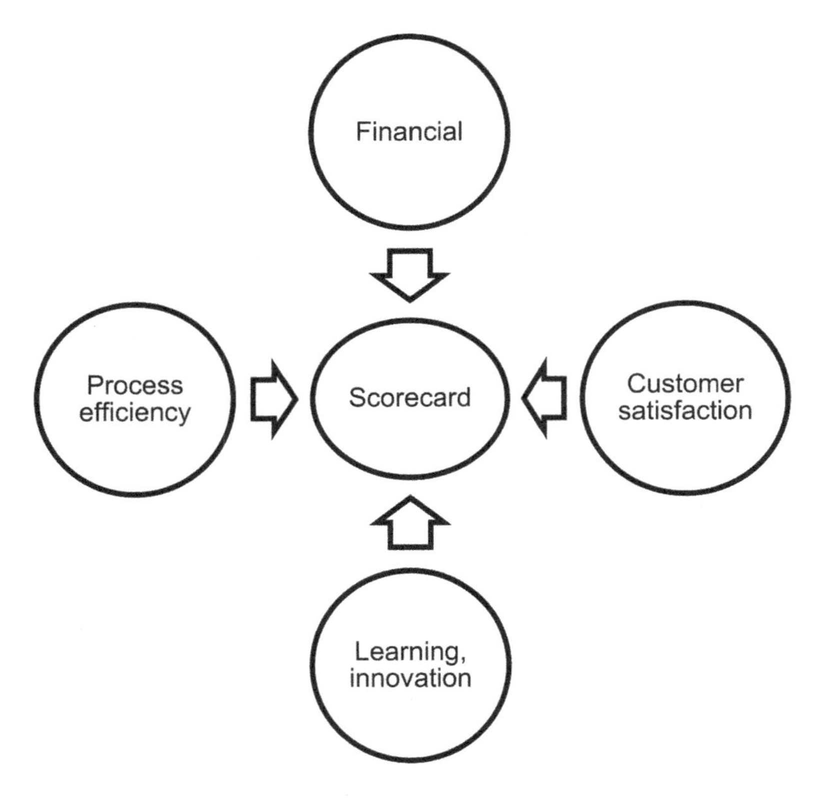

FIGURE 10.4 Balanced project portfolio concept.

He argues that none of these areas can be slighted without damaging the organization overall.

Where you fit in: As with the other items in this chapter, these are business issues even though understanding them requires some grasp of technology

If an information system that you propose fits into an underserved category, pointing out its potential to balance the project portfolio can improve its chances of acceptance.

Infrastructure Considerations

Any organization large enough to have a procedure for selecting new information systems has an existing technology infrastructure and architecture. New information systems must take these into account. They fall into three categories:

Proposals That Fit into the Existing Infrastructure

An organization's existing hardware infrastructure—servers, networks, and clients—can support some new applications. It may be necessary to add a few pieces, but these do not change it in a significant way. Any such changes are a cost in the financial evaluation of a new system.

Proposals That Extend the Existing Infrastructure

A new information system may require extending the existing infrastructure. Suppose a new ERP system requires a new enterprise database. That, in turn, argues for a storage-area network (SAN). While the cost of the SAN, if assigned in full to the ERP project, could make it difficult to justify that system, the organization expects to use the SAN for other systems later. A fraction of the cost can be allocated to the ERP system, with the rest considered an infrastructure modification.

A good architecture is designed to be extendible. Such extensions usually fit the existing system architecture. When this is impossible, the architecture must be replanned.

Proposals Whose Purpose Is to Modify the Existing Infrastructure

Sometimes the infrastructure must be extended as a foundation for future systems, even if few of those are under consideration. An example is how networks were once extended to connect to the Internet. Few people imagined what the Internet would be, but they knew it would be important and wanted to be ready. Another example: a data warehouse for future business intelligence use.

Infrastructure proposals often come from an organization's information systems group. They are the ones who first notice that a system component is reaching capacity or that new technologies will facilitate future information systems.

It can be hard to justify infrastructure projects on financial grounds. How much an organization is willing to invest in infrastructure depends on the value the organization places on information systems. Organizations that see them as a support function with little strategic value will tend not to invest in infrastructure that is not clearly required. Those that try to use information systems creatively to improve their competitive position want to make sure their computing infrastructure is ready for whatever comes. There is no "right" attitude. The choice depends on the industry an organization is in, its position in that industry, and its strategy for maintaining or improving that position. What is important is that managers should have a clear idea of how they feel in that regard and act consistently with that position. Any position is better than randomness. (The discussion of the Linking Strategy concept in Section 12.1 will expand on this theme.)

Where you fit in: When you propose a new system, keep the steering committee's priorities in mind. Your system is more likely to be developed if you can explain why the organization must develop it, if it is consistent with corporate strategy, and if you can make a case that it is in an area that has been underfunded in the past.

10.3 MAKE OR BUY?

You don't have to write a computer program to benefit from it. You have a word processor and a web browser, and you didn't write either. That's true even if you're a programmer and could have if you had to. You made a decision: There are lots of word processors, your WP needs are much the same as anyone else's, and you value your time, so you took advantage of what others have done.

In business, this is known as a *make-or-buy decision*. Such decisions appear often: Should we build electric motors for our vacuum cleaners, or buy them from an electric motor manufacturer? Should we manage our cafeteria, or pay a food service company to run it? The same applies to information system development.

The major options and two important selection factors are shown in Figure 10.5. The factors are:

1 How unusual is the application? If it's standard, like word processing, there's a large market for packages to handle it. A large market means choices. You can probably find a package that meets your needs. If your application is unique the market is smaller, so it's less likely that a software company has developed a package for it.

In addition, if an application is standard, again like word processing, your organization will not get a competitive advantage by using its own software. The value of a word-processed document is in its words, not in how they were processed. A company does not give

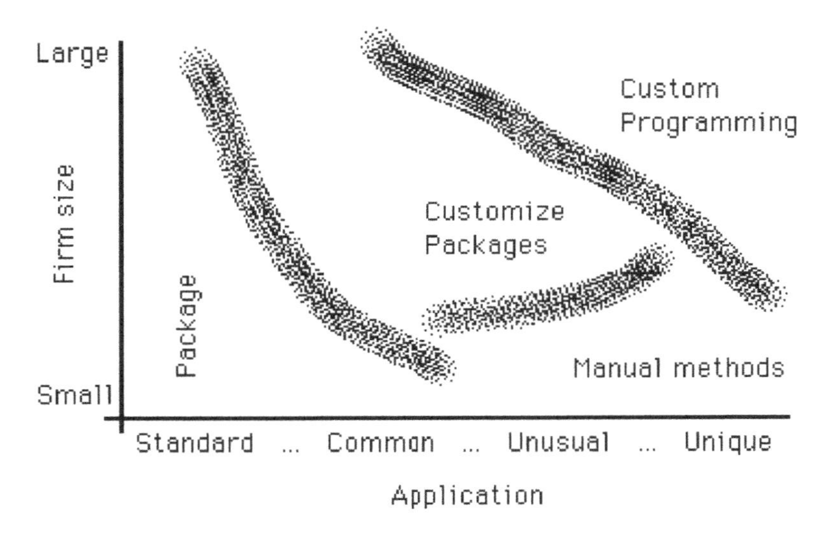

FIGURE 10.5 Factors affecting IS make/buy decision.

ground to competitors by using the same word processor. When an application is unique, though, its potential strategic value is large.

2 How big is your company? The benefits of any information system are roughly proportional to company size, but the cost of developing it rises much more slowly.

COMPANY SIZE AND INFORMATION SYSTEM VALUE

At the end of 2018, General Motors' inventory was valued at about $9.8 billion. The inventory of a local bicycle shop might be worth about $1 million. By this measure, GM is about 10,000 times larger. If each can reduce its inventory by 5% by managing it with a custom information system, that inventory reduction is worth 10,000 times as much to GM.

Development costs, however, are closer. GM's inventory needs are more complex. It has more product lines, more complex products, more locations, a need to supply dealers and repair shops in still more locations, many people using the system at the same time, and a host of additional complicating factors. However, its needs are not 10,000 times as complicated. They're probably more than ten times as complicated, but less than a hundred times as complicated. Let's say 50 times.

So, GM will pay 50 times more than the bike shop for custom inventory management software, but will benefit 10,000 times more. We might expect GM to get custom software for this system, while the bike shop uses a commercial package or manages inventory the same way it always did: The owner looks around the back room and decides to order road bike frames and FSA cranksets.

Figure 10.6 illustrates this. The benefits of a new information system, or of improving an existing one, are proportional to the size of the organization that uses it. It will always have at least some minimum cost. This cost will increase with organizational size more slowly than benefits increase. At some point the lines cross. Above this point a custom system is justified. Below this point, it isn't. That's why large organizations develop custom software, but small ones don't.

When an organization chooses custom development, it has another choice: have its staff develop it, or to use a specialized software development firm? This is a make-or-buy decision. It is based on cost, availability of skills within the organization, and, if the required skills aren't available, how useful they'll be in the future. Engaging an outside firm to develop software (or do anything else)

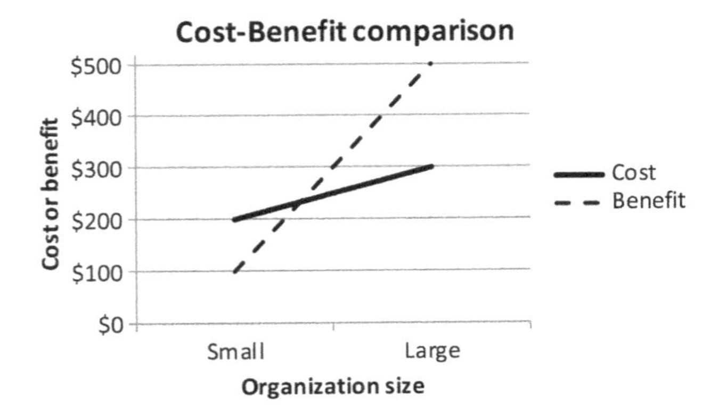

FIGURE 10.6 Costs and benefits of improved inventory management.

is *outsourcing*. If that firm is in a different country, it's *offshoring*. Offshoring offers savings when developers are in a low labor cost region, but increases the difficulty of managing a project.

Customized packages lie between custom software and standard packages. Your school probably uses one for registration. Course registration is a common need, shared by most institutions of higher education and many secondary schools. It is a large market, but not as large as that for word processors. The basic concepts are common, but there is enough variation to make a fully standard package impractical. Your school probably bought a package, spent months customizing it (far easier than writing a registration system from scratch), and is satisfied with the results.

If customization calls for specialized skills, or for resources that the organization doesn't have, customization can be outsourced as well. Software suppliers often offer this service. Customers can use supplier resources to train their own programmers, to work alongside them, or to take on the entire job. The choice is a business decision based on cost, schedule, and how important an organization feels it is to be self-sufficient.

The lower right corner of Figure 10.5 is labeled *manual methods*. If a small company has a rare need, an information system may not be justified. Consider sales forecasting for a sports arena pretzel vendor. One could develop a system that considers day of the week, time of the game, weather, holidays, league standings, and more. The cost of that system would exceed the vendor's annual revenue. Instead of hiring a programmer to develop that system, our vendor makes a best guess based on years of experience. It may not be the best possible forecast, but it's good enough.

Where you fit in: Consider this framework when you propose a new system. If your organization is small, you might do online research to see if there are packages to do what you need. The cost savings from using a commercial package, versus needing to develop a new information system from scratch, can make the difference between accepting and rejecting a suggestion.

10.4 SELECTING SOFTWARE

When your organization's IS steering committee considers a proposal, it decides if the proposed system will use an off-the-shelf package, a customized package, or custom software. This choice is based on information in the proposal, general knowledge of the area, research by the IS department, and discussions with knowledgeable people. Cost differences among approaches can affect whether or not developing a new information system is justified. Most companies, most of the time, opt for either a standard or a customized package.

If custom development will be used, the committee begins the process you'll read about in Chapter 11.

Many organizations have a procedure for purchasing software. Its nature depends on the size of the organization and type of system. A billion-dollar company selecting an ERP system will use a more formal process than a startup picking a web design app. This section describes the high end of the spectrum. Each step has a purpose, so small organizations should think about how they will accomplish that purpose even if they don't follow the full process.

The IS steering committee is responsible for software procurement. For important systems, it may manage the procurement itself with support from the IS department. For less critical systems, it will delegate the work to a subcommittee or another group, while exercising general supervision. The group that handles the selection is referred to as the *selection committee* below.

Where you fit in: Being part of a selection committee is one way that potential members of the IS steering committee are evaluated with little risk to the organization if they don't work out.

Establish Requirements

The first step in obtaining software is to establish what it should do. A student registration system should enable students to register for courses—but such a general statement is not enough. Every requirement, every desirable feature, must be spelled out. This is important for software vendors who will respond to the committee's solicitation and for the committee in evaluating responses.

User requirements should be:

1. **Verifiable:** Stated in such a way that conformance to the requirement can be demonstrated.
2. **Clear and concise:** Easily read and understood, unambiguous, and free of irrelevant information such as reasons for a need.
3. **Complete:** Contains all required information, including amounts and measurement units.
4. **Consistent:** One part doesn't conflict with another, all use the same terminology.
5. **Traceable:** Identifiable to be followed through development and testing.
6. **Viable:** Can be met within technology, budget, and schedule constraints. (If this is not possible and the requirement is not flexible, project feasibility must be reevaluated.)
7. **Necessary, not optional:** Must be present for the system to meet its objectives.
8. **Free of technical detail:** Define what must be done without telling developers how to do it.

Figure 10.7 shows a small piece of a requirements document, to suggest the level of detail that is required, from a proposed curriculum information system for a university. This section deals with changing the status of a course proposal. If it goes into this much detail about such a tiny part of curriculum management, imagine what the entire document is like!

It is important to establish requirements before considering possible suppliers. Otherwise, there is danger of biasing the requirements based on appealing features of one candidate product that are not really important.

Develop a List of Potential Suppliers

Next, the selection committee uses all available sources to come up with a *long list* of potential vendors. Little or no screening is done at this stage. It is better to include a vendor who will be ruled out later than exclude one who might have a suitable product.

4.24 Change Status of a New Course Proposal

4.24.1 Description

Proposals go through status codes 'New In-work' to 'Submitted' to 'Approved.'

Once approved, a proposal is kept permanently, but new versions can be undertaken that, when approved, supplant the currently approved version. The exact set of status codes and allowed transitions is detailed in CISWorkflow.doc. Status changes are effected by user request and are subject to authorization checks. The system displays only allowed status changes to the user.

The CIS tracks editorial changes not as status code changes as revision code changes (most of which are automatically determined by comparing the current version with the prior approved version) – see 'Automatic Setting of Proposal Revision Codes.'

4.24.2 Functions

Determine user's authority to perform requested proposal status change.

If allowed, change the status code in the proposal database.

If not a valid status change, display an error message and do not change the status.

After changing a status code, the user's authorizations, menus, etc. must be recalculated to take into consideration the new proposal status.

4.24.3 Data

Retain the ID of the user who changed the status code and the date/time of the change. Also retain any comments made at the time of the status code change.

FIGURE 10.7 Sample software requirements document excerpt.

Reduce the Long List to a Short List

A long list contains too many vendors to evaluate carefully. The selection committee screens them to narrow its options to a *short list*, typically consisting of three to five potential suppliers.

The first step in reducing the list consists of online research to confirm that each vendor has a product that seems to meet the need. That need not take long. The purpose is not to evaluate products. It is to weed out vendors who shouldn't have been on the list in the first place.

Next, the selection committee sends remaining vendors a *request for information* (RFI). An RFI describes the organization, states its requirements and any constraints (such as compatibility with existing infrastructure), and asks if the vendor has a product that meets those needs. It also states requirements for a reply: deadline, format requirements (a PDF file or four copies of a paper document), and often a maximum length (and a minimum type size).

Pricing may or may not be requested at this point. A reason not to request it is that the short list should be unbiased by cost considerations. The argument for it is that sometimes a price is so high or so low that cost becomes a valid reason to include or exclude a vendor.

The selection committee then evaluates responses to its RFI. It may use outside help by engaging consultants who are familiar with the application area, or reading reports of industry analysts who know about many vendors' offerings. It does not attempt to rank the offerings. Rather, it divides them into two groups: those that will advance to the next stage and those who won't. Sometimes it uses a waiting list: firms to be added to the short list if it drops below minimum desirable size.

An organization might not receive any acceptable responses to its RFI. This doesn't happen often, if it did its preliminary research well, but it can. If that happens, it must figure out why.

1. If no packages meet the need, the system will require custom development. In that case its cost will be higher, so the desirability of proceeding must be revisited. If the organization is still interested, it proceeds with custom development as discussed in the next chapter.

2 If the selection committee remains convinced that suitable packages exist, it should revisit its RFI to understand what scared off potential vendors. Perhaps requirements were too specific to allow vendors to meet the business need in the best way. Perhaps the schedule wasn't realistic. Perhaps the terms were too far from how vendors do business. Discussions with vendors can help figure this out. Once the reason for lack of replies is understood, the organization can issue an RFI that is more likely to obtain responses—or decide not to.

Request for Proposals

Once the selection committee has a short list, it writes, again perhaps using outside resources, a *request for proposals* (RFP). The RFP spells out in detail what the organization wishes to obtain. Figure 10.8 shows the content of a typical RFP. Its main content, in terms of what the new system will do, is in its Section 3.2.

Vendors have a few weeks to respond to the RFP. Selection committee members, most of whom have been borrowed from other jobs in the organization, use this time to catch up on their work.

It is possible that no vendors will submit acceptable responses to an RFP, but since it only goes to those who have already replied to the RFI—thus indicating that they intend to reply to the RFP when it is issued—this almost never happens.

RFP Response Evaluation

Once responses to the RFP are in, the selection committee evaluates them. This includes a functional evaluation, financial evaluation, and sometimes a benchmark test. It leads to a tentative selection: the vendor with whom the organization will try to negotiate a contract.

Request for Proposals
Table of Contents

1. Introduction
1.1. Company Profile
1.2. Project Background
1.3. Objective
1.4. Overview of Requested System
1.5. Anticipated Project Schedule

2. Administrative
2.1. Proposal submission process
2.2. Confidentiality
2.3. Future requirements (including finalist presentations)

3. Vendor Question Matrix
3.1. Vendor History/Profile
3.2. Application Functionality
3.3. Technology
3.4. Security
3.5. Reporting
3.6. Testing
3.7. Support
3.8. Service Level Agreements
3.9. Costs/Fees

4. Selection and Evaluation and Selection Criteria

5. References

FIGURE 10.8 Typical RFP table of contents.

Functional Evaluation

The most important part of RFP evaluation is *functional evaluation*: How well does each proposal meet our goals for the new system? How close is each to what we want? Do all proposed systems meet the absolute requirements (*musts*)? How many of the desirable features (*wants*) does each one have? If some factors are measured numerically, how does each proposed system rate on those, and what is each one's overall score?

Different groups use different decision-making methods to combine all these factors into a ranking of alternatives. One common method starts by screening all the alternatives for all the *must* factors. Any that fail to meet at least one of those are ruled out. (If a system can be acceptable without meeting a must factor, that factor wasn't really a must in the first place.)

Next, *want* factors are given weights that reflect how important each one is. This weight can be a number on a scale such as 1–10, or a fraction of the total score such as a percentage. Weights should be assigned before evaluating alternatives. (This process was used in Section 9.4 to select a corporate jet.)

Proposals are then evaluated on each factor. A 1–10 scale is one approach. Proposals that meet a capability best, or completely, score 10 on it. The others score lower, based on the committee's perception of how close they are to the best proposal or to an ideal solution. A system that is close to the best might receive a 9. One that doesn't have that capability at all will get a 0 for that item.

Finally, the scores of each alternative on each factor are multiplied by the factor weight, and the products added up to give the overall score of each alternative.

Figure 10.9 shows part of a functional evaluation scoring grid for a CRM package. (It is based on an example from Axia Consulting of the U.K. Their full grid has many more lines that are not shown.) This grid uses two scales, which are not exactly the same as those discussed above:

Importance weights ("W" column):
- 3: Important (but not mandatory)
- 2: Desirable
- 1: Nice to have
- 0: Not needed

Vendor scores ("S" column):
- 3: Requirement exceeded
- 2: Requirement fully met

FUNCTIONAL EVALUATION OF CRM SOFTWARE				
Vendor:	XYZ CRM Inc.			
Date:	June 23, 2014			
Section	Functional Requirement	W	S	WS
3.11	***Pipeline management***			
3.11.1	View, monitor sales pipeline/opportunities	3	3	9
3.11.2	Predict probability of successful close	2	2	4
3.11.3	Real-time update of pipeline figures	3	3	9
3.11.4	Identify top opportunities and required actions	3	3	9
3.11.5	Measure sales process effectiveness	2	0	0
3.11.6	Identify sales cycle bottlenecks	2	1	2
3.12	***Lead management and routing***			
3.12.1	Lead capture	3	3	9
3.12.2	Fast access to new sales leads	3	3	9
3.12.3	Automatically evaluates/merges sales leads	2	2	4
3.12.4	User-defined rules for lead prioritization	2	0	0

FIGURE 10.9 Functional evaluation scoring example.

1: Requirement partially met
0: Requirement not met

This vendor scored 55 total weighted points ("WS" column) in these two areas. Other vendors have their scores, and many areas aren't shown in this excerpt from the full evaluation form.

One could also score vendors in different columns on one sheet. That isn't basic to the concept.

Financial Evaluation

Vendors may propose different financial arrangements. The usual approach for personal software, a one-time payment, is rare with enterprise software. An initial cost plus a regular maintenance fee is more common. There can be periodic fees with no initial payment, often with a minimum term. The cost of enterprise software can also be based on actual or potential usage.

Measures based on *actual* usage include:

- Number of people who use the software in a given period of time, such as a month
- Number of hours people are logged into the software, regardless of how much they use it during this time—or whether they even use it at all
- Size of the application's database, in GB or a quantity such as number of students
- Volume of data entering and/or leaving the system
- Number of transactions processed, such as the number of students registered
- Another usage measure, such as the number of queries made or reports produced

Measures based on *potential* usage include:

- Number of authorized users, regardless of how many actually use it
- Number of different geographic sites at which the software may be used
- Maximum number of concurrent users
- Number of servers on which it runs, or the computing power of those servers
- Any other capacity measure, such as maximum database size

Such measures are often *tiered*: divided into ranges or tiers, where any quantity within a tier pays the same price. For example, a software license for computers belonging to one organization might have one price for 2–9 computers, a higher price for 10–99, a still higher one for 100–999, and one even higher with no limit. (Your university probably has such a *site license* for much of the software in its computer labs.) A site with 15 computers pays the same price as one with 95. If two packages are comparable, the site with 15 might do better to select a vendor that has a tier break at 25—unless it expects to have 30 computers next year.

Comparing alternatives requires estimating how the system will be used. Suppose one system is priced at $10 per month per authorized user, another at $100 per month per concurrent user. If 1,000 employees will be authorized to use the system, the break-even point is 100 users on the system at a time. The choice requires knowing if users will stay logged on all day, or if they'll log on for a few minutes when they need to. With the first pattern, paying per authorized user is the better approach. With the second, paying for concurrent users is less expensive.

Figure 10.10 compares three different options: a one-time license fee with annual maintenance fees assumed to rise at 5% per year, leasing at a cost that is assumed to rise at the same rate, and a per-seat charge assuming the number of users will increase by 10% per year. The calculations assume that the organization's cost of capital is 15% per year and use a six-year planning horizon.

In this case, both leasing and paying per user are more economical than buying a license and paying for maintenance. If these are different payment options for one system, the customer can choose between B and C on the basis of confidence in its growth estimates or by flipping a mental coin. If they represent three different systems, B and C are more desirable on financial grounds.

System Acquisition Financial Comparison							
Year:		1	2	3	4	5	6
Option A:	Annual cost:	100.0	5.0	5.3	5.5	5.8	6.1
	PV:	87.0	3.8	3.5	3.2	2.9	2.6
	NPV of cost:	$102,846					
Option B:	Annual cost:	20.0	21.0	22.1	23.2	24.3	25.5
	PV:	17.4	15.9	14.5	13.2	12.1	11.0
	NPV of cost:	$84,128					
Option C:	Annual cost:	18.0	19.8	21.8	24.0	26.4	29.0
	PV:	15.7	15.0	14.3	13.7	13.1	12.5
	NPV of cost:	$84,278					
Annual figures in thousands of U.S. dollars							
Annual figures rounded to nearest $100 for display purposes							

FIGURE 10.10 Financial evaluation example.

Organizations differ in how they combine cost with other factors in the final selection. Some treat cost as a want factor in scoring the alternatives. Thus, low cost can offset functional advantages. Others evaluate the two separately, then look for the best combination of capability and cost.

Check References

While most technology vendor representatives will not knowingly lie, they will put things in the best light and are not motivated to disclose problems. System selection advisory firm Infor of Alpharetta, Ga., urges clients to get a list of references from vendors. While there are unlikely to be unhappy customers on these lists, comparing the satisfied customer experience across vendors can be enlightening. Infor recommends asking questions such as these:

- What type of business is the reference? (Not providing references in your industry, unless a vendor is very small or is moving into your industry for the first time, is a negative point.)
- Explain the top three to five goals you want to achieve. Did the reference have similar goals? Were they achieved? If not, was the software at fault, or was it an internal problem?
- What are the greatest benefits the reference firm saw after the software was implemented?
- Were there any surprises?
- How was vendor support? Were cost and timeframe in line with what was discussed and quoted? If not, did the scope of the project change, or did the reference feel that the vendor underestimated the work required?
- If the reference firm had to repeat the process, would they choose the same package?

Benchmark Test

A *benchmark test* is a controlled evaluation of a computer system as it performs realistic tasks. Benchmark tests, or *benchmarks* for short, can be used to confirm that a system can carry out its functions and will perform adequately under the expected load.

Personal computer reviewers report benchmark results. They may find, for example, that adding blur to a test image in Photoshop takes 15.9 seconds on one computer, 19.3 on another. Most users wouldn't notice, but for those who spend a lot of time in Photoshop, small differences add up. Benchmark tests can also show how this changes if RAM is upgraded from 8 to 16 GB, if a disk drive is replaced by solid-state storage, how long it takes in a different package such as Gimp, or how a new version of Photoshop compares with its predecessor in this respect.

A benchmark test for an enterprise system is more complicated. It requires a database: perhaps not the full database the system will eventually have, but large enough to expose issues that may arise with it. It requires users, perhaps a lot of them, or a computer to mimic the activity of many users. It may require software customization. It requires instrumenting the software to measure how well it is performing. It requires multiple runs, with precise changes from one to the next, to understand how workload variations affect performance.

Benchmark tests are a lot of work! Customizing even part of a large system, installing it on the customer's servers or arranging to use the vendor's, finding and training enough "users" for a realistic workload, running it, measuring its performance, and analyzing the data are a major investment of time, cost, and effort by both parties. As a result, benchmarks are rare. When they are used, they are usually run only for a system that has been tentatively selected. Planning and carrying out more than one benchmark is seldom worthwhile. A benchmark may be used to confirm that the preliminary choice is up to the task. Both parties understand that if the system passes—criteria for that having been agreed to in advance—contract negotiations will follow.

Demonstration

While it may be impractical to run more than one benchmark, that's not true of demonstrations. Each vendor who submits a response to an RFP should demonstrate their system. Demonstrations usually take place at vendor facilities or at an existing customer's. The intent is to show how the system feels in use and how it will help the potential new user meet its needs.

Buyers vary in their approaches to vendor demos. Some provide data for the vendor to use and even a script for it to follow. They do this to make demos directly comparable. However, this limits vendors' ability to showcase their product's strengths. Infor recommends telling the vendor the goals you hope to achieve. Then, let the vendor show how its product will help achieve them.

Negotiation

Personal computer software is offered on a take-it-or-leave-it basis. If we don't like the contract that we accept when we break the seal on a CD envelope, or when we check "accept" on a web site, we're free to get a different package with better license terms—if we can find one.

With enterprise systems there is almost always room for negotiation. The software and a vendor's general business policies won't change, but the buyer has the power of the checkbook. Discounts from the asking price are common. Extra support may be available for the asking now, but not later. Concessions you can ask for, and may receive, include:

- A lower price. Software prices are like car prices: There is a level below which a seller won't go, but the initial asking price is higher. Reductions can apply to any price element. Advisors can often tell a prospective purchaser how much customers of that vendor actually paid. That is useful information, just as knowing how much a car costs dealers is useful to car buyers.
- A vendor may provide free or low-cost training on a new system. This can be a win-win: The vendor has training staff and can provide it at little incremental cost, but the new user doesn't have such people and would have to pay a great deal to get them.
- A vendor may make experts available to help with customization or to answer questions that come up during it. Here, again, the vendor already has resources that the user needs and that would be expensive to find elsewhere.
- A vendor may agree to waive upgrade fees if a new version comes out within, say, six months. This costs it nothing and can motivate a user to buy now rather than wait for the new version that's been rumored in the press—or buy a competitor's system instead.

As with writing RFIs and RFPs, outside advisors can help in negotiations. A user organization is probably involved in information system negotiations once every few years. Vendors negotiate with customers daily. Every sales rep handles several contracts a year. As with cars, the seller has an edge. Using a consultant who negotiates contracts regularly can offset that edge, saving many times the consultant's fee.

Where you fit in: If you're on a selection committee, these steps will be part of your job. Be sure your manager understands the time commitment that your committee assignment will involve, in both duration and level of activity.

10.5 AFTER THE CONTRACT IS SIGNED

It may be hard to believe after all the work that goes into selecting and purchasing a new system, but most of the activity still lies ahead. This is when a general-purpose package from a vendor is turned into a customized version that fits your company, is deployed to users' desks or mobile devices, users are trained in it, and they're persuaded that changing their habits to use the new system is a good idea. Since the customization stage resembles the development stage of custom software, and the other stages are the same for software that is developed internally and software that is purchased from an outside firm, we'll cover those activities in the next chapter.

KEY POINT RECAP

Most organizations, and almost all large ones, have an IS steering committee: *a group of managers from all areas who are collectively responsible for its information systems choices.*

As a businessperson, you may be assigned to your organization's IS steering committee and will surely be affected by its actions. The better you understand how it works, the more you will be able to contribute as a member, and the more you will be able to get its support when you need it.

Organizations are regularly faced with a decision whether to buy a commercial software package or have a program written specifically for them.

As a businessperson, you must make sure that these "make or buy" decisions are made for good business reasons, not solely on technical grounds (though technical factors are important).

Almost every business need for which a software package exists can be met by more than one existing package. *Organizations must choose one of the many on the market.*

A structured approach to software selection usually leads to better results than would be obtained otherwise. As a businessperson, you will be in a position to propose and use such a process.

KEY TERMS

Balanced portfolio: A set of development projects chosen as a group for balance in important respects, such as risk and potential reward.

Benchmark test: Controlled measurement of an information system as it performs realistic tasks.

Chief Information Officer (CIO): Senior executive responsible for an organization's information systems.

Corporate strategy: Overall approach that an organization takes in order to succeed.

Custom software: Software written specifically for the needs of one organization. Compare *standard package, customized package.*

Customized package: Software purchased from a supplier that must be modified in significant ways before it can be placed into use. Compare *standard package, custom software.*

Discount rate: The annual rate at which an organization reduces expected future costs or benefits to arrive at their *present value.*

Feasibility study: Preliminary study of an information system concept to determine if it is feasible within organizational constraints.

Financial evaluation: Comparing proposed systems on the basis of expected cost.

Functional evaluation: Comparing proposed systems on the basis of what they do.

Hurdle rate: Minimum *internal rate of return* that justifies proceeding with a project.

Information systems steering committee (IS steering committee, steering committee): Group of senior managers charged with making decisions about information systems.

Internal Rate of Return (IRR): The rate of return that a financial investment would have to offer in order to equal the profitability of a proposed project. (Negative if the project will lose money.)

Long list: Initial list of potential suppliers before it is narrowed down.

Make-or-buy decision: A decision as to whether a need, such as a new information system, should be met internally or via purchase from an outside supplier.

Must (decision factor): A characteristic that, if an alternative does not have it, renders that alternative unacceptable. Compare *want.*

Net Present Value (NPV): The sum of the *present values* for a project in every period over its expected life or until the *planning horizon.*

Offshoring: *Outsourcing* to a company located outside one's home country, often to take advantage of lower labor costs in another part of the world.

Outsourcing: The business process of obtaining a needed capability from another company. See also *make-or-buy decision*, *offshoring.*

Planning horizon: How far into the future an organization considers the costs and benefits of a potential new information system.

Present value: The amount of money which, were it in hand today, an organization considers financially equivalent to an expected cost or benefit at a specific time in the future.

Request For Information (RFI): Screening questions used to reduce a *long list* to a *short list.*

Request For Proposals (RFP): Document that allows vendors to make specific offers, including cost and schedule, to meet a user organization's needs.

Selection committee: Group of managers and senior professionals charged with selecting a new information system or part of one.

Short list: A few (typically three to five) vendors who are most likely to have a suitable product.

Site license: License that allows software to be used on an unlimited number of computers at a specified location, such as a university campus.

Standard package: Software purchased from a supplier and not modified in a significant way before use. Compare *custom software*, *customized package.*

Want (decision factor): A characteristic that, if an alternative does not have it, makes that alternative less desirable though it is still acceptable. Compare *must.*

REVIEW QUESTIONS

1. What is an information systems *steering committee*?
2. Why should a company care about an IS steering committee if it's too small to have one?
3. What does a steering committee do with suggestions it receives for new information systems?
4. What is a preliminary investigation? What sort of document does it produce?
5. Give an example of a mandatory new information system.
6. Why is having a dollar today worth more than expecting to have a dollar in five years?
7. Define *hurdle rate.*
8. Why should an information system be consistent with an organization's corporate strategy?
9. Suggest a few ways in which an organization's system development projects can be balanced.
10. What are two factors that affect whether an organization should make or buy a new system?
11. What is the category of new systems between custom development and standard packages?
12. What is the difference between an RFI and an RFP?
13. How can an evaluation of several "want" requirements be combined into a single score for a software package?

14. Give a few measures on which the monthly cost of an information system might be based.
15. Define *benchmark test.*
16. Why are benchmark tests not widely used for large-scale information systems?

DISCUSSION QUESTIONS

1. Discuss when and why your school might consider replacing its current student information system. In your discussion, suggest at least three reasons why it might want to do this at some future time, and what approach it will probably take to its new one.
2. Discuss who, by job title or area of responsibility, you think would comprise a good MIS steering committee for the following organizations. (There are many possible right answers.) Make any assumptions you need, but be sure to state them. If any of the organizations don't need an MIS steering committee, explain why.
   a. Your school.
   b. A hotel chain with 300+ hotels on four continents.
   c. A corner barber shop.
   d. The government of a large state/province of the U.S., Canada, or Mexico. (If you're not in North America, use the government of a small country in your home continent.)
3. Real requirements documents do not always meet the eight requirements in Section 10.4. Which of them are the curriculum information system requirements in Figure 10.7 consistent with? Which are they not consistent with? For which is there insufficient information to tell?
4. Your employer is considering two vendors for a new system. Expected benefits and costs for each are shown in the following table for a seven-year planning horizon. (All figures are in thousands of dollars.) Vendor A offers on-premises software with an initial license fee, annual maintenance fees, and a major upgrade expected in Year 4. Vendor B offers SaaS for an annual fee. It will take a year and a half to customize and implement either system; that cost is included in the costs for each vendor. You have been asked to carry out the financial evaluation on behalf of the IS steering committee. Write your report in the form of a memo.

	Year 1	Year 2	Year 3	Year 4	Year 5	Year 6	Year 7
Benefit	0	50	100	105	110	115	120
Cost: A	100	100	20	20	50	20	20
Cost: B	50	70	70	70	70	70	70

5. Interview one or more people from your school's information systems department to find out how the registration system you use was chosen. Compare your findings with the description of a selection process in Section 10.3.

   It may not be practical for everyone to approach the IS department individually with this question. Your instructor may assign this question to groups, ask someone from there to visit the class and discuss what they did, or ask students to interview people in several firms.
6. Your company's IS steering committee is considering requests for three systems. All sound good, but there are only enough funds for two. Write a memo to the committee from the vice president in charge of each of the following areas, explaining why that system should be chosen. Make any assumptions you want, but state them.
   a. For the marketing department: a system that will track and analyze mentions on social media, enabling the company to anticipate trends and improve its sales campaigns.

b. For the manufacturing department, a system that will tie your company more closely to its suppliers, enabling them to supply its needs more accurately and with less lead time.
c. For the engineering department, a computer-aided design system that will shorten product development times and result in lower production costs.

7. Near the end of Chapter 9 it says "Homegrown dashboards are seldom necessary or cost-effective." Explain why, using concepts from this chapter.
8. An airline needs to track aircraft maintenance. (An airplane may not carry passengers if it has not been maintained on schedule.) In addition to recording what has been done, this system will plan what should be done, when, and where. This is not simple: If maintenance is done earlier than necessary, money is wasted; if later than necessary, the aircraft must be flown empty to a suitable facility. No other airline has a system that plans maintenance effectively.
   a. Would you expect to use custom programming or a customized package for this? Why?
   b. If this program works well, it can give the airline an edge over its competitors. Explain why, using concepts in Chapter 2.
9. Apply the stages of the software selection process to selecting a word processor for yourself.
10. You are performing a financial evaluation of three competing software packages. Package A has a fixed annual license fee of $10,000 with no usage limitations. Package B has an annual license fee of $50 per authorized user regardless of how much they use it. Package C has an annual license fee of $1,000 plus $200 per concurrent user allowed, with no limit on how many users are authorized. Draw a diagram that covers 0–100 concurrent users and 0–500 authorized users, showing the regions in which each of the three is least expensive.
11. Jira team-oriented project management software (*www.atlassian.com/jira*) was priced as follows for cloud usage a few years ago. (Their 2019 pricing is, thankfully, simpler.)

No. Users	10	15	25	50	100	500	2,000
Cost/mo.	$10	$50	$100	$200	$300	$500	$1,000

    Plot the cost per user at the tier breakpoints. As a businessperson, how do you explain this pricing approach?
12. Consider Oracle software prices described in Section 4.6. Its August 2019 price list is 15 pages (including five pages of definitions and footnotes), in small print, with 4–6 columns of numbers ranging from 44¢ (per-user annual support of a small add-on program for identity management) to $300,000.00 (processor license for a complete business intelligence suite).
    a. If you are on a selection committee that is considering Oracle software, how might you approach the negotiations to minimize your cost?
    b. If you are an Oracle competitor, how could you use this complexity to your advantage?
    c. If you are an Oracle business planner, what are the advantages of this complexity?

KHOURY CANDY DISTRIBUTORS LOOKS AT TRUCK ROUTING SOFTWARE

The following week Jason's daughter Jennifer, manager of KCS's Atlanta distribution center, was in town for the quarterly KCD distribution center managers' meeting. These meetings rotated through the centers, and it was Springfield's turn. Jason made sure there was a gap in the meeting schedule at the time of Jake and Isabella's usual weekly visit. Such gaps were standard procedure; there was never enough time during Springfield meetings for the distribution center managers to sit down with all the headquarters people they wanted to meet with but couldn't justify a special trip to see.

Jennifer greeted the two students warmly in the KCD reception area. The light blue background of her employee badge indicated that her normal workplace was Atlanta. The three went to the MIS conference room. The candy bowl had hardly any attraction for Jake and Isabella by now, though they had to admit that Armand's Coffee Crisp from Canada (see Chapter 6 episode) was special.

After the usual pleasantries, during which the students congratulated Jennifer on her upcoming promotion (which she waved off, saying that being the next generation of the founding family helped a bit), Jennifer said that KCD was looking at software to select optimal routes for their trucks. "Picking the best route among a bunch of known points is old hat by now," she conceded. "What we want to do is a step beyond that. Given a bunch of shipments going to a bunch of places, which together will fill several trucks, we want to assign shipments to trucks and then route those trucks."

"That does sound a lot more complicated," Isabella agreed. "Does anyone do that?"

"A few companies claim to," Jennifer replied. "There's a company called Descartes,* one called EVOS, another is TruckLogics, and a few more whose names I forget. The concept is called 'LTL routing,' for 'Less Than Truckload.' Most of them look for clusters of shipments going in the same direction. That tends to give good routes, but not always the best routes, and our shipping managers can usually get close to that on their own. If we can't do better than that, we won't bother."

"Don't most companies have competent shipping people like that?" asked Jake. "If they do, why would anyone go to the trouble of getting software to do the same thing?"

"One reason is that these packages do more than pick routes," Jennifer explained. "They keep you informed where your trucks are, they handle driver payments when they're paid by the mile, they allocate fuel tax when trucks cross state lines, do expense reporting, and so on. Each one's feature list is different. Thing is, most of these don't matter to us. Either we don't care about them, perhaps because our drivers are on our payroll, or we already do them another way and we're happy with it."

"So," Isabella mused, "it becomes a matter of not paying for a lot of stuff you don't need and that makes the package cost more than it's worth to you—even if it could be worth that much to someone else. Are there any other big decision issues?"

"Yes, actually. I heard you were at the meeting where we decided to stay with SAP instead of going to NetSuite. (See the Chapter 7 episode.) The same question of running the software ourselves versus letting someone run it for us in the cloud comes up here too. The big reason we stayed with SAP, we were already using it and had gone to a lot of trouble to get there, doesn't apply here, but the cloud issue still does. We already own computers, they have enough capacity, we have people to run them, so we have to justify paying someone else to do that for us. There are good reasons to use the cloud besides cost, we know that, but they'd have to be pretty strong."

"Where are you with this decision now?" Jake asked.

"That's what I was meeting with Lakshmi about just before," Jennifer answered. "We have a short list of three vendors. Two have cloud software they'd run for us, one sells software we'd run ourselves. We've contacted all three and told them what we want to do. Now we want to meet with all three of them and see how close they come to matching what we need. That should happen in the next couple of weeks."

"Have you read any analyst reports on any of these applications?" asked Isabella.

"Hmmm," said Jennifer as she thought this over. "I haven't personally, no. I'm not sure I'd understand them anyhow. Still, that's a good suggestion. If Lakshmi hasn't, I'll make sure she does. They probably won't be free, but an informed, objective opinion is worth something. It might save us from an expensive mistake."

---

* As elsewhere in this book, mention of specific companies should not be taken as endorsements.

Questions

1. You are returning home for the winter holidays. Many of your classmates don't have cars. You have offered rides to three whose families live in the same general area as yours—about 200 miles/300 km from campus, all within a 20 mile/30 km radius. You don't care about the order in which you drop off your friends. How might you select an optimal route using commonly available software? Do you think there would be a market for an application that found such routes?
2. You have asked three LTL routing software suppliers to demonstrate their products. You want to focus on features KCD cares about, and not waste time on features that are not important to it. Write an email to one supplier's sales representative saying what you want them to demonstrate and what they shouldn't waste everyone's time with. (You may want to research the features that such software can have.)
3. Relate this case to the Milhench example in Section 11.2. (You don't need to study the whole chapter.)

CASE 1: MUDDLE THROUGH OR REPLACE?

North America is crisscrossed by telecommunications links: wires of various types, fiber optic cables, and more. These all involve physical devices: the connections themselves for wired and fiber optic links, but even wireless links have switches and connections of many types. Physical devices, however well they are designed, sometimes fail. A company that operates a network might not have the skills to fix them.

Enter Ledcor Technical Services. This company, which grew out of the Alberta oil fields in Canada, has 800 field technicians across North America, positioned to be within 4 hours of nearly every telecommunications link in the U.S. and Canada in order to provide front-line maintenance of those links. This is a valuable resource. It was also a scheduling and reporting nightmare.

"The spreadsheets, Word documents, and PDF files the company was using just weren't up to the task," says Ledcor Technical Services CIO Greg Sieg. So they began a search for software to handle their scheduling and reporting needs. They had several Must requirements:

- The selected package had to bring immediate improvements to those tasks.
- It had to *scale* to meet increasing demands: that is, accommodate more employees and more activities without running up against internal limitations. Such expansion might require more hardware, of course, but the software had to be able to take advantage of that hardware.
- It had to be able to handle all of Ledcor's resource planning tasks, from scheduling workers to collecting reports on completed assignments.
- It had to work with the wide range of mobile devices that Ledcor employees used.
- It had to deliver clear financial benefits.

In addition, they had two Wants:

- Ledcor did not want to invest in new hardware.
- Ledcor did not want to add responsibilities to its information systems staff.

Evaluating several products against the Must requirements led Ledcor to a Mobile Workforce Management software from ClickSoftware. This software was available in two versions: on-premise, which Ledcor would operate, and cloud-based, Software-as-a-Service. Their Want requirements led them to the cloud-based version. Had it been significantly more expensive than the on-premises solution, they might have chosen differently.

Specific benefits that Ledcor has seen include these three:

- Average work done per day is up 22%.
- Billing time was cut in half, because the required information is collected automatically.
- Dispatchers can manage 50% more technicians because much of the dispatch process is automated.

QUESTIONS

1. What does it mean for spreadsheets, Word documents, and PDF files not to be "up to the task?" Specifically, in what ways do you think they would not have been up to the task? What business needs do you think they didn't meet? What is the value of meeting those needs?
2. Many companies besides ClickSoftware sell workforce scheduling and reporting programs. Others might be better for other firms in other situations. Find two more such packages in an online search. What advantages do their suppliers claim? What types of companies might value these advantages?
3. Jeanine Sterling, an analyst at Frost and Sullivan, says about a third of the companies that use mobile workforce-management suites choose cloud-based software; a third, on-premise software; and a third, a combination of both. What factors, other than the two that led Ledcor to choose the cloud, do you think might influence that decision?

CASE 2: UNIVERSITY OF WÜRZBURG

The University of Würzburg (JMU for short, for its full name) traces its roots to 1402. Wilhelm Röntgen was chair of its physics department when he discovered X-rays in 1895—and he is just one of the 14 Nobel Laureates who taught or did research there, most recently Harald zur Hausen in 2003 (Medicine). Today, it is home to over 28,000 students in fields as diverse as Human Factors in Computing, Games Engineering, and Nanostructure Technology. Its motto, "Science for Society," drives both its teaching and its research.

The University of Würzburg's data center is its central IT service provider. Its 50 employees provide IT services, including system operation, network and communication, consultation, and training and multimedia services, to the university's students plus its 4,000 research and teaching staff across all its faculties and institutes.

By 2015 JMU's existing storage environment, two disk arrays and two tape libraries, had reached its limits. It held 300 TB of data in its storage-area network (SAN). "Over the past ten years, our data volumes have increased by a factor of 12. And we're expecting it to grow even more in the coming years," says deputy center director Dr. Matthias Reichling. "We can't tell exactly how much, as it depends on the projects our faculties have planned, and we often have no idea when they're coming."

To make things worse, many research applications generate larger and larger volumes of data. "The resolutions of the cameras, microscope, and measuring devices that many of our faculties use have increased dramatically in recent years," continues Reichling. Histological tissue samples, for example, easily generate images of 100,000 × 100,000 pixels. As they are often in color, file sizes can be 30 GB or more. Experts predict that data volumes produced by imaging software will approximately double every three years.

It was time to move forward. JMU consulted with IT services provider Bechtle Hosting & Operations of Neckarsulm, part of the Bechtle Group and Germany's largest systems integrator. Since most of JMU's existing infrastructure was from Hewlett-Packard Enterprise (HPE), they began by evaluating that firm's offerings to see if they suited the new requirements. The university took part in workshops with them and visited HPE's technology center in Böblingen before concluding that they did.

The new JMU storage complex includes two storage arrays with an initial capacity of 800 TB, expandable as need arises. Critical data is duplicated on both. It also includes two automated tape libraries with an initial capacity of 6,000 slots, up from the previous 1,400, and expandable up to a total capacity of 75 PB. Regular backups from disk to tape ensure data protection. Tape is also used for archival storage of data for ten years after a research project ends. The tape libraries are located in rooms 700 meters (about 2,300') apart for disaster protection, as well.

"Whether we'll actually ever need [the maximum tape capacity of] 12,000 slots, I don't know. But it was important for us to have the ability to expand [them] over the coming years should we need to," explains Reichling. He never wants to have to remove second copies from the library again just to be able to cope with the backup data volumes, as he had to do in the final months before the new solution was installed.

Managing the data center's storage environment is now much easier than it was before. JMU students and researchers have ample space for their work. The IT staff feels that the initial storage capacities will suffice for two or three years, with its expansion capability enough for several more.

Questions

1. Find two other vendors of storage systems that could support what JMU needed, considering only storage. Summarize their relevant offerings in a brief report.
2. JMU's new system will handle needs for 24–36 months, after which it can be expanded. There is a cost for this expansion capability. If you were buying a system such as this, how would you compare the benefits of a lower initial price with the benefits of this ability to expand later without another conversion? Discuss, with an example using data you invent.
3. JMU skipped the stages of developing a long list, whittling it down, and so on. They did so in order to stay with an existing, satisfactory, vendor.
   a. Do you agree or disagree with their reasoning? Give your reasons.
   b. Whichever way you answered the previous part, describe a hypothetical situation in which you would feel the other way. Explain why the two situations are different.
   c. JMU is not unusual in this respect. Many firms (and individuals) feel likewise. From the vendor's point of view, this suggests that they should make concessions on the initial sale, as repeat business over the years is likely. However, users who know this can play an incumbent vendor off against proposals from competitors who want to get a "foot in the door." As an incumbent vendor, how would you counter this?

BIBLIOGRAPHY

Bechtle Hosting & Operations web site, www.bechtle.com/de-en/about-bechtle/company/locations/bechtle-hosting-and-operations-bochum, accessed September 21, 2019.

ClickSoftware web site, www.clicksoftware.com, accessed September 21, 2019.

Crumrine, R., "Curricular information system software requirements document," 2000, web.archive.org/web/20180724224532/web.mit.edu/ssit/cis/CISRequirements.html, accessed September 23, 2019.

Hewlett-Packard Enterprise, "University of Würzburg brings scalability to its data centre," case study, 2016, cc.cnetcontent.com/vcs/hp-ent/inline-content/SY/E/8/E8F562FBAAF10F10CA5C0C1008690B8DD48F17BF_source.PDF, accessed September 21, 2019.

Infor, "Selecting an ERP solution: A guide," 2008, http://www.btasystems.com/Downloads/Whitepaper_SelectingAnERPsolution.pdf, accessed September 21, 2019.

Julius-Maximillians-Universität Würzburg web site, www.uni-wuerzburg.de/en/home, accessed September 21, 2019.

Karlsson, G., "8 characteristics of good user requirements," January 18, 2009, www.slideshare.net/guest24d72f/8-characteristics-of-good-user-requirements-presentation, accessed September 21, 2019.

Ledcor Technical Services web sites, www.ledcor.com and www.ledcor.com/what-we-do/communications September 21, 2019.

M.I.T. Center for Information Systems Management, "IT portfolio management," cisr.mit.edu/research/research-overview/classic-topics/it-portfolio-management, accessed September 21, 2019.

Oracle Corporation, "Oracle technology global price list," August 12, 2019, www.oracle.com/assets/technology-price-list-070617.pdf, accessed August 17, 2019.

Pratt, M.K., "Construction company looks to the cloud for workforce scheduling," *CIO*, July 29, 2013, https://www.cio.com/article/2383498/construction-company-looks-to-the-cloud-for-workforce-scheduling.html, accessed September 22, 2019.

Software Advice, "Ten steps to selecting the right ERP software," www.softwareadvice.com/imglib/lightbox-download-assets/Ten_Steps_to_Selecting_the_Right_ERP_Software.pdf, accessed September 22, 2019.

Warnock, R. "How to select the right ERP system solution," *CIO*, April 6, 2018, www.cio.com/article/3268128/how-to-select-the-right-erp-system-solution.html, accessed September 22, 2019.

Weill, P. and S. Aral, "Managing the IT portfolio: Returns from the different IT asset classes," *M.I.T. Center for Information Systems Research*, Research Briefings, Vol. IV, no. 1A (March 2004), cisr.mit.edu/blog/documents/2004/03/12/2004_03_1a_itportrtrndifcls.pdf (free registration required), accessed September 22, 2019.

11 Developing Information Systems

CHAPTER OUTLINE

WHY THIS CHAPTER MATTERS

Most information systems use commercially available software, as is or customized, but most is not all. Organizations can gain a competitive advantage by using information systems that nobody else has. Getting those means custom software development.

Businesspeople should know about custom software development because obtaining its benefits depends on the active participation of those who will use it. Parts of development must be done by people who will use the resulting information system. If those people (you, in a few years) don't do their part, they will get a system that doesn't meet their needs or doesn't yield the hoped-for benefits. They will have only themselves to blame.

This chapter will tell you what you must do during software development, when you will do it, and why doing it can't be passed off. That will prepare you for being called upon to participate in information system development at work.

CHAPTER TAKE-AWAYS

As you read this chapter, focus on these key concepts to use on the job:

1 Organizations can choose from different software development processes.
2 Each process suits certain business situations and technologies.
3 Each process has stages during which specific people carry out specific tasks. These involve the user side of the organization (that's you) at different times in different ways.
4 The quality of the software that emerges from any process depends on all its participants. Knowing what is expected of you will help you contribute and thus improve the outcome.

11.1 OVERVIEW OF SOFTWARE DEVELOPMENT

As you know, *software* refers to instructions that tell a computer how to perform a job. Software consists of *programs* for specific tasks. A short program like Figure 4.1 might be assigned in the first week of a programming course. An application is probably tens of thousands of lines long. Enterprise-level information systems have millions. Figure 11.1 shows a tiny part of the Android operating system module that splits a task over multiple processing units.

FACTOID

SAP's popular ERP system has been estimated to contain over 250 million lines of code. Every one of those lines was written by a programmer.

```
setContentView(R.layout.main_layout);
mBitmapIn = loadBitmap(R.drawable.data);
mBitmapsOut = new Bitmap[NUM_BITMAPS];
for (int i = 0; i < NUM_BITMAPS; ++i) {
        mBitmapsOut[i] = Bitmap.createBitmap(mBitmapIn.getWidth(),
                mBitmapIn.getHeight(), mBitmapIn.getConfig());
}
mImageView = (ImageView) findViewById(R.id.imageView);
mImageView.setImageBitmap(mBitmapsOut[mCurrentBitmap]);
mCurrentBitmap += (mCurrentBitmap + 1) % NUM_BITMAPS;
SeekBar seekbar = (SeekBar) findViewById(R.id.seekBar1);
seekbar.setProgress(50);
seekbar.setOnSeekBarChangeListener(new OnSeekBarChangeListener() {
        public void onProgressChanged(SeekBar seekBar, int progress,
                    boolean fromUser) {
                float max = 2.0f;
                float min = 0.0f;
                float f = (float) ((max - min) * (progress / 100.0) + min);
                updateImage(f);
        }
        @Override
        public void onStartTrackingTouch(SeekBar seekBar) {            }
        @Override
        public void onStopTrackingTouch(SeekBar seekBar) {
        }
});
createScript();
updateImage(1.0f);
}
private void createScript() {
mRS = RenderScript.create(this);
mInAllocation = Allocation.createFromBitmap(mRS, mBitmapIn);
mOutAllocations = new Allocation[NUM_BITMAPS];
for (int i = 0; i < NUM_BITMAPS; ++i) {
        mOutAllocations[i] = Allocation.createFromBitmap(mRS, mBitmapsOut[i]);
}
mScript = new ScriptC_saturation(mRS);
}
private class RenderScriptTask extends AsyncTask<Float, Integer, Integer> {
Boolean issued = false;
protected Integer doInBackground(Float... values) {
        int index = -1;
        if (isCancelled() == false) {
                issued = true;
                index = mCurrentBitmap;
                mScript.set_saturationValue(values[0]);
                mScript.forEach_saturation(mInAllocation, mOutAllocations[index]);
                mOutAllocations[index].copyTo(mBitmapsOut[index]);
                mCurrentBitmap = (mCurrentBitmap + 1) % NUM_BITMAPS;
        }
        return index;
        }
}
```

FIGURE 11.1 Sample code fragment.

Conceptually, the process of developing a new custom system is simple:

1 An organization decides to develop custom software for a new system.
2 Someone figures out, and writes down, what that software should do.
3 If the system is large enough for more than one person to work on (most are), someone divides it into modules for different programmers.
4 Those programmers write sequences of instructions that guide the computer, step by tiny step, to carry out the tasks of their modules.
5 People test those modules to confirm that each works properly. Then they combine them and confirm that the entire system works properly.

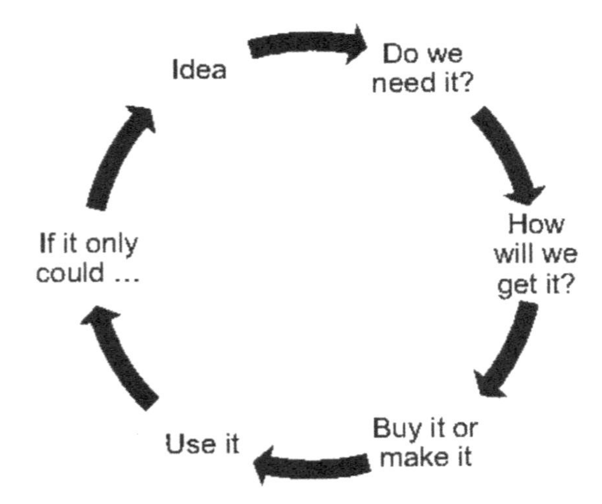

FIGURE 11.2 Cyclical diagram of this process.

6 The new system is installed on the organization's computers.
7 Members of the organization use it.
8 Over time, changes to the system become desirable. Programmers modify its code. The modified programs are tested, installed, and used.
9 Eventually, the system can no longer be modified enough to meet new requirements. A new information system is needed. The process repeats from Step 1 (Figure 11.2).

There are choices in how each step will be carried out. The next section describes the traditional approach, called the *System (or Software) Development Life Cycle* (SDLC). The SDLC has drawbacks. Other methods have been developed to avoid them. Section 11.3 discusses some of those other methods.

11.2 THE SYSTEM DEVELOPMENT LIFE CYCLE APPROACH

The SDLC (where the S may stand for "System" or "Software") approach was first described in 1970. While we recognize its drawbacks today, it was a huge improvement over the earlier anarchy. With it, software development could be approached as a systematic process. People use other software development approaches today, but it's still worth learning about the SDLC. It has a place, there's a great deal of overlap between the SDLC and newer methods, and it's easier to understand those methods after looking at their roots.

We show the SDLC with seven stages. There is no official list of SDLC stages, or even general agreement about how many there are. Different people give the SDLC five to nine stages, and give some stages different names. Stages may be called steps or phases. Still, the concept stays the same. If you understand the ideas behind one version of the SDLC, you can adapt to another. This understanding will also help you understand other development approaches.

How the SDLC Works

The SDLC consists of the stages shown in Figure 11.3. Because this chart resembles a riverbed under a waterfall, the SDLC is sometimes called the *waterfall method* of system development. (The stages are sometimes drawn as a "V," working down from a general view of the system into technical details and up the other side to its use as a whole, and relating stages at the same height on opposite sides of the V to each other.)

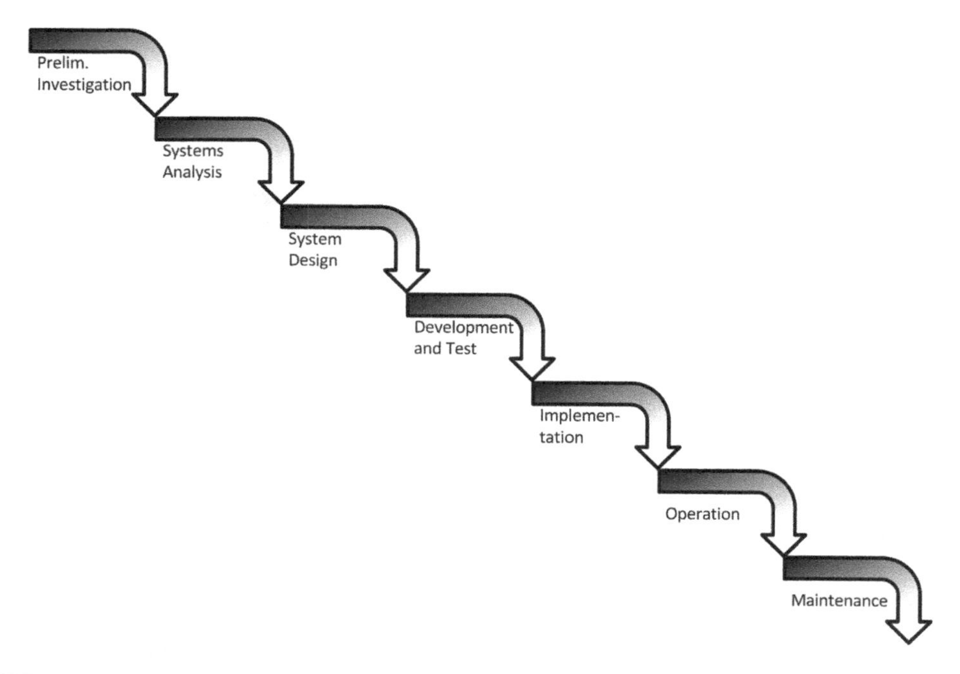

FIGURE 11.3 SDLC stage "waterfall" diagram.

Each SDLC stage produces a *deliverable*, often a document. No stage is finished until its deliverable is ready. Once it is, management conducts a *stage review*. This review results in one of the outcomes in Figure 11.4:

1 The project team is authorized to proceed with the next stage.
2 The team is told to make changes to the last stage, then proceed with the next.
3 The team is told to make changes to the last stage, then go through a re-review.
4 The team is told to go back to an earlier stage, make changes there, and review that stage before continuing.
5 The project is cancelled.

Outcomes 2 and 3 are most common. It is rare to find that nothing needs fixing. It is not common, though less rare, to find a problem more than one stage back. Cancellations are usually due to external circumstances or recognizing that a project is more difficult than anyone thought.

> ***Where you fit in:*** As a potential user of a new system, you may be asked to join a review board. You'll spend time before the review learning about the system you'll review. You'll then spend from half a day to a week meeting with members of the project team, going over what has been done and how well it meets the requirements. The review board will then write up its findings and discuss them with management.

The SDLC Stages

1. Preliminary Investigation

The first thing a steering committee (see Chapter 10) does in considering a software project is to initiate a *preliminary investigation*. It gives the steering committee information to decide whether or not to fund this project. The systems analysts assigned to this job ask if the proposed system is feasible from three standpoints:

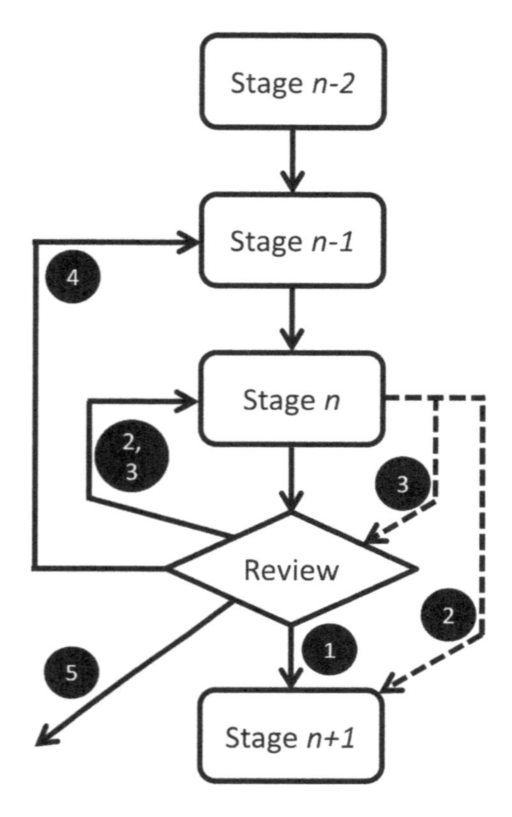

FIGURE 11.4 Possible outcomes of SDLC stage review.

- **Technical:** Can it be built with technology that we have, or can get?
- **Economic:** Does it make economic sense?
- **Operational:** Does it fit our processes and culture? Will our people accept it?

Developing a system is feasible only if it passes all three tests.

THE SYSTEM MUST FIT THE COMPANY!

In the early 2000s, Milhench Supply Company,* a New Bedford, Mass., distributor of cleaning, packaging, and material handling supplies (*www.milhench.com*), acquired truck routing software. The software had been developed by a large logistics firm for its own use. That firm then offered it to other firms that had to assign cargoes to vehicles and route those vehicles through multiple delivery points. Despite that firm's success with the package, it didn't work at Milhench.

The reason: The firm that developed the package had standard delivery vehicles. Milhench had a variety, from box trucks through semi-trailers. Some destinations had constraints on vehicles that could be used. For example, semi-trailers couldn't navigate the narrow roads to some customers or turn around at their locations. The software wasn't designed to deal with vehicle constraints.

Fortunately, Milhench Supply operates in a limited geographic area: eastern Massachusetts and Rhode Island. Its delivery points don't change much from week to week. Its drivers and shipping personnel know the area. They can choose good routes, even if not always optimal

* The author is grateful to Heike Milhench, president, for this information and for permission to identify her company.

ones. The cost of developing software to optimize the routes would exceed the value of its benefits.

This is an example of a system that was not *operationally feasible*. The package was technically feasible (it ran under Windows on standard hardware), economically feasible (it didn't cost much), but not operationally feasible. It wouldn't work at Milhench, regardless of its technical and economic feasibility.

Deliverable: The deliverable of the preliminary investigation stage is called a *feasibility study*.

Where you fit in: Analysts who carry out a preliminary investigation can't answer these questions themselves. They need help from the user side of the organization. In a small company, they may talk with everyone affected by the proposed system. In a larger one, they'll ask people they think will be able to help them. If you are called on to lend your time and expertise to a preliminary investigation, your participation will contribute directly to the quality of the system you use later. It can also increase your visibility to management.

Once it receives the feasibility study, the steering committee decides whether or not to proceed with this project. A system may be feasible in every respect, but not of high enough priority to justify developing it. For example, the *internal rate of return* (IRR) of a proposed system may be above the organization's *hurdle rate*, making it financially feasible, but lower than that of another proposed system. If there are insufficient funds to develop both systems, all else being equal, the one with the higher ROI will be developed. Only a fraction of proposed development projects eventually become working systems, as shown in Figure 11.5.

While a project can be called off at any point, a change in mindset takes place at the dotted line. Until then, people are deciding if a new information system *should* be built. After it, absent any surprises, it *will* be built.

Where you fit in: Some lists of SDLC steps don't include the preliminary investigation. The logic is that system development doesn't really start until it's been decided to develop a system. The point is not that one or the other is right, but that you should expect to see this sort of difference.

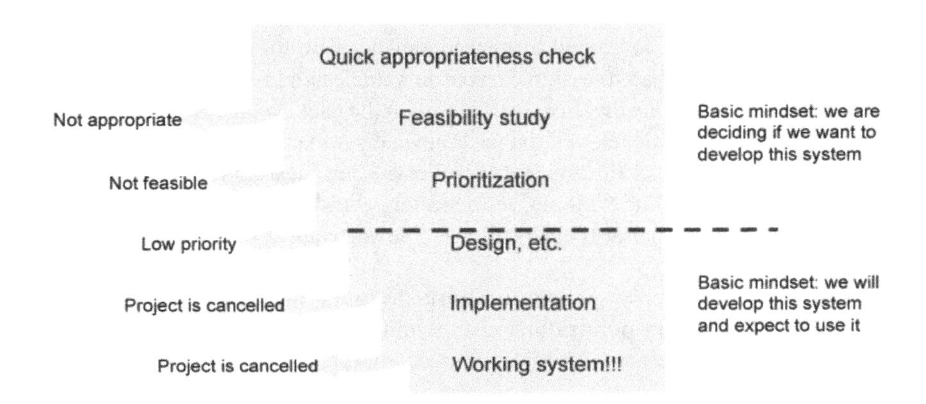

FIGURE 11.5 Diagram of project attrition.

2. Systems Analysis

If the proposed system is approved for development, it is necessary to define what it will do in more detail than the general description that sufficed for the preliminary investigation.

Suppose your university hires a programmer to develop a new student information system. That programmer knows them from the student side, at the school he or she attended, when he or she was there. He or she knows nothing about their use by instructors, the registrar's office, or anyone else; doesn't know what has changed in them since he or she was in school; and has no idea how your school's needs differ from those of his or her alma mater.

This programmer—or *systems analyst*, a specialist in this aspect of system development—must learn what everyone at your university requires of this system. Systems analysts start by learning how the existing system works, using methods such as:

- *Reading:* Documents, on paper or online, acquaint the systems analyst with how existing systems work.
- *Observation:* Watching people use an existing system helps learn about many systems. Here, a systems analyst might (with permission) watch students sign up for courses.
- *Interviews:* Asking questions helps learn how an existing system works and what its users need in a new one. Systems analysts use interviews after learning what they can from reading and watching. They use structured interviews, with planned questions; unstructured interviews, which range broadly over the area of the new system; or a combination.
- *Questionnaires:* In all but the smallest organizations, it isn't practical to interview everyone affected by a new system. Systems analysts interview a few from every affected area. They then use questionnaires to give others a chance to weigh in on the new IS.

These require user cooperation. If users don't cooperate, analysis will be deficient. If analysis is deficient, a system based on it will also be deficient. Users will suffer then, not systems analysts.

WHAT'S THE REAL REQUIREMENT?

This dialogue is modified from U.S. Federal Highway Administration training materials:

User: I need liquid drainage channels on my keyboard.
Systems Analyst: Why?
User: I sometimes spill coffee on my keyboard.
Systems Analyst: Why?
User: I need to have three or four manuals open to use the system. Sometimes one of them knocks my coffee over.
Systems Analyst: Why do you need to have three or four manuals open?
User: Because the list of commands is in one manual, the explanation of what the function keys do is in another manual, the list of valid parameter values is in a third manual, and sometimes I have to refer to the mathematical methods in a fourth manual.

The dialogue continues, with the systems analyst asking "why?" until the underlying needs are discovered. (The real need may not be drainage channels.)

Around now the development team may begin to use *Computer-Aided Software [or Systems] Engineering* (CASE) *tools*. These are computer applications whose purpose is to help people develop computer programs more quickly, systematically, and reliably.

Different CASE tools are used in different aspects of software development. The tools used in systems analysis help analysts understand the relationships among parts of an information system and how it interacts with its users, and help system designers transform these relationships into

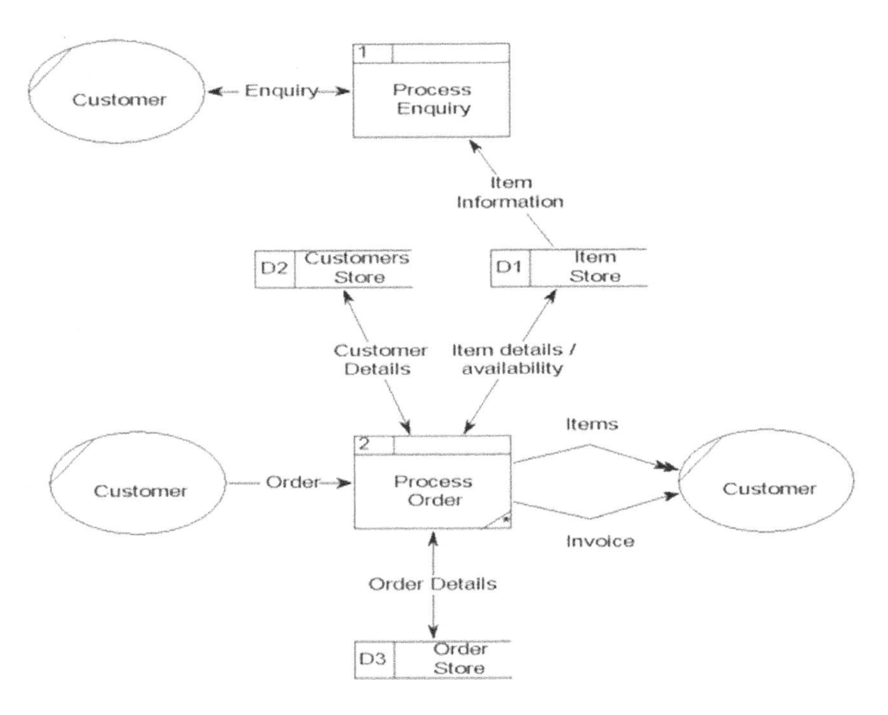

FIGURE 11.6 DFD produced by CASE tool (Source: Wikimedia Commons).

software designs for the next SDLC stage. Such CASE tools, in a pun on the term for capital letters, are called *Upper CASE*. Figure 11.6 shows a *data flow diagram*, which shows how information moves through a system, as an Upper CASE tool might draw it. Other CASE tools, called *Lower CASE*, are used in later stages of developing a new system.

The definition of a new system must include its functions and its user interface. Even if two systems do the same thing, one can be easier to use than the other. This affects how well a new system will be accepted, how much time users will need to carry out their tasks with it (time costs money), and how often they will have to call for support (support costs money). A programmer might prefer the dialogue box of Figure 11.7(a), but most users would prefer (b).

Once information gathering is complete, the systems analyst documents it. He or she then reviews the write-up with those who helped earlier. It's easy to make changes at this point, harder later on. Figure 11.8 shows one expert's opinion of the relative cost of changing a system at various times. This isn't a precise curve, and it differs from system to system, but the concept is universal.

Deliverable: The deliverable of the systems analysis stage is called a *functional specification.*

WRITING A CLEAR FUNCTIONAL SPECIFICATION

A requirement from U.S. Federal Highway Administration training materials: "The system shall turn off the alarm when the user presses the F6 key." This was interpreted two ways:

- The user wanted the key to silence the alarm if it was sounding.
- The programmer had the key turn the alarm off, so it wouldn't sound at all in the future.

It may be impossible to write requirements that can't be misinterpreted. Misunderstandings *will* happen. Review specifications carefully with this in mind. Be sure

(a)

UserForm1

TV Store Order Optimization

Enter Parameters 20,30,20,40,100,Y

Run it!

(b)

UserForm1

TV Store Order Optimization

	Lower Limit	Upper Limit
Reorder Point	20	30
Reorder Amount	20	40
Number of Iterations per Point		100

☑ Watch Graph Build

Run it!

FIGURE 11.7 (a) and (b) User interface examples.

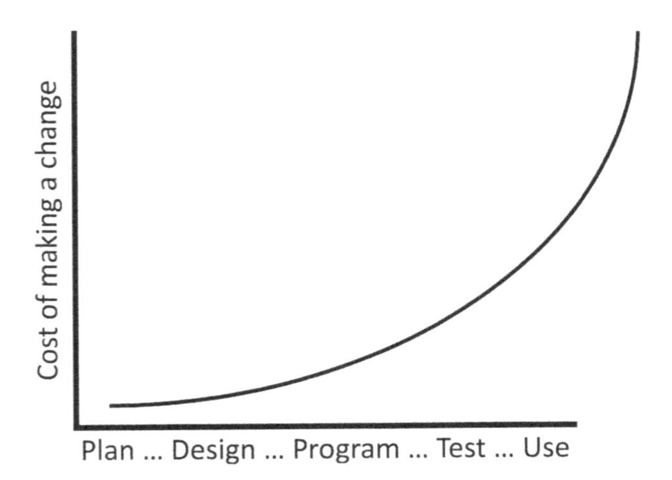

FIGURE 11.8 Chart of increasing cost of making changes over a system's life.

everything is tested to make sure it's what users meant, not what a programmer mistakenly thought they meant.

Where you fit in: As with the preliminary investigation, participation in systems analysis can be a good career move. It can improve your knowledge of the organization, enhance your relationship with its information systems department, and earn you positive attention from management.

When you tell a systems analyst what you need, be crystal-clear about how important that need is. Systems analysts are used to users who ask for features on a whim. It's hard for a systems analyst to tell the difference between "it would be nice if it did X" and "it *must* do X." If it really *must* do X, make sure the systems analyst understands why it's that important so that requirement isn't dropped later.

3. System Design

Systems analysis defines *what* a new system will do. System design defines *how* it will do it.

In this stage, experienced programmers decide what programs and databases will be needed for the new system. They specify them in sufficient detail for less experienced people to write those programs and create those databases.

CASE tools can be used to create database design specifications. Many such tools can then write commands to define those data elements to a computer. This ensures that the entire system shares one set of data definitions, removing a common problem in module interactions.

CASE tools can also define what the programs will do. Some are able to create parts of programs automatically. Other parts will still need to be coded by hand. In this case, the boundary between design and programming is fuzzy. When design is complete, a mouse click will create part of the program. Still, a formal design document is needed for review at the end of this stage.

Deliverable: The deliverable of this stage is called a *design specification* or *system specification*.

Where you fit in: Users don't do much during system design. If you helped in systems analysis, someone may need to follow up on something you said earlier. That won't take much time.

4. System Development and Test

Once the design specification is approved, programmers write programs and create databases that will do what the specifications call for. This stage may take as much time as the others combined.

CASE software can automate parts of system development. That saves time and reduces errors. Other kinds of software may also be used now, such as a *source code control system* to keep track of modules as they are written, to ensure that team members use the latest version of each one, and to facilitate reverting to an earlier version if a change introduces problems.

No matter how careful programmers are, programs seldom work correctly the first time. (You've probably encountered this with spreadsheet formulas.) Testing, to locate programming errors or *bugs* so they can be corrected, is intended to prevent defects from reaching users where they would waste time and potentially harm the business. Systems are tested in stages:

1 *Unit test.* Programmers send inputs to each module and inspect its output. Some test inputs are correct. For those, the output should also be correct. Some are deliberately incorrect. The module should detect these errors and respond appropriately. If any inputs don't produce the proper output, the reason must be found, the program corrected, and the test repeated. Unit test comes before modules are released to the entire project team.

2 *Subsystem test.* Tested modules are combined into subsystems, which are then tested in much the same way. Subsystem testing catches errors in communication among modules. Perhaps one module sends order data as "quantity, product ID" while another expects "product ID, quantity." Each module may operate correctly, but they don't communicate correctly.
3 *System test.* When all subsystems work, programmers test the system as a whole. They use it as a user would. They make mistakes they expect users to make and fix problems they find.
4 *User test.* Programmers test for mistakes they expect. Users will make different mistakes. The programmers didn't expect those, so they didn't program the system to catch them. Testing by users helps find these missing error checks.

Deliverable: A system that, as far as its developers can tell, works. Anything known not to work can't interfere with the system's ability to carry out its mandatory functions. All such deficiencies must be documented. *Work-around procedures* (alternative ways to achieve the same end) should be identified, if possible, and should also be documented.

Where you fit in: As with system design, programming and testing are technical activities. You can let the professionals handle them. You may be called on for user testing, but that's all.

5 Implementation

The *implementation* stage of the SDLC prepares the new system for *production use:* use in the course of business. This includes *training*, *change management*, and *system conversion*.

Training: In all likelihood, a new system doesn't work exactly like the old one. There may be new business processes. The users of a new system must learn how to use it effectively.

Training is a profession. It may be easier to teach a professional trainer about a new system than to turn a specialist in that system into an effective trainer. In any case, trainers must recognize that people who are used to the older system have a mental model of their information world based on how that older system works, and will have to change it.

***Change management*:** Moving to a new information system is organizational change. Change is difficult, often threatening. As Machiavelli pointed out over five centuries ago:

> *There is nothing more difficult … or more uncertain in its success than to take the lead in the introduction of a new order of things. For the reformer has enemies in all those who profit by the old order, and only lukewarm defenders in all those who would profit by the new order … [This comes] partly from the incredulity of mankind, who do not truly believe in anything new until they have had actual experience of it. Thus it arises that on every opportunity for attacking the reformer, his opponents do so with the zeal of partisans, the others only defend him half-heartedly, so that between them he runs great danger.*

Change must be managed. Today's theory behind *change management* is by Kurt Lewin. With additions by Edgar Schein, it is known as the *Lewin-Schein change model*. It says that effective change requires *unfreezing* people from what they're used to, *moving* them to a new place, and *refreezing* them there. All three must take place if change is to succeed. If change is not managed properly, employees will resent the new system. Those who are used to the old one will resist the new—openly, covertly, perhaps to the point of sabotage.

Where you fit in: Change cannot be managed by the technical staff. They don't have personal credibility with users. Only someone from the user community, a manager or a respected individual contributor, can manage this change. That may be you.

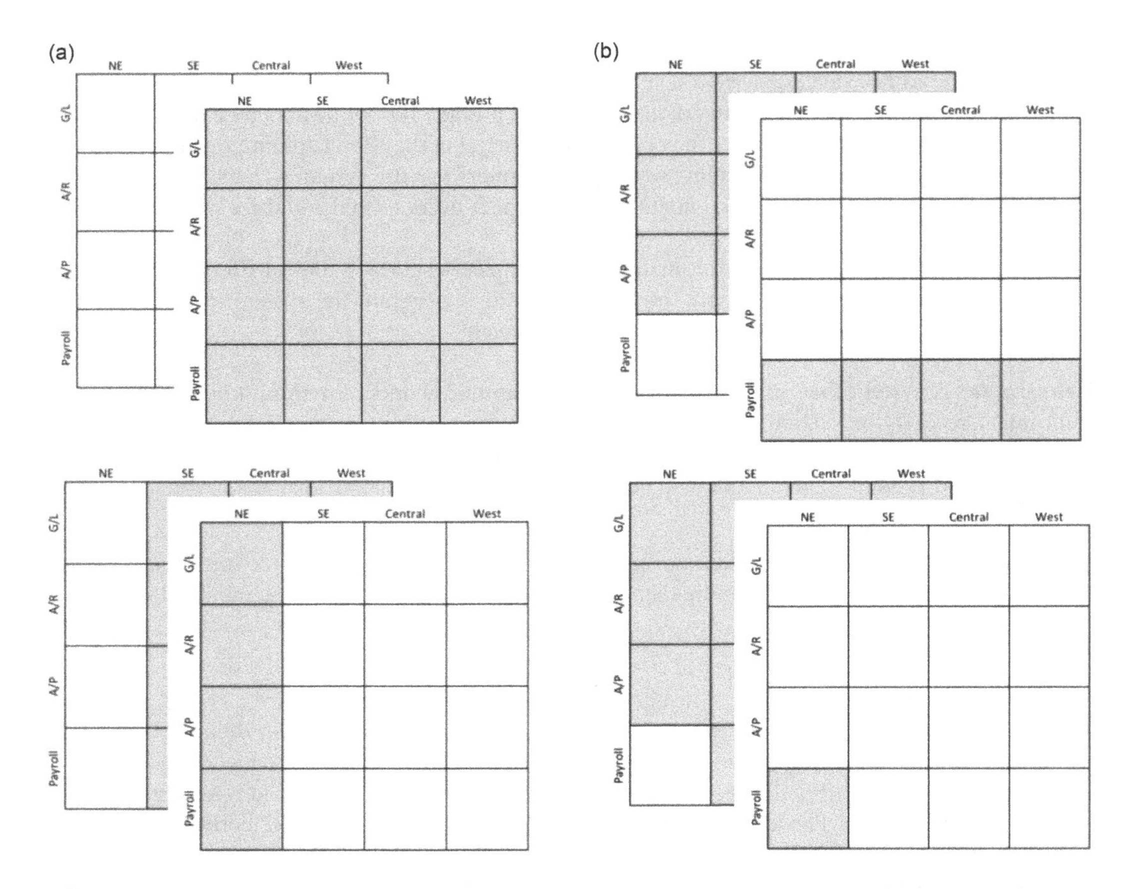

FIGURE 11.9 Conversion method diagrams: (a) direct cutover, (b) pilot, (c) phased, (d) pilot-phased.

System conversion: Installing a new word processor on your personal computer is simple. Pop in a CD-ROM or download a program, click a few buttons, accept the license terms, and it's there. This approach is risky with enterprise-wide information systems. Too much can go wrong. Other conversion methods reduce this technical risk. These methods, shown in Figure 11.9, are:

- *Direct cutover* (how you installed that word processor). The entire organization stops using the old system and starts using the new one. This has high technical risk. If anything goes wrong, the organization may not be able to function until the issue is resolved. However, this approach is best from a human standpoint since change can be managed at one time for everyone.
- *Pilot conversion.* Part of the organization switches to the new system while the rest of it uses the old one. Problems are confined to the pilot group. Problem-solving resources can be focused on it, while the rest of the organization operates as before.

  This reduces technical risk. However, it requires isolating part of the organization from the rest as regards the new system's tasks. (A university can't use a new registration system for math but not English, since many students take both.) It calls for managing change carefully so those outside the pilot group won't become dissatisfied with the old system too early.
- *Phased conversion.* If the new system can be broken down into parts that handle different functions, it may be possible to introduce them one at a time. Once the first part is working properly, a second can be introduced, and so on until the entire new system is in use.

This also reduces technical risk. (It may require temporary programs to move data between parts of the new system and parts of the old one.) The change management issue is unfreezing users from the old system while they must keep using some parts of it.
- *Pilot-phased conversion.* This combination of pilot and phased is used for complex systems in large organizations. Part of the organization starts to use part of the new system. Once it works properly there, the rest of the organization adopts it, and the same or another group starts to pilot another module. This continues until the whole new system is in general use.
- *Parallel conversion.* Here, the same data are input into both systems and their outputs are compared. Any differences are presumably due to an error in the new system. After that is confirmed (the old system might have had an undetected error), the reason is found and the problem is fixed. Once the new system provides reliable results, the old one is retired.

  Parallel conversion isn't keeping old software for backup. When you install a new word processor, you may keep the old one in case the new one has a problem. That's not parallel conversion. You don't use them both and compare printouts. It's an intelligent precaution.

  Parallel conversion was popular when it involved running a deck of punched cards through a computer twice. Output consisted of paper that could be easily compared. Today's systems are online. A conversion method that makes users enter data twice isn't viable. Differences in input timing can cause two online systems to produce different results even if both operate correctly. Parallel conversion is only used for a few financial reporting systems today.

Once users have been trained, the entire system has been converted, and the refreezing stage of change management is complete, implementation is over.

Deliverable: A system that the organization uses in its work. The stage review should confirm that implementation was successful.

Where you fit in: Technical professionals in charge of conversion may consider only technical risks. This may lead them to approaches such as pilot and pilot-phased conversion, which reduce technical risk but increase organizational risk. One must balance both types of risk in choosing a conversion method. Businesspeople should provide this balance. If a method such as pilot conversion is used, managers must understand its change management implications.

6 Operation

Today, most systems large enough for custom software run in client/server mode. Users access them through clients: desktop, laptop, tablet, or phone. They may use client software or browsers. The rest of the system, including its database, is on its servers. Professional operators make sure those servers operate properly, keep the database secure and reliably backed up, plan for sufficient capacity as the database and the number of users grow, and guard against intrusions.

Deliverable: Ongoing, reliable, and secure access to the new system. This stage isn't followed by a stage review because it has no endpoint. Management may conduct periodic *system audits* to make sure everything continues to work as it should.

Where you fit in: The IS department can keep a new system running, but the responsibility for getting business value from that system lies with its users on the business side.

- If it does what you need, use it. Your work, or your colleagues', during its development has paid off in a useful information system.

- If it doesn't do what you need, identify the issues to your management. Be specific. Don't blame or point fingers, unless you want to take responsibility for not having been clear about your needs earlier. Don't complain; suggest improvements.
- If the system is basically OK but could be better, read the next section.

7 Maintenance

A system that meets its users' needs perfectly on delivery will not meet them perfectly forever. Software reflects its developers' best efforts to understand what users needed at the time, and to express those needs in computer programs within the limits of their abilities, their resources, schedule constraints, and the available technology. People will find things to improve. Such *maintenance requests* come from several sources:

- Programming errors: bugs that were documented in the deficiency list or found later.
- Requests for features that were deferred due to time and resource constraints.
- Requests for features whose value was recognized only after using the system for a while.
- New requirements: These can come from many sources—legislation, a need to support users with mobile devices, customer requests for access to inventory data, a merger, etc.

Maintenance requests are collected and discussed by the steering committee, or a group to which the steering committee has delegated this task. It obtains estimates of what's needed to meet each request. Then, considering the cost and importance of each, it decides which will be in the next version of the system. Those that don't make the cut may be kept to be reconsidered later.

Programmers then go through a streamlined version of design, development, and test. The new version of the system is then installed. It is usually close enough to the previous version that little training and no change management are required. Direct cutover is common when installing a new version, though a pilot approach can be used if the changes were large.

Deliverable: An improved system that satisfies the approved maintenance requests.

Where you fit in: Most maintenance requests originate with users. (Exceptions: deferred items and the deficiency list.) Try to notice opportunities to improve the system, and submit them via the appropriate procedure.

The User's Job: The Overall Picture

The "where you fit in" paragraphs in the above sections had a lot for you to do at the beginning and the end, but not much in the middle. That is typical. User involvement in the SDLC follows a curve much like Figure 11.10. Understanding it will give you a good idea of when your participation and that of your colleagues will be needed and when it won't be.

What If We Bought Our Software?

The previous sections were about custom software. As you read in Chapter 10, most organizations buy most of their applications. That changes the SDLC:

Preliminary Investigation

An organization must still decide if a proposed system is feasible. Economic feasibility reflects the cost of purchasing and customizing the application rather than the cost of writing it.

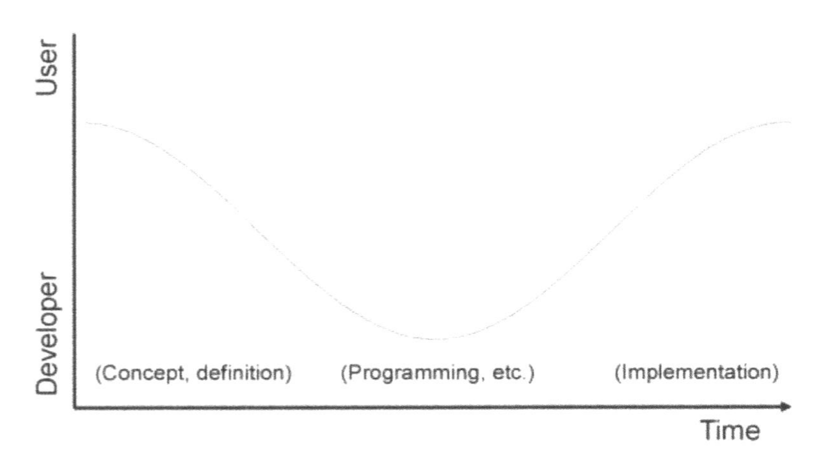

FIGURE 11.10 User involvement curve.

Systems Analysis

If purchased software is to be customized, that customization must be planned. If an organization selects such a system, it must document its own processes carefully.

System Design

Customization must be designed just as programs are. It is often necessary to specify database details, even if the general structure of the database won't change. For example, the student ID field in a university database must be set for the length and format of its ID numbers. If procedures will be added to a package, those must be designed just as for custom software.

System Development and Test

Testing is as important for customization as it is for development. All the testing steps are needed for every customized module. System and user tests are the same as for custom software.

Implementation

No change here. Implementation doesn't depend on where software came from.

Operation

No change here either.

Maintenance

Changes that involve customized modules are handled by the organization's staff.

Changes to the basic package are made by the software supplier. Even if an organization's staff could make them, modifying a package can interfere with moving to its next release. Therefore, the organization tells the vendor that it would like certain changes or enhancements. It's up to the vendor to make them. Your requests will not be the only ones it gets. The vendor must prioritize them. There is no guarantee that yours will be selected. The *roadmap* (the plan for future versions) of purchased software is out of your hands.

> ***Where you fit in:*** The channel for vendor requests probably passes through your company's MIS department. If you have an enhancement idea, contact them to find out how to proceed. How you couch your request, your business reasons for it, and how persuasively you show that all the vendor's customers can benefit from it will improve the odds that they'll do what you ask.

How the SDLC Differs for Software Products

The above sections describe the SDLC in organizations that use computers in their business. Software is also developed by *software vendors* for others to use. Software vendors also need a systematic development process. Understanding their process will help you select software. The stages for vendors are:

Preliminary Investigation

This is market analysis for a proposed product. A decision to develop a new system is based on how well it is expected to sell and whether the vendor can charge enough for it to earn a profit.

Systems Analysis

Vendors meet with potential users of the proposed product to understand their needs. Their systems analysts then document, in a functional specification, what the new system should do. They document what can be customized and the process for customizing it.

Where you fit in: A vendor could approach your employer for this purpose. You could be invited to such a discussion. What you learn in it can help you. You'll also get a chance to influence the development of a product that you may, down the road, use. The better you understand where this meeting fits into their development process, the better the chances that your opinions will count.

System Design

Once a new system has been specified at the functional level, programmers create its design specification exactly as they would if a company were developing it for its own use.

System Development and Test

Programming is the same whether a company develops software for itself or for resale. Testing, however, is different. After system test, user testing has two phases:

- ***Alpha test:*** Users in the vendor organization try to use the system as an outside user would. When Microsoft develops a new release of Word, its employees use it. They document problems they encounter. Those are fixed before anyone outside Microsoft sees this release.

  Sometimes alpha test is impractical. Microsoft has employees who use word processors, but a software firm is unlikely to own a university or an airline. In that case, this test step may be skipped.
- ***Beta test:*** After alpha testing, users outside the organization try the software. Continuing the example, Microsoft could ask companies that use the current release of Word if they would like to test the next one. Some will: Beta testing gives users early access to new features, may improve their relationship with their software supplier, and in some cases may come with extra support or price concessions. In return, they agree to provide feedback on the new software, with details of any problems. The vendor can fix these problems, document work-around procedures, or remove the affected features from the product.

Following successful beta test and the correction of any problems that were found, the software goes to *general release*. It is now available to anyone with a sufficiently large checkbook.

Implementation

A software product is implemented by the organization that purchases or licenses it, not by its developer. Training, change management, and conversion are the same whether software is developed in-house or purchased. An outside developer may make experts, such as experienced trainers or customization consultants, available to customers—usually for a fee.

Operation

Operation is also independent of where software comes from. An organization that uses software operates it, unless it contracts for SaaS or another arrangement. As with implementation, vendor help is usually available. Some may be included with a license. If not, or if more is needed, it may be available for a fee. User groups and online forums also help with using commercial software.

Maintenance

Software vendors fix bugs and enhance their products regularly. Enhancements are a competitive necessity: If one vendor develops a feature that customers like, others must offer similar features or lose sales. Ideas for new features that will appeal to customers come from users, competitive products, and the vendor's staff. Once fixes and enhancements have been tested, a new release of the system is made available. Over time, support for old releases is phased out. This avoids diluting support resources and encourages users to upgrade to the current release (for a fee).

> ***Where you fit in:*** Moving to a new version of software is a business decision. An organization must weigh the benefits of new features and ongoing support versus the cost of the move. New software may require a newer operating system, hardware upgrades, and repeating customization work in addition to the vendor's license fee. As a businessperson, you must ensure that all these factors are considered before choosing to upgrade software.

SDLC Strengths and Weaknesses

Strengths

SDLC offers control of software development. Management knows what is being done, who is doing it, when they should finish doing it, and what the organization will have when they do. If system requirements must be defined carefully, such as for a system with financial or legal implications or one to be outsourced under a fixed-price contract, this is important.

Weaknesses

The SDLC does not handle dynamic needs well. It is difficult or impossible to change anything without reverting to an earlier stage of development. In fast-moving fields, an old functional specification can mean a useless system. If a user goes to programmers and says "This isn't what we need now," the programmers will reply "Too bad. We have to write what the spec says."

Formal specifications are a poor communication vehicle. Those who approve specifications seldom have time to study them, and might not understand them anyhow. They may approve a spec, assuming that its writers knew what they were doing, and be surprised later to find that the system they approved is far from what's needed.

Problems are often discovered late. Winston Royce, one of the originators of the SDLC concept in 1970, wrote at the time:

> *The basic framework described in the waterfall model is risky, and invites failure. The testing phase ... at the end of the development cycle is the first event for which timing, storage, input/output transfers, etc., are experienced as distinguished from analyzed. ... The resulting design changes are likely to be*

> *so disruptive that the software requirements upon which the design is based and which provide the rationale for everything are violated. Either the requirements must be modified, or a substantial design change is warranted.*

This may be overly pessimistic. Successful SDLC use in thousands of projects over decades proves that disasters are far from necessary. Still, his basic point—that experience with the software comes only well after design is complete, and cannot influence that design—is valid.

Stage reviews, which ought to be a vehicle for open exchange of information, can put developers on the defensive and lead to confrontation instead. Development teams may hide problems to make a project look as good as possible.

11.3 OTHER SYSTEM DEVELOPMENT APPROACHES

Other software development approaches accept change and improve communication by involving users more. Some alternative approaches are described here. Their common premise: One can't know all the requirements ahead of time, so software developers must embrace change and minimize its cost. SDLC says the opposite: Enough analysis and design will eliminate the need for changes. If changes are necessary anyhow, as they often are in today's world, they'll be costly.

The important thing is not the specifics of each method, but how they approach the development process. New approaches appear often. The methods described here are not mutually exclusive. Parts of different methods can be used together. Some methods your employer will use probably haven't been invented yet. If you understand the concepts behind these methods, you'll be able to adapt to any method you see in the future and understand your role in it.

The Prototype Concept

A *prototype*, in software, is a preliminary version of an information system that demonstrates important features that the finished item will have.

A prototype is not a finished system, though it may look like one. It is developed using tools that focus on its user interface and lend themselves to quick changes. It may lack error checks, a full database, security, and other behind-the-scenes aspects. Their absence may not be obvious. People who see a prototype may think the system is almost finished. In fact, most of the work—and the most difficult work—remains. An information system is like an iceberg (Figure 11.11): We see only its tip. Users who work with a prototype must understand that unseen parts are absent.

FIGURE 11.11 Sketch of iceberg mostly below the surface.

Prototypes can be used in two ways:

- In one approach, a prototype replaces some or all of the functional specification. It's better than a document because it's dynamic. Users can try it out and see how the system responds to their actions. Once they approve the prototype, programmers reproduce its behavior in their usual programming languages and/or CASE tools.
- In another approach, often called *evolutionary prototyping* or *rapid application development* (RAD), features are added until the prototype becomes a usable system. This requires building the prototype with software that can be extended to become an operational system.

Where you fit in: The first approach creates a more efficient system. The second produces it sooner. The tradeoff—efficient or sooner?—is a business decision. It cannot be made on technical grounds. Transaction processing systems must be efficient. Delay in getting them is acceptable. Systems intended to give a company a competitive edge are needed as quickly as possible. Worse performance, which can often be offset by spending more on hardware, is usually a small price to pay for the business benefits of getting them sooner.

Alternative Development Methods

Many development approaches based on prototyping have emerged since the drawbacks of SDLC were first recognized. They differ in how prototyping is used, in the people who participate in the process, and in their sequence of activities. A few such methods are discussed here.

Each development approach has unique features and differences from the SDLC. An organization need not adopt one approach in its entirety. For example, Agile development calls for writing a module's tests before programming that module. This can be done without adopting Agile in full.

Where you fit in: These development methods change the curve in Figure 11.8. The middle doesn't dip as far. Users stay involved as a new system is developed and tested.

You may have little say in your employer's choice of a system development method. Whether it uses SDLC, one of the methods outlined below, a method not discussed here (no book can cover them all!), or a mix of several methods, users will have parts to play. You may be selected to play one. Find out what is expected of you and when. You may need time to prepare for your task.

Joint Application Design

Joint Application Design (JAD) was conceived in the late 1970s as a way to involve users in the design of an information system. Its intent is to develop functional and design specifications as a collaborative process involving both users and technical staff.

In JAD, joint activities are organized into *workshops*. Each workshop, led by a trained facilitator, lasts from a day to a week and has a defined goal: for example, a complete design for a software component. To make this happen, the JAD group includes a *modeler*: a person experienced in the use of CASE tools to document a design idea on the spot for the group to discuss and agree on.

At the time JAD was introduced, software that could evolve a prototype to an operational system didn't exist. If it had, it would have been prohibitively slow on computers of that era. Now that such software is available, JAD concepts can be extended into the development stage. The name then becomes *joint application development*, keeping the acronym JAD.

In JAD of the 1970s, group meetings were always face-to-face. Today, technology supports virtual meetings and can also improve the productivity of face-to-face meetings.

Rapid Application Development

Rapid Application Development (RAD) begins with JAD, or something much like it, to develop the design of an information system. This design is embodied in a prototype, which is then extended until it becomes the complete system. Users continue to be involved throughout this construction stage, suggesting changes or improvements as they watch the system take shape.

Agile Development

Agile development methods break information systems into small pieces that can be developed quickly. The result is a useful new version, with new capabilities, every few weeks. The new version might not be released to its users (or to the market, if it is for a software product), but after several such iterations, the system will have changed significantly.

Agile development methods involve *user stories*, *sprints*, and *scrums*.

A *user story* is a description of something a user needs to be able to do with a system. In a student registration system, a user story might be "As a student, I want to be stopped if I try to register for a course for which I don't have the prerequisites." User stories follow the INVEST rules: They must be Independent, Negotiable (their priority can be negotiated), Valuable, Estimable (the effort of programming them can be estimated), Small, and Testable.

A *sprint* is the basic unit of development. At the beginning of a sprint, user stories are chosen from a list of requirements called a *backlog*. The development team sets a schedule, a week to a month, to work on those elements. At the end, a new version of the system is released.

A *scrum*, a term borrowed from the sport of rugby, is a daily status meeting that lasts no more than 15 minutes. Scrum meetings are often held standing up (Figure 11.12) to keep them short. They can result in changes to the requirements backlog. Thus, the functions of the information system (other than what is being worked on in the current sprint or sprints) are flexible during agile development.

Agile methods prioritize collaboration with a system's prospective users, getting software to work, and flexibility to change. Other items—written specifications, documentation of how it works,

FIGURE 11.12 Photo of agile development scrum meeting (Source: Wikimedia Commons).

efficient use of computer resources—take a back seat. Is this the right set of priorities for a specific system? That is a business decision. It must be made before development starts.

Agile development has implications beyond development. The concept of creating functionality increments in sprints is different from developing a major release with many changes. This has implications in moving new versions of applications from development to use. The operations team must accept a new version every few weeks rather than a few times per year. Alternatively, the operations team can roll out selected development releases to users, skipping intermediate ones until enough changes have been made to justify moving users to a new version.

Where you fit in: Frequent rollout of new system versions impacts all its users. It can be easier to keep up with an evolving application than one that has less frequent but more drastic changes, but users may need to be trained to work this way. Agile methods may make this part of your job.

DevOps

After successful testing and review, the development team traditionally passes a major system enhancement to an operations group for deployment. This time-tested approach has drawbacks. Separating the two groups can cause problems when things don't work as expected. Calling developers off new projects to fix problems leads to rushed fixes, poor testing, delays in their new projects, and mistrust between development and operations groups. In addition, separating developers from users can perpetuate problems from release to release when Operations copes with these problems and developers therefore never learn of them.

The DevOps approach (**DEV**elopment/**OP**eration**S**) addresses these concerns by delivering small incremental improvements on a more frequent basis and using the same team that developed them to deploy them. The team includes specialists in testing, implementation, and so on—adopting DevOps doesn't eliminate the need for specialized expertise—but they are under one manager, not in separate parts of the organization.

Any organization can try out DevOps on a small scale. The main concern with using it only in part of the organization is that it requires some aspects of matrix management (see Section 12.2). It can be hard to institute matrix management in part of an organization but not all of it. Still, for a limited pilot project, this difficulty is not insurmountable. Once it has been used successfully on a small scale, the organization can—if it so wishes—extend it to all its software projects.

You may have experienced one aspect of DevOps, frequent deployment of small enhancements, if you use Chrome or Firefox. These use a "rapid release" cycle, with releases every few weeks. Other software suppliers use it as well.

In Summary

Whatever approach is used to develop information systems, they always follow a life cycle such as Figure 11.2. That is their nature.

11.4 MANAGING THE DEVELOPMENT PROCESS

System development approaches don't just happen. These processes must be managed.

Project Management

The *project manager* (PM) is central to system development. The PM works with users to gather requirements, in testing, and in implementation; communicates project status to the IS steering committee, supervises the development team, and is ultimately responsible for the value that the organization receives—or doesn't receive—from the system.

Project managers trade off resources (people, money, and time) against system capability. That includes what the system does, its efficiency, its security, and any other measures that may apply.

- If fewer people are used, it will take longer to develop a given capability.
- If fewer people are used for the same time, the system will have less capability.
- If the same number of people have less time, the system will also have less capability.

Project managers develop plans that (ideally) lead to an agreed result, at a scheduled time, using the planned resources. To do this, they break projects into steps as called for by the development method to be used, estimate the time each will take, and analyze the order in which steps must be carried out. Once the plan is accepted by the IS steering committee, PMs monitor progress against it, report project status to the steering committee, take corrective actions if the project deviates from its plan, and alert the steering committee if the needed corrective actions exceed the PM's authority or if no corrective action seems feasible.

The earliest project management tools were charts. A *Gantt chart* (first used by Henry Gantt in 1910) shows tasks vertically and time on the horizontal axis. Figure 11.13 shows a Gantt chart for a public relations project. This chart was created in a spreadsheet program. Project management software can position tasks automatically on a Gantt chart, based on dates and durations.

Network charts show task relationships. Such a chart for building a house (Figure 11.14) shows that the foundation must be completed first, then the framing, then the roof. They can't be done in parallel. If each takes a week, the house can't be ready for exterior painting in less than three weeks. Wiring and inside wall construction can be done in parallel with roofing, so if they take no longer than a week each, they do not add time.

Month:	1	2	3	4	5	6	7	8	9	10	11	12	13	14	15	16	17	18
Activity:																		
Strategizing	■																	
Reality Check	■	■																
Planning		■	■	■	■													
Approval					■	■												
Enlisting support							■											
Staffing							■	■										
Database setup							■											
Initial database entry								■										
Newsletter plan, write									■	■								
Newletter appears											■	■	■	■	■	■	■	■
Seminars, events														■			■	
Visits to analysts													■		■		■	

FIGURE 11.13 Gantt chart for a PR project.

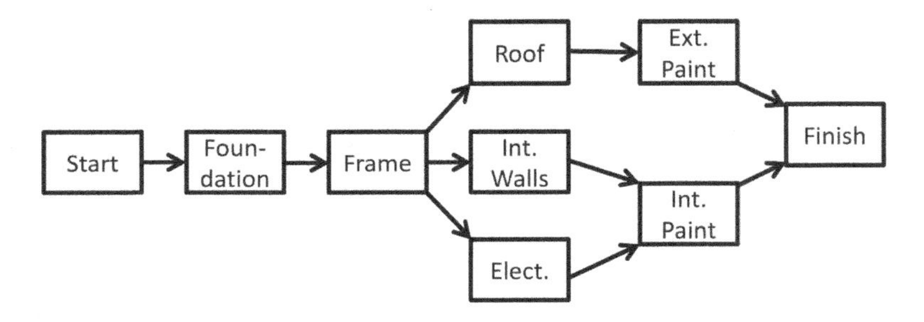

FIGURE 11.14 Network chart of building a house.

The *Critical Path Method* (CPM) formalizes network chart concepts. A CPM chart shows which tasks are on the *critical path*. If any of them take longer, project completion is delayed. Tasks off the critical path have *slack*. Those tasks can take longer, up to the amount of slack, without delaying project completion. In the house project, if exterior painting will take a week but interior painting will take only three days, interior painting has two days' slack.

Program Evaluation and Review Technique (PERT) evolved from CPM. It uses three durations for each task: optimistic, normal (most likely), and pessimistic. This permits statistical calculation of expected project duration. For example, if a project has six parallel tasks, and each is expected to take two weeks, CPM will show the entire project as taking two weeks. PERT recognizes that some tasks will end up taking more than two weeks, some less, because of unpredictable variability. It will therefore predict a later date for completing all six and finishing the project.

PERT and CPM ignore resource constraints. A network chart may show that two tasks can be completed in parallel, but won't show that they share resources. In Figure 11.14, if there is just one paint crew, the house can't be done in four weeks even though neither painting job depends on the other. *Critical Chain Project Management* (CCPM) integrates resource dependencies. It also makes other changes to PERT/CPM, such as reducing task time estimates and putting half the reductions into an overall project buffer. It is difficult to use part of CCPM without using all of it, and it can be hard to use it for one project unless it is used for all the projects that share its resources, so a firm must consider CCPM carefully before adopting it.

Project management software helps track activities, people, and resources. It can produce a variety of charts and reports. Figure 11.15 shows project management software in action. This screen shows a project as a Gantt chart, but clicking the icons at the upper left will show it in other ways.

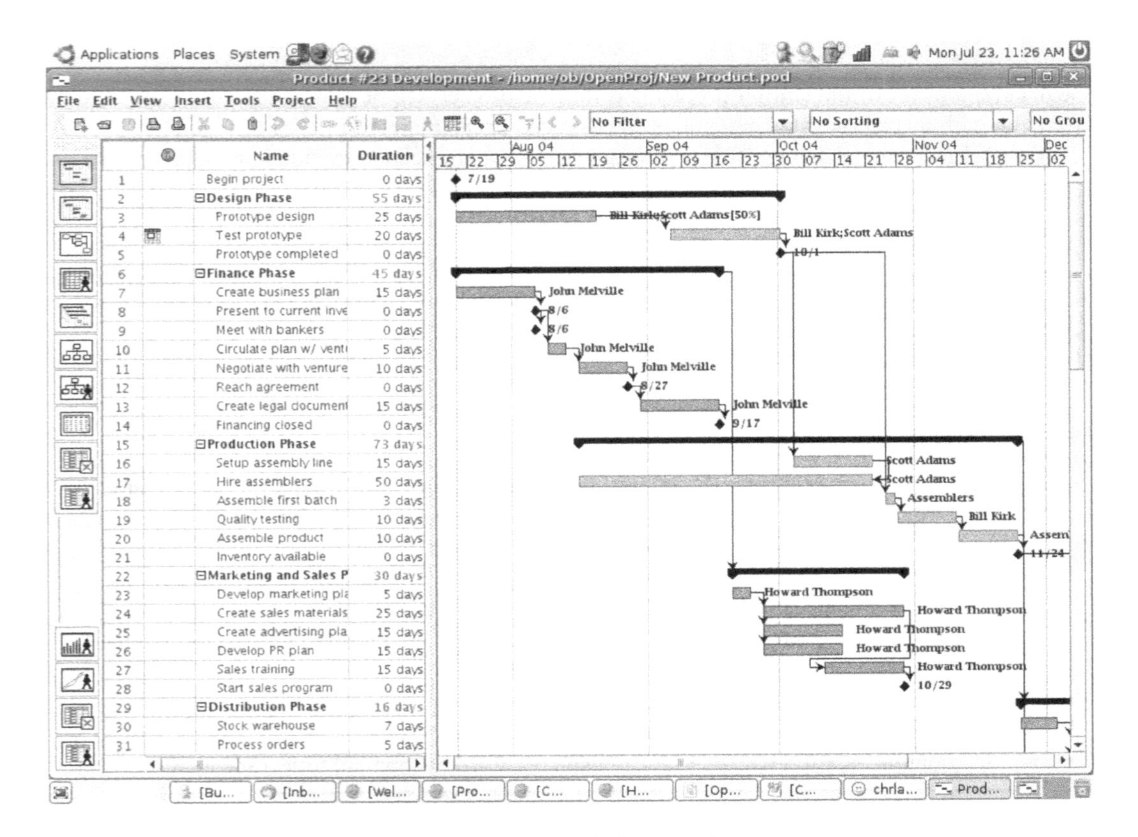

FIGURE 11.15 PM software screen (Source: Wikimedia Commons).

Where you fit in: The project manager is normally the person with whom to discuss software that is under development. If you need to do that, find out who the PM is and start there.

Late Projects and Brooks's Law

Fred Brooks argues in his classic *The Mythical Man-Month** that "adding staff to a late project makes it later." This effect, now known as *Brooks's Law*, is not what we'd expect. Common sense tells us that, if we put more people on a job, it will be finished sooner.

This "common sense" does not apply to software. If a three-person software project is behind schedule and two more programmers are assigned to speed it up, this happens:

- Modules are reallocated from three people to five. This takes time. During this time, nobody knows what to work on, so no work gets done.
- Everything that was done before is explained to the new participants. This takes time. During this time, the old team members are busy explaining, so they can't do anything else.
- New team members have ideas. Their ideas are discussed. That takes time. Some of their ideas seem good and are adopted. That means scrapping some of what was done before. That will delay the project, though it might also improve the finished product.
- While new team members are coming up to speed, their contributions are limited. During this time, they also make demands on other team members' time for explanations and help.
- Instead of communicating with two team members, each member must now communicate with four. Team communication takes longer, leaving less time for programming.

The net effect of these and similar problems is that the project will be completed later than it would have been with its original staffing, not earlier—and at greater expense.

Brooks's Law is not absolute. If staff is added early in a project, before much work has been done or much of the schedule has been used up, negative effects can be minimized.

Where you fit in: When you see a software project fall behind schedule, it can be tempting to add staff to bring it back on track. Remember that this will usually not achieve the desired effect.

Points to Remember

Limoncelli lists "top 10 [actually 13] things executives should know about software" in a 2019 paper. Several you should know follow, with his original numbering. His paper is in the chapter bibliography. You can find all 13, with more discussion of each, there.

1 Software is not magic. It must be designed, built, and operated.
2 Software is never done, any more than marketing is ever done. The world changes, competitors offer new features, people have new ideas.
3 Software is a team effort. Nobody can do it all.
5 Security is everyone's responsibility. If you think you are not a target, you are wrong.
6 Feature content, as perceived by users, doesn't predict development time. (See next section.)

* Despite a title that comes across as sexist today, this book contains much timeless wisdom.

9 Software doesn't run itself. Someone has to turn it on and tend to it. Operation must be planned for, budgeted, and managed.

Estimating and Pitfalls

Managing time, people, and results for system development requires estimating the effort it will take to achieve a result. Systems analysts develop preliminary estimates to assess economic feasibility during system investigation. Those are refined during a project. If they grow (shrinking is rare), the IS steering committee must be informed, since that may affect decisions to proceed.

Estimating system development effort is difficult. There are no reliable methods, though many have been proposed. Managers compare new projects with earlier ones of known duration, use formulas involving programmer experience and project complexity, use other formulas based on how many basic functions a program performs, and more. None of these works well in practice. They also apply rules of thumb such as "take your best estimate and triple it," or even "take your best estimate, double it, then use the next larger time unit." In other words, if a programmer thinks it will take two days to write a program, it will be four weeks until that program is ready for user deployment. This may seem extreme, but history suggests it's not far off.

The purpose of a program also affects development effort. One rule of thumb states that, if it takes X effort to write a program, it will take 3X to make it into a usable system. Not recognizing this has led companies to find that they have a program that works, but don't have a usable business system. It will take three times that—9X—to make it into a marketable package. Firms that say "We have a great program, let's sell it" can lose a lot of money by not understanding this.

Estimating for Agile and similar methods differs. Agile managers fix team size and schedule, prioritize system capabilities, and let the team loose on the backlog. This approach can be scary to managers who are used to knowing, in advance, what will be available at each milestone if all goes well. The unfortunate fact is that all often does *not* go well. Agile, if system requirements are prioritized properly, can produce good results.

Pressure to get information systems developed quickly can backfire. Consider this dialogue between a new project manager (NPM) and a senior vice president (SVP):

NPM: "This system will take six months to develop."
SVP: "We don't have six months. We need it in three."
NPM: "We can do it in three if we defer features A, B, and C."
SVP: "We can't defer those. We need all of it in three months."
NPM: "It's impossible."
SVP: "If you can't do it in three months, you can forget about your promotion to project manager and your raise. I'll find a project manager who can."
NPM: "Okay. I'll figure something out. You'll have it in three months."
SVP: "Thanks."

NPM then updates his résumé, emails it to his LinkedIn connections, and submits it to three job openings on Indeed.com. He has a new job in two months. His replacement tells the senior VP that the project always needed six months, two of which have gone by, and that NPM couldn't have been serious when he promised it in three. The SVP can't do a thing about it.

Where you fit in: At some point in your career you will be a senior vice president. Don't be *that* senior vice president.

KEY POINT RECAP

Software is best developed using a systematic approach. *The right approach increases the odds that it will be completed on schedule and within its budget, and will meet organizational needs.*

As a businessperson, you will often depend on custom software. The quality of your input during development affects how useful that software will be. Understanding the development process will improve the quality of that input.

The System Development Life Cycle (SDLC) approach is appropriate some of the time and is the basis for many other methods.
As a businessperson, you should understand its principles of deliverables and stage reviews. You should know what to do in each of its stages if you are called on to contribute.

Other approaches, most based on the prototyping concept and/or taking small steps toward a big goal, can be more responsive to changing user needs.
Businesspeople know needs change. The pace of change is increasing. It may not be practical to freeze requirements early in a project. You may have to advocate for a responsive development method rather than the SDLC when the business requires responsiveness.

Implementation is essential to having an effective system.
If a system is not rolled out to users in a way that they can accept, it can never succeed. Some aspects of implementation can only be handled by businesspeople such as yourself. You must understand why and be ready to do this.

Project management is essential to successful information systems development.
You may find yourself managing an information system project even if you are not an IS professional. You should be familiar with project management concepts and tools so you can brush up on them quickly at that time. They'll come in handy in anything else you manage, too!

KEY TERMS

Agile development: System development approach in which an information system is divided into small parts that provide new capabilities and can be developed quickly.

Alpha test: Preliminary *user testing* of a new software product, by people who work for the firm that developed it.

Beta test: Final *user testing* of a new software product, by people who don't work for the firm that developed it.

Brooks's Law: "Adding staff to a late project makes it later."

Bug: Informal term for a programming error that causes a program to behave incorrectly.

Change management: The process of motivating users to want to use a new system.

Computer-Aided Software [or Systems] Engineering (CASE): Software tools designed to help with development. See *lower CASE*, *upper CASE*.

Critical Chain Project Management (CCPM) chart: Project management approach which integrates task and resource dependencies.

Critical path: The sequence of activities which, if delayed, will delay project completion.

Critical Path Method (CPM) chart: Network chart in which the critical path is identified.

Deliverable: Tangible item that proves completion of an SDLC stage.

Design specification: Deliverable of the system design SDLC stage.

Direct cutover: System conversion method in which an older system is replaced in its entirety by a new one for an entire organization at the same time. Compare *pilot conversion*, *phased conversion*, *pilot-phased conversion*, *parallel conversion*.

Economic feasibility: Of a project, providing economic benefits that exceed its costs. Compare *operational feasibility*, *technical feasibility*.

Evolutionary prototyping: See rapid application development.
Feasibility study: Deliverable of the preliminary investigation SDLC stage.
Functional specification: Deliverable of the systems analysis SDLC stage.
Gantt chart: Project control chart in which time is shown on the horizontal axis and resources or activities along the vertical axis.
General release: Decision by a firm that develops and sells a software product that testing is complete and the product should be available for anyone to purchase.
Implementation: Fifth SDLC stage, in which the system is delivered to its users.
Interviews: System analysis method, consisting of speaking to users of an existing system prior to writing functional specifications of a new one. Compare *observation*, *questionnaires*, *reading*.
Joint Application Design (JAD): System development method in which users collaborate with technical staff in developing system specifications and in extending them to a finished system.
Joint Application Development (JAD): System development method in which users collaborate with technical staff in developing system specifications.
Lower CASE: Software tools designed to help with the later stages of system development, including software/database design, programming, and testing. Contrast with *upper CASE*.
Maintenance: Seventh SDLC stage, in which a system is modified to meet new requirements.
Modeler: Participant in *JAD workshops* who is experienced in the use of CASE in JAD.
Moving: Stage of *change management* in which users are taken from an existing system or manual procedure to a new information system. Compare *unfreezing*, *refreezing*.
Network chart: Project control chart in which activities form a network that shows their precedence relationships.
Observation: System analysis method, consisting of watching people use an existing system prior to writing functional specifications of a new one. Compare *interviews*, *questionnaires*, *reading*.
Operation: Sixth SDLC stage, in which a new system is used on a regular basis.
Operational feasibility: Of a project, able to work with an organization's culture and processes. Compare *economic feasibility*, *technical feasibility*.
Parallel conversion: System conversion method in which both old and new systems process the same inputs, and their outputs are compared, until the new system is seen to operate correctly. Compare *direct cutover*, *pilot conversion*, *phased conversion*, *pilot-phased conversion*.
Phased conversion: System conversion method in which an organization as a whole uses part of a new system, along with the rest of the existing one, until that part is seen to operate correctly. Compare *direct cutover*, *pilot conversion*, *pilot-phased conversion*, *parallel conversion*.
Pilot conversion: System conversion method in which part of an organization uses a new system while the rest continues to use the existing one until the new system is seen to operate correctly. Compare *direct cutover*, *phased conversion*, *pilot-phased conversion*, *parallel conversion*.
Pilot-phased conversion: System conversion method in which part of an organization uses part of a new system while the rest continues to use the existing one, until that part of the new system is seen to operate correctly. Compare *direct cutover*, *pilot conversion*, *phased conversion*, *parallel conversion*.
Preliminary investigation: First SDLC stage, in which it is determined whether or not a proposed system is feasible.
Production use: Regular use of an information system in the normal course of business.
Program: A group of computer instructions that is intended to carry out a specific task.
Program Evaluation and Review Technique (PERT) chart: Variation of CPM chart in which project durations are described statistically rather than being fixed.
Project Manager (PM): Person responsible for all activities related to a development project.
Prototyping: Any system development method that uses one or more *prototypes*.
Prototype: Preliminary version of an information system that demonstrates some of its features but is not suitable for production use.

Questionnaires: System analysis method, consisting of submitting written questions to users of an existing system prior to writing functional specifications of a new one. Compare *interviews, observation, reading.*

Rapid Application Development (RAD): System development approach in which features are added to a prototype until it has become the desired system.

Reading: System analysis method, consisting of examining written material about an existing system prior to writing functional specifications of a new one. Compare *interviews, observation, questionnaires.*

Refreezing: Stage of *change management* in which users are motivated to want to continue using a new information system. Compare *unfreezing, moving.*

Scrum: Short status and planning meeting in *agile development.*

Software: A collective term for computer programs.

Source code control system: Computer program that keeps track of modules that comprise an information system, their versions, and the status of those versions.

Sprint: Unit of activity in *agile development,* in which software to support one or more *user stories* is developed.

Stage review: Formal project review in the SDLC method,

Subsystem test: Testing part of a system to confirm that it operates correctly with artificial inputs designed to test it under a variety of conditions. Compare *unit test, system test, user test.*

System conversion: The process of replacing an existing information system or manual procedure with a new information system.

System design: Third SDLC stage, in which the way a new system will work is determined.

System development and test: Fourth SDLC stage, in which a new system is put into technical form and determined to operate correctly.

System [or Software] Development Life Cycle (SDLC): A formal method of software development, based on milestones with deliverables and review of each deliverable before proceeding.

System specification: See *design specification.*

System test: Testing a complete system to confirm that it operates correctly with realistic inputs designed to test it under a variety of conditions. Compare *unit test, subsystem test, user test.*

Systems analysis: Second SDLC stage, in which detailed functional requirements of a new system are determined.

Systems analyst: Person who specializes in the *systems analysis* stage of the SDLC.

Technical feasibility: Of a project, able to work with an organization's existing technology or with technology it can obtain. Compare *economic feasibility, operational feasibility.*

Training: The process of teaching users what they must know to use a new system.

Unfreezing: Stage of *change management* in which users are motivated to want to stop using an existing system or manual procedure. Compare *moving, refreezing.*

Unit test: Testing an individual module of a system to confirm that it operates correctly with artificial inputs designed to test it under a variety of conditions. Compare *subsystem test, system test, user test.*

Upper CASE: Software tools designed to help with the earlier stages of system development, including systems analysis and specification. Contrast with *lower CASE.*

User story: In *agile development,* a description of an activity that a user should be able to perform with a new information system.

User test: Testing a complete system to confirm that it operates correctly with inputs produced by those who will eventually use it. Compare *unit test, subsystem test, system test.*

Waterfall method: Alternate, informal term for the *SDLC method.*

Work-around procedure: An alternate way to accomplish an objective that cannot be achieved as planned because of a flaw in an information system.

Workshop: Unit of JAD activity leading to a previously defined goal.

REVIEW QUESTIONS

1. What are the stages of the SDLC approach?
2. What happens after each stage in the SDLC approach?
3. List the deliverables of each stage in the SDLC approach.
4. What can happen as the result of an SDLC stage review?
5. What are the three types of feasibility that are evaluated in a preliminary investigation?
6. List the four information gathering methods that systems analysts use.
7. During what SDLC stages might CASE tools be used?
8. What are the four levels of testing for custom software development?
9. What three types of activities are carried out in system implementation?
10. What situations give rise to maintenance requests?
11. During what parts of the SDLC are users most involved? Least involved?
12. How does maintenance of purchased packages differ from maintenance of custom software?
13. What types of testing do software vendors perform on new software?
14. List the strengths and weaknesses of the SDLC method.
15. What is a *prototype*?
16. Describe, in general terms, two fundamentally different approaches to using prototypes.
17. What types of resources does a project manager have to trade off among?
18. What is the difference between a CPM chart and a PERT chart?
19. What is Brooks's Law?

DISCUSSION QUESTIONS

1. Your company's IS department asked you to represent your department on a team that will plan a new information system your department will use next year, to provide input on its needs. This would require 8–10 hours a week over the next two months. Your boss turned down their request, saying "I don't pay you to do that, and your work here is too important." Write a memo to your boss explaining why he should support your participation.
2. The "factoid" at the beginning of Section 11.1 gives the size of today's SAP software. SAP software is nearly 50 years old. A lot of code was scrapped and replaced over that time, so assume twice that many lines were written in total. Assume, also, the widely used figure of a programmer producing ten lines of tested code per day. (This figure is flawed, but will do for this question.) How many people, on the average, worked on this program over that time? If the project team was quite small in 1973, how big do you think it is now? What conclusions can you draw from this as regards the software, working at SAP, the cost of enterprise information systems, the likelihood of errors in them, or anything else?
3. Write a two-page feasibility study for a new information system for your school. It will take in information about applicants and follow them through the rest of their association with the school, including academic records, through their participation in alumni activities. Make and state any necessary assumptions.
4. Consider the conversation between a user and a systems analyst in Section 11.2, where the user opens by stating a need for drainage channels in a keyboard.
   a. What is the *real* problem here?
   b. Come up with five solutions for keeping this user's coffee from damaging the keyboard. (Be creative. Don't just pick obvious ones.) Then write down four criteria, one of them cost, for evaluating solutions. Assign a weight to each so the weights total 100%. Rate each solution from 1 to 10 on each criterion and calculate a total weighted score for each. Would you recommend the highest-scoring solution? Why or why not?
5. Are system design and maintenance related? If so, how?

6. You and three other students are working on a term project that counts for half your grade in a senior course. It will use calculations and graphs from spreadsheets in a written report and a presentation. The spreadsheets must be done first so their data can be used in the other parts. Your team assigned parts of the report to each team member. After each member completes his or her part, the team will review that part as a group and agree on changes. That member will make those changes. The entire report will then be assembled. At the end of the term, you will submit the report to your instructor and present your project to the class.
   a. Compare this process to the SDLC.
   b. Draw a network chart for this project.
   c. Find out more about *source code control systems*. Do you think one could be useful on this project? If so, how would you use it? If not, what changes could make it useful?
7. An SDLC development project has about six stage reviews. Each review occupies the development team (if it's small) or parts of it, and the review board, for the review's duration. The development team must prepare pre-review materials, and the review board must study them. The review board then writes up its findings and presents them to management. The development team responds to those findings. None of this moves the project itself forward.
   a. A project is estimated to require an average team size of ten people for a year, 120 staff-months, not including these reviews. Estimate the time that people (both team and review board) will spend preparing for, in, and following up on stage reviews for this project. What percentage of the project effort is that? Discuss the implications of your findings.
   b. Is this level of effort for reviewing justified? Why or why not?
8. Recommend a conversion method for each of the following. Justify your recommendation:
   a. A new airline airport passenger operations system to be used by check-in and gate agents.
   b. A new release of a serious amateur photographer's favorite photo editing program.
   c. A new income tax preparation package for a CPA firm with five accountants.
   d. A new student information system for a small liberal arts college.
9. Automobile designers use two types of prototypes: styling prototypes, traditionally made of clay, to gather feedback on proposed designs; and functional prototypes, drivable but with bodies disguised to avoid premature disclosure of styling, to test engines, transmissions, and suspensions. How does a software prototype resemble each of these? How does it differ?
10. Using an online search, identify three project management software packages. Write a report comparing them and recommending which one your company should purchase. (You don't really have a company, so make and state any necessary assumptions.) Software vendor sites aren't objective, so try to find independent reviews, user case histories, or industry analyst reports to support your evaluations.

KHOURY CANDY DISTRIBUTORS PLANS ERP CONVERSION

Jake and Isabella came back the following week to discuss the upcoming ERP conversion with Brian Greenwood, who was assigned to that project after meeting with Prof. Acton from Standish and deciding not to pursue an AI-based program to forecast retail sales for KCD customers. (The new SAP release was mentioned in the Chapter 7 episode. The forecasting application was discussed in the Chapter 9 episode.) When Brian escorted them to the familiar MIS conference room, another man was already sitting there.

"This is Nikau Taumata," said Brian as he introduced the two students to him. "Nikau studied information systems at Massey University in Auckland, New Zealand; met an American who spent a year there studying Maori language and culture; and the rest is history. Nikau's going to be the lead on our move to the new version of SAP. Right now he's planning conversion strategy. I thought you might be interested in that."

"We certainly are," agreed Isabella. "Our professor said a bit about it, just enough to make us want to hear more."

"Did he talk about the basic approaches?" Nikau asked.

"It's *she,* actually," Jake corrected him gently, "and yes, she did. She talked about four basic approaches. She also said that most of the time big organizations use a combination of pilot and phased conversion."

"I can't say what most organizations do," Nikau said, "but that's what we're doing. The folks from SAP who came out here to support us said that's what most of their customers do. It's what Massey did when it went through this with its ERP software a few years ago, and it worked well there."

"Why did you pick that method?" asked Jake.

"For two reasons," Nikau replied. "One, SAP is too big to bring up all at once. It has fingers in every part of the pie. There's one piece that makes sure we follow the correct procedures when we hire someone and another piece that handles travel expenses when a distribution center manager goes to a meeting. They're all sort of connected through the main database, but you can start with some of the pieces and then add more over time.

"Two, there are going to be problems. We don't know what they'll be, but there always are. Say we put all KCD on the time reporting system for hourly employees. When you put in your time, it starts as "Needs Approval." Then it goes to "Approved," which means you'll get paid, and finally "Taken by Payroll." Someone has to customize that for our pay periods and so on. If that person does anything wrong—not that anyone did, but *if*—it would affect lots of people and be really hard to deal with. If we put that module into just one distribution center and it has that problem, we can cut their paychecks by hand while we find the problem.

"There could still be a problem we don't catch in the pilot group, though. For instance, there's a screen where someone can see what absences an employee has. What if nobody happens to look at absences during the pilot period? If there's a problem with absence checking, it won't get caught at the pilot stage. We can't be sure nothing will surface later on, but this will improve our odds."

"Is headquarters involved in this too, or just the distribution centers?" Elizabeth asked.

"Yes. They use SAP for a lot of administrative work. They'll be the pilot group for the human resources hiring modules. They hire almost every kind of employee we have. The big hole is hourly workers: they don't have a warehouse, do their own custodial work, or anything else. We'll have to test that part in the Springfield distribution center."

"When is this going to start? It would be fascinating to see how it goes in."

"Not for a few weeks, I'm afraid," Nikau replied. "Your semester will be over by then. Still, I don't think anyone would mind if you showed up to watch."

Questions

1. Do you feel that pilot-phased conversion is a good choice for KCD? If you do, explain why. If you don't, identify a method that you think would be better, and explain why it would be.
2. Discuss change management issues you think might arise during this conversion. Include suggestions as to what KCD should do about them.
3. What implementation issues should Brian, Nikau, and other KCD staff members be concerned with, other than those discussed in the case? What, if anything, should they do about them at this time?

CASE 1: THE GENDARMES MOVE TO LINUX

The Gendarmerie Nationale, the French national police force, traces its history to the 12th century. Today it is a modern force of about 100,000, with 21st-century responsibilities such as airport security and crowd control, and a Facebook page with 787,500 Likes in December 2019. Naturally, it couldn't function without information systems.

With over 70,000 desktop computers, networks, servers, and more, information technology uses a considerable amount of the Gendarmerie's €7.7 billion annual budget (about US $8.5 billion at the late 2019 exchange rate). Since every public agency is constrained by its budget, economizing is essential. Its command staff saw opportunities for savings in desktop computers, but the conversion would not be simple.

This conversion involved the operating system. (Part of the motivation for the move was that Microsoft was ending Windows XP development, so some operating system conversion would have been required even if the Gendarmerie Nationale had stayed with the Windows family.) It also included replacing Microsoft Office with the open-source OpenOffice suite, standardizing on the Firefox browser, adopting Thunderbird for email, using the Gimp image-editing software, and the VLC multimedia application.

Prior to the conversion, only about 20,000 of the force's desktop computers had Microsoft Office, primarily for cost reasons. They were able to install OpenOffice on every computer, conforming to a mandate to expand the use of standard productivity tools. At that point, the force also moved to Open Document Format, the native format of OpenOffice, for internal reporting. It continues to use Office formats to exchange files with other organizations.

Major Stéphane Dumond of the Gendarmerie Nationale recognizes that there was risk in this conversion. "It is a risk, but a controlled risk, counterbalanced by the lower service costs," he says. He estimates the reduction in the total cost of ownership at about 40%. This comes from several sources: savings in license fees, reduced need for local technical intervention via central administration of desktop computers, and, in the future, hardware savings due to the lower resource requirements of Linux compared to Windows.

This conversion did not take place overnight. The first users converted over a four-year period. First, OpenOffice replaced Microsoft Office. Two years later, Firefox and Thunderbird were phased in. All these still ran under Windows, but had Linux versions as well. After Firefox and Thunderbird had been in use for two years, and with other applications such as Gimp now installed and running, an initial 5,000 users moved to a customized version of the popular Ubuntu Linux package, called GendBuntu, without any further application software changes. In mid-2013, 37,000 of the Gendarmerie's desktop systems used Linux. By June 2018, 82% were.

In addition to the direct cost benefits, the Gendarmerie Nationale values its new independence from commercial software vendors. "This is priceless," says Major Dumond.

The Gendarmerie Nationale is not the only public organization moving to Linux, though it may be the most successful large one as this is being written. Several Italian provinces (Perugia, Cremona, Macerata, Bolzano, and Trento) have moved to LibreOffice, an offshoot of OpenOffice, and to Linux. Based in part on their experiences, the Italian Ministry of Defence announced in late 2015 that it would follow suit. The government of South Korea is poised, as of mid-2019, to make a similar move. The governments of Brazil and the U.K. are using more and more open-source software. These experiences show that converting systems with common horizontal applications to open-source software is practical, and can bring significant savings.

(The city of Munich also moved to Linux and documented substantial savings, but returned to Windows. It has been asserted that this return was related to Microsoft's promising to move its German headquarters to Munich, which it did in late 2016. The city gave different reasons. Linux supporters claimed to refute them. The truth cannot be proven either way.)

Questions

1. What conversion method did Gendarmerie Nationale use here?
2. Do you consider this an appropriate choice for the Gendarmerie Nationale? Why or why not?
3. Suppose the Gendarmerie Nationale's desktop computers also had applications that required the Windows operating system. How might that have affected this conversion?

CASE 2: MOODY'S AND AARON'S GO AGILE

It would be hard to find two businesses less alike than Moody's and Aaron's. Moody's sells financial information services. Aaron's is a chain of rent-to-own stores selling furniture, electronics, and appliances. Yet, like every other business, both need software. Like other large organizations, both use custom software for critical applications. And, like more and more businesses of all types, both adopted Agile methods to develop it.

Aaron's needed to replace an aging point-of-sale system. They built its replacement, CustomerCore, over a period of several years. It is used by more than 19,000 employees to process about three billion dollars of business every year.

More than 15 teams of five to ten people each worked on CustomerCore over six years. Each team worked on a different business problem. Each had its own work style. All fit the Agile philosophy, though. The teams were coordinated by VersionOne* project management software.

VersionOne project management software was designed for the Agile methodology. It uses Agile terms such as *story* and *epic* (a group of related stories). It helps a team prioritize its requirements based on parameters such as scope, value, and risk. It also provides management visibility into the development process. This can be a concern with Agile methods, since they do not provide the confidence that some managers get from milestones, deliverables, and project reviews.

John Trainor, Aaron's CIO, was pleased with VersionOne's flexibility to match the way each team wanted to work. "VersionOne gently reminds you of some of the better practices of Agile project management," said Trainor. "It gives a framework that helps guide you toward best practices. You can choose not to use them if they aren't applicable to your team, but it helps guide teams and helps shape the process."

Moody's KMV, a subsidiary of Moody's Corporation, uses financial and statistical analysis to help businesses decide when and how to extend credit. They don't just say "this borrower has X percent probability of making its payments." They also predict how much lenders can expect to recover if a borrower defaults, using a database of thousands of situations where lenders recovered something after a default. This is just one of Moody's KMV's many analytical systems.

Moody's project management requirements differed from those of Aaron's. Since they used Agile in many projects, they needed to support hundreds or thousands of users, not just dozens. Their motivation for using Agile was to speed product development in a competitive market where the first with a new capability gets the lion's share. They chose Rally for several reasons:

1. It could support large numbers of users.
2. They offered a great deal of support, including on-site training.
3. They were pleased with its *roadmap*, the vendor's plan for future versions.

Darren Stovel, Senior Director of Program Management at Moody's KMV, is pleased with Agile. Referring to one project, he says "A project of that scope would likely have taken two to three years using our older waterfall-style process; we released version 1.0 of the redesigned product in twelve months." As for Rally, Stovel likes how "team members could look at a newly requested feature and see everything related to that feature – user stories, test cases, tasks, defects, notes, and discussions were all linked to the feature. We especially liked that you could see a rollup of the story hierarchy, with tasks and test cases all tied into the dashboard views."

Lessons we can learn from these experiences include:

- Agile methods can produce useful software, often more quickly than traditional methods.
- Project management tools are just as important with Agile as with the "waterfall" method.
- Companies that use Agile methods should use a project management package designed for it.

---

* Mentions of software in this case, as elsewhere in this book, should not be taken as product endorsements.

Questions

1. Do you agree with Aaron's and Moody's choice of Agile development for their systems? Explain why or why not.
2. Other than number of users, what differences do you see between the two firms' project management needs? How do you think those differences affected their choice of a solution?
3. One reason Moody's KMV chose Rally was a report by industry analysts at Forrester Research, which took it from "never heard of them" to the short list. Moody's KMV still wouldn't have chosen Rally had it not won their evaluations, but without that report they probably wouldn't have looked at it at all. What does this suggest for software marketing?

BIBLIOGRAPHY

Arin, K., "S. Korean government to switch to Linux: Ministry," *The Korea Herald*, May 17, 2019, www.koreaherald.com/view.php?ud=20190517000378, accessed September 22, 2019.

Bowers, T., "Why ignoring the end-user makes you seem incompetent," *TechRepublic*, November 7, 2013, www.techrepublic.com/blog/career-management/why-ignoring-the-end-user-makes-you-seem-incompetent, accessed August 29, 2019.

Brooks, F.P. Jr., *The Mythical Man-Month: Essays on Software Engineering*, Addison-Wesley, 1975.

Craig, J., "DevOps for the new millennium: A lifecycle approach enabled by automation," Enterprise Management Associates "Analyst's Corner" blog, June 2013, www.enterprisemanagement.com/web/ema_ac0613.php, accessed August 29, 2019.

Federal Highway Administration, U.S. Department of Transportation, "Systems engineering for intelligent transportation systems," Publication FHWA-HOP-07-069, January 2007, ops.fhwa.dot.gov/publications/seitsguide/index.htm, accessed August 29, 2019.

Finley, K., "French national police switch 37,000 desktop PCs to Linux," *Wired*, September 30, 2013, www.wired.com/wiredenterprise/2013/09/gendarmerie_linux, accessed December 11, 2019.

Gagne, C., "A practical guide to DevOps: It's not that scary," *InformationWeek*, July 5, 2019, informationweek.com/devops/a-practical-guide-to-devops-its-not-that-scary/a/d-id/1335046, accessed August 21, 2019; see also the articles it links to.

Gendarmerie Nationale web site (in French), www.gendarmerie.interieur.gouv.fr, accessed September 22, 2019.

Goldratt, E., *Critical Chain*, North River Press, 1997.

Hillenius, G., "French Gendarmerie: 'Open source desktop lowers TCO by 40%'," *European Commission JoinUp*, September 30, 2013, joinup.ec.europa.eu/community/osor/news/french-gendarmerie-open-source-desktop-lowers-tco-40, accessed September 22, 2019.

Hillenius, G., "Italian military to switch to LibreOffice and ODF," *European Commission JoinUp*, September 15, 2015, joinup.ec.europa.eu/collection/open-source-observatory-osor/news/italian-military-switch, accessed September 22, 2019.

Keogh, L., "The real cost of change," January 30, 2012, lizkeogh.com/2012/01/30/the-real-cost-of-change, accessed August 29, 2019.

Limoncelli, T., "The Top 10 things executives should know about software," *Communications of the ACM*, vol. 62, no. 7 (July 2019), pp. 34–40.

Palmquist, M.S., et al., "Parallel worlds: Agile and waterfall differences and similarities," Carnegie Mellon University, Technical Note CMU/SEI-2013-TN-021, October 2013, resources.sei.cmu.edu/asset_files/TechnicalNote/2013_004_001_62918.pdf, accessed September 22, 2019.

"Pjotr," Re: South Korea says goodbye to Microsoft, Linx Mint forums post, May 19, 2019, forums.linuxmint.com/viewtopic.php?t=294401, accessed September 22, 2019.

Rally Software case study, "Rally provides Moody's with real-time visibility and metrics to measure success," web.archive.org/web/20130908173402/https://www.rallydev.com/sites/default/files/moodys_03.pdf, accessed September 23, 2019.

Rally Software web site, www.ca.com/us/products/ca-agile-central.html, accessed September 22, 2019.

Royce, W., "Managing the development of large software systems," Technical Papers, WesCon, August 1970, web.archive.org/web/20171212205352/http://www.cs.umd.edu/class/spring2003/cmsc838p/Process/waterfall.pdf, accessed September 22, 2019.

Vaughan-Nichols, S., "Italian Ministry of Defense moves to LibreOffice," *ZDnet*, September 16, 2015, www.zdnet.com/article/italian-ministry-of-defense-moves-to-libreoffice, accessed September 22, 2019.

Vaughan-Nichols, S., "Why Munich should stick with Linux," *ZDnet*, February 14, 2017, www.zdnet.com/article/why-munich-should-stick-with-linux, accessed September 22, 2019.

VersionOne case study, "Customer success: Agile made easier," 2013, www.versionone.com/pdf/case_study/Aarons_Customer_Success_Story.pdf, accessed September 22, 2019.

VersionOne web site, www.collab.net, accessed September 22, 2019.

Wiedemann, A., et al., "Research for practice: The DevOps phenomenon," *Communications of the ACM*, vol. 62, no. 8 (August 2019), pp. 44–49.

12 Managing Information Systems

CHAPTER OUTLINE

WHY THIS CHAPTER MATTERS

Information systems are a vital organizational resource. They can make the difference between business success and failure and require a substantial budget. Many IS-related decisions must be made by businesspeople, not by technical specialists. Managing these resources is therefore a concern of all managers, not only of IS specialists and their supervisors.

In some companies, information systems management evolved without much thought. It can be good to step back, look at the options and at the factors that affect the choice, and decide if they are appropriate or should be changed. This chapter will give you some tools to do that.

In others, decisions that should be in the realm of management are left to the technical staff. No matter how competent those people are in their professions, they may not be qualified to make business decisions. This chapter will cover one such area, information systems security, to show you why you, as a manager, must take an active role in decisions that affect it.

Finally, knowing how your company works is essential to your career success. This chapter will show you some things to look for.

CHAPTER TAKE-AWAYS

As you read this chapter, focus on these key concepts to use on the job:

1 The best way to manage information systems varies with an organization's dependence on its IS, now and in the future. There is no single best answer for all organizations.
2 Information systems can provide interesting, fulfilling, and well-paid work. The demand for information systems professionals is expected to continue to grow.
3 Keeping information secure and private requires everyone's attention. The most important precautions, and those that cost the least, involve people more than technology.

12.1 THE LINKING STRATEGY

Amazon couldn't function without its information systems. They are vital to its business concept. It can't stay ahead of its competition unless it keeps innovating with them.

A local heating oil supplier could operate without its information systems. If it couldn't use its trucks' computerized oil delivery system, drivers could record each delivery on paper. Clerks could then figure out bills and mail them. That would be annoying but not disabling. Heating oil suppliers also use computers to plan customer deliveries on the basis of temperatures and previous oil demand, but those applications can be down for several hours, even a day or two, without affecting the business.

The difference between Amazon and a heating oil supplier isn't just size. They are in different businesses. Their profitability is determined by different factors. Their customers choose them for different reasons. As a result, their dependence on information systems is different. Should the way they manage those systems be different as well? Of course it should. The approach an organization should take to managing its information resources is influenced by two factors:

- How dependent is the organization on reliable, well-performing information systems? In 2019 all organizations need them, but does it depend on them more or less than average?
- How much will the organization depend on information systems to maintain or enhance its competitive position in the future? Again, this is relative to typical organizations.

Researchers created the grid of Figure 12.1 from these factors. Don't read too much into the quadrant names. For example, a firm in the Factory quadrant (depends heavily on reliable, well-performing information systems today, won't gain a competitive edge from IS in the future) can be in any industry. The name of the quadrant doesn't mean that the firm has a factory.

Separately, the researchers determined that the way an organization manages its information systems can be described by one of six *linking strategies*. Those are shown in Table 12.1.

- The Centrally Planned strategy requires a strong commitment from top management, in terms of their time and financial support when opportunities are recognized.
- The Leading Edge strategy requires willingness to experiment. The culture must not penalize failure, though it may penalize being slow to recognize it.
- The Free Market strategy requires an information systems department that is willing to compete with outside vendors.
- In the Monopoly strategy the IS department controls the organization's information systems, has enough funding for good results, and commits to meeting user needs, but might not be particularly innovative.
- The Scarce Resource strategy focuses on cost control, recognizing that spending more won't help the business.
- The Necessary Evil strategy meets basic requirements but little more.

Every organization has a linking strategy, even if its managers aren't aware of it and couldn't articulate it. In some, it's carefully thought out. In others, it just grew. People who work in those would say "it's how we do things."

How do these two concepts, the grid and the six linking strategies, come together? *Certain linking strategies fit organizations in certain quadrants of the grid better than others.* Figure 12.2

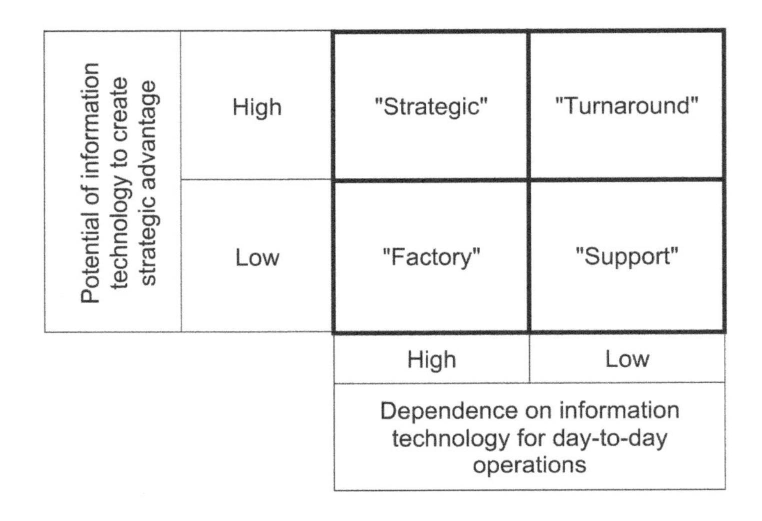

FIGURE 12.1 Linking Strategy grid.

TABLE 12.1
Linking Strategies

	Centrally Planned	Leading Edge	Free Market	Monopoly	Scarce Resource	Necessary Evil
Best fits quadrant	Strategic or Turnaround	Turnaround or Strategic	Turnaround if users are sophisticated, Support if not	Factory or Support (more expensive than necessary there)	Support or Factory (risk of underspending there)	Support (low level only)
Requires	• Unit with responsibility, authority for IT/business strategy • Knowledgeable, involved senior management	• Resource commitment • Innovative, aggressive IT management • Interface to users • Strong technical skills	• Knowledgeable users if TA • Autonomous users • Removal of IT budget controls • Willingness to duplicate effort	• User acceptance • Policies to enforce single source • Usage forecasting methods • Excess capacity	• Budget control on IT • Standards, monitoring procedures • Policies for controlling users	• Tight IT control • Meeting basic needs
Management logic	Central administration makes best decisions	Technology will create business opportunities	Market makes best decisions	Information is a corporate good	Information is a limited resource	Information is not important for business
Internal IT role	Link to the business at multiple levels	Push technological boundaries on all fronts	Compete against outside vendors; develop, market, and supply profitable services	Sole-source utility; satisfy users	Maintain systems, control costs, maximize resource usage	Manage conservatively within limited scope and abilities
Users' role	Identify IT opportunities at all levels	Use the new technology that's been developed	Identify and execute IT opportunities at all levels	Go to internal information utility when needs are realized	Identify cost-justified projects, be fairly passive	Very few users, cannot influence IT

Potential of information technology to create strategic advantage		Dependence on information technology for day-to-day operations: High	Dependence on information technology for day-to-day operations: Low
	High	"Strategic" 1. Cent. Planned 2. Leading Edge	"Turnaround" 1. Leading Edge 2. Free Market, Cent. Planned
	Low	"Factory" 1. Monopoly 2. Scarce Resource	"Support" 1. Scarce Res. 2. Monopoly, Necessary Evil

FIGURE 12.2 Linking Strategy grid with strategies.

shows which strategies fit each quadrant. The strategies shown as 1 are usually best, but those shown as 2 also work if they fit an organization's culture better.

Using a linking strategy that matches a firm poorly will cause poor organizational performance, missed opportunities, or wasted money. This is just as true of a strategy that overemphasizes IS as it is of one that underemphasizes it. A firm that doesn't stand to gain a competitive advantage from innovative systems shouldn't use a linking strategy that demands much management attention. Every hour a CEO spends on information systems is an hour that CEO does not spend on sales, product development, or other important areas. Top management time should be used where it will do the most good. That may not mean using a lot of it on information systems.

Where you fit in: Any organization you work for will have a linking strategy. If you know what it is, because it's stated or because you figured it out by watching how things are done, you'll know what you can expect and what's expected of you. If you feel it's inappropriate to the organization, you may be able to come up with suggestions to improve matters.

12.2 THE INFORMATION SYSTEMS DEPARTMENT

People form groups to accomplish tasks that one person can't accomplish alone. This was as true of hunters stalking a wooly mammoth as it is of today's startups.

Once an organization grows beyond a certain size, it is divided into parts. The top manager then leaves the internal workings of each part to its manager. This, too, has been so throughout history. In the information age, one of these parts is the information systems department.

This section focuses on the IS departments of large organizations. Smaller organizations have simpler structures or, in the smallest, no structure at all.

The CIO

The head of the information systems department goes by any of several titles, perhaps Director or Vice President of Information Systems. One common title is *Chief Information Officer* (CIO). This title arose by analogy to Chief Financial Officer (CFO).

An organization's CFO is responsible for the financial aspects of the organization, even if they're in another department. The CFO of a construction company doesn't know that a dump truck must be replaced and has no say in what it will be replaced with. The CFO doesn't decide what it will carry, where, or when; or make sure its oil is changed. Still, the CFO must find funds to pay for it, report its depreciation, and pay property taxes on its value.

So it is with the CIO. The CIO is responsible for the information resources of the organization, even if he or she has no direct say in how they are used. A university CIO doesn't decide how many sections of English 101 to offer next term, but is responsible for making sure this number is stored securely and reliably, can be retrieved, and can be updated.

As another example, the CIO is responsible for security policies. Those policies can annoy users: Why must I enter a network password if I just want to check Wikipedia? If a security breach occurs, though, the CIO will get the blame. "I didn't want to annoy users" won't serve as an excuse. This is an aspect of being responsible for information system use without direct control.

Where the IS Department Fits

Everyone except the head of an organization reports to someone else. The position of one's boss signals how the organization regards one's own position. In a university there is significance in how colleges and departments are set up. If a university has a College of Music, it values music education. If there's a Music Department in the College of Liberal Arts, it's less central. If music is in a Department of Visual and Performing Arts, it's even less important than that. Potential music students and faculty members look at those signals in deciding where to study or work.

The importance an organization attaches to its information systems can be seen in where the CIO reports. When computers just automated accounting calculations, its head (probably called Data Processing Manager then) was in the accounting department, under the corporate controller or lower. As the role of computers grew and their strategic importance was recognized, CIOs moved up the organization chart. Today they are usually on a par with the heads of other key functions, directly under the CEO or the Chief Operating Officer, as shown in Figure 12.3.

Where you fit in: The reporting position of your employer's CIO is a clue to approaching your information systems' needs. If this position is part of top management, it's important to work with the IS department. If it's way down, working with your own management may be more effective.

IS Department Structure

The developers you read about in Chapter 11 are in the IS department. This department also has people who plan the organization's future systems; troubleshoot, maintain, and expand its internal networks and external connections; operate its servers; design and maintain its databases; train and support desktop, laptop, tablet, and phone users; monitor information privacy and security; and carry out all the other behind-the-scenes tasks it takes to run today's systems.

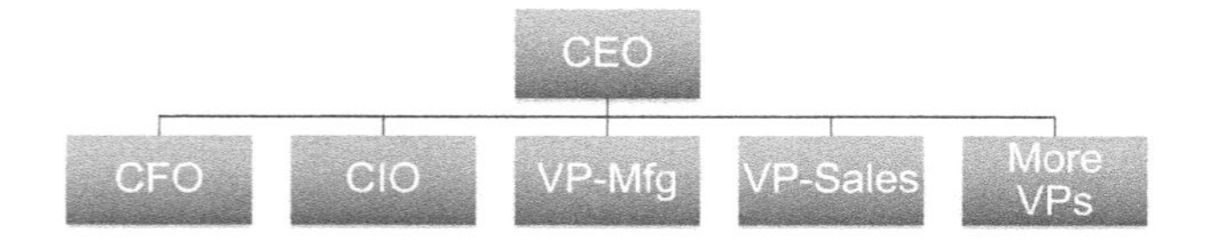

FIGURE 12.3 Corporate organization chart showing CIO on executive level.

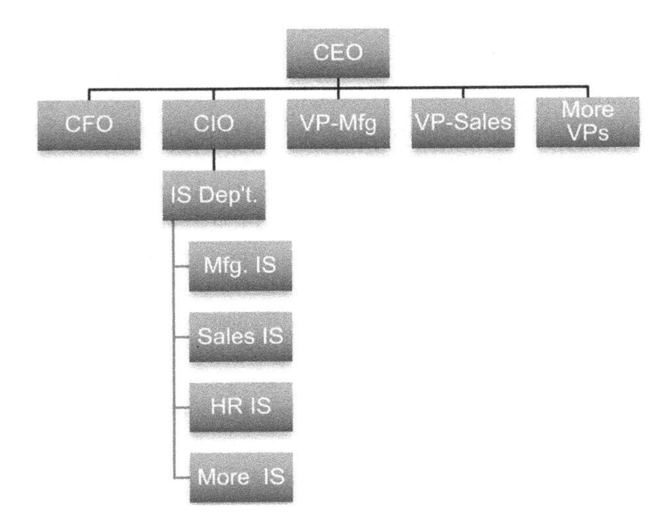

FIGURE 12.4 Organization diagram of centralized IS structure.

The best way to organize these people varies from company to company. An organization is, after all, a way for people to get work done. Since people are different, the best way to organize them will also be different. Information systems departments can be organized in many ways:

Centralized IS Department Structure

One way to structure a large organization's information systems function is to put IS personnel in one place. This department can be organized internally along functional lines (by its members' professional skills), by the part of the business that each part supports as in Figure 12.4, or some other way. In a functional IS organization, the help desk will be part of the IS department. In a business-oriented IS organization, the marketing information systems section will have a help desk to support users in Marketing. Other sections will also have their own help desks.

A centralized structure provides central control over resources. They can be allocated where the need is greatest. In addition, department members feel they belong to a professional community and have a visible career path. Its main disadvantage is that they are organizationally distant from those they support. They want to help the IS department succeed, not help Manufacturing or Sales succeed.

Decentralized IS Department Structure

Decentralizing the IS function, as in Figure 12.5, shows the real meaning of the title CIO. The firm's information resources are not under the CIO, but the CIO is still responsible for them.

The plusses and minuses of this structure reverse those of a centralized structure. IS people who report to business unit management tend to prioritize the needs of that business unit. They may feel out of touch with their professional peers, though, and it may be hard for them to take

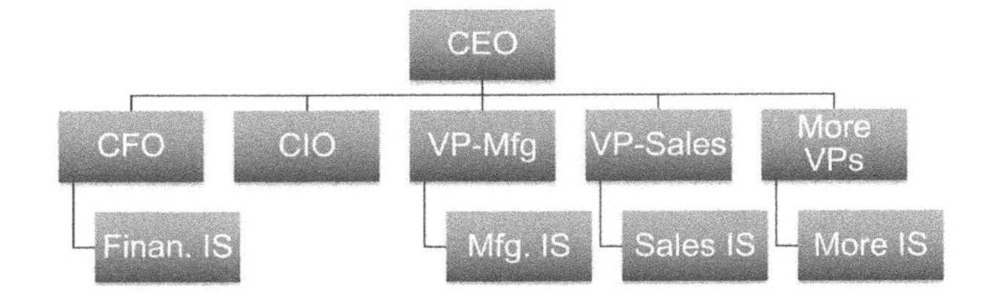

FIGURE 12.5 Organization diagram of distributed IS structure.

advantage of advancement opportunities in the IS department that supports a different part of the business.

Matrix IS Department Structure

In the *matrix organization* shown in Figure 12.6, each IS staffer has two managers: one from a business function, one from IS. That can offset some issues with centralized and decentralized structures, but this approach has its own. For one, it can be difficult to get managers from different parts of the organization, with different objectives and different criteria, to define good job performance or to agree on raises, promotions, and future assignments.

Hybrid IS Department Structure

Some firms decentralize part of the IS function while keeping part of it centralized. Application development might be distributed to business units, because people who develop applications should understand the needs of people who will use those applications, but databases, security, network management, and anything related to the corporate infrastructure would be centralized.

> ***Where you fit in:*** You probably won't have much say in how your employer's IS department is organized until you're well into your career. However, if you know how it works, you'll be in a better position to work with it so it can meet your needs. You'll save time, by knowing who to go to with a problem, and you'll get better results by approaching the right person or group first.

12.3 INFORMATION SYSTEMS JOBS AND CAREERS

Information systems offer an excellent career for intelligent, motivated professionals. The need is expected to grow for years to come. Figure 12.7 shows expected U.S. job growth, 2018–2028, for major categories of information systems personnel. Few other large groups show similar growth. (The apparent drop in the "Computer Programmers" category is more than offset by growth in categories such as application developers, who are also programmers.) It is fair to say that no comparably large category offers such growth along with attractive pay and working conditions.

Information systems jobs that one can obtain after graduating from a university or with a few years' experience include the following. This is just a sampling, not a complete list:

- **Programmer:** Writes the instructions that tell a computer how to carry out a process. This job requires the ability to conceptualize how a process can be broken down into elementary steps and attention to detail in writing out those steps precisely.

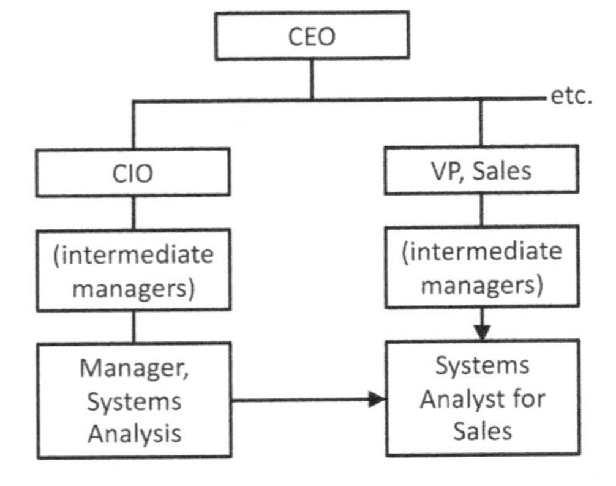

FIGURE 12.6 Organization diagram of matrix IS structure.

Job category	Employment		Change, 2018-28		Openings, 2018-28 annual average
	2018	2028	Number	Percent	
All computer occupations	4,490.1	5,036.3	546.2	12.2	403.5
Computer and information research scientists	31.7	37.0	5.2	16.5	3.2
Computer and information analysts	746.2	837.6	91.5	12.3	66.2
Computer systems analysts	633.9	689.9	56.0	8.8	53.4
Information security analysts	112.3	147.7	35.5	31.6	12.8
Software developers and programmers	1,776.3	2,063.3	287.1	16.2	164.8
Computer programmers	250.3	232.3	-17.9	-7.2	15.1
Software developers, applications	944.2	1,185.7	241.5	25.6	99.2
Software developers, systems software	421.3	463.9	42.6	10.1	35.4
Web developers	160.5	181.4	20.9	13.0	15.1
Database and systems administrators and network architects	660.2	697.3	37.1	5.6	51.2
Database administrators	116.9	127.4	10.5	9.0	9.7
Network and computer systems administrators	383.9	402.1	18.2	4.7	29.3
Computer network architects	159.3	167.7	8.4	5.3	12.2
Computer support specialists	863.1	946.2	83.1	9.6	82.5
Computer user support specialists	671.8	742.7	70.9	10.6	65.1
Computer network support specialists	191.3	203.4	12.2	6.4	17.4
Computer occupations, all other	412.8	455.0	42.2	10.2	35.7

FIGURE 12.7 IS job growth, 2010–2020 (figures in thousands except percentages).

- **Systems analyst:** Studies business processes to develop information systems that will carry them out. This job requires the ability to draw out the people who will use a new system and to write clearly. When the focus is more on the business process and less on the information system, this person's job title may be *business analyst.*
- **User support:** Who users turn to when something doesn't work. This job calls for patience: You may have heard a question 50 times, but it's new to the person who asks it!
- **Web developer:** Writes the instructions that tell a computer how to display web content. The skills resemble those of a programmer, especially for complex sites that access and update databases, but a flair for design helps too.
- **Database administrator:** Designs databases, sets them up, and keeps them running efficiently and securely. The work resembles that of a programmer, but focuses on data rather than its processing.
- **Network administrator:** Plans local area networks, Internet connections, intranets, and extranets for the organization; selects the equipment they need, supervises their installation, and keeps them running securely. (See the Chapter 6 episode of the Khoury Candy Distributors case for Armand Rocher's take on this type of work.)

People who want to work at the leading edge should consider careers with IS vendors. Vendors need people in all the above categories. They also have all the jobs of any company: accounting, sales, human resources, and so on. Some of their positions, such as computer design, require specialized technical training, but vendors have jobs for business school graduates as well.

Where you fit in: If you're reading this book, you're probably in business school. Most business schools offer majors with titles such as Computer Information Systems that can prepare you for a career in IS, especially as a systems or business analyst.

If your interests run more to building information systems than using them, a computer science program may be for you. Graduates of such programs can "hit the ground running" in database or programming jobs, but may lack the perspective to see how their work fits the broader picture.

12.4 MANAGING INFORMATION SYSTEMS SECURITY

Information systems security is ensuring that information resources are available for authorized uses, and only for those. More formally, it is *protection of information against unauthorized access, modification, or destruction.* Like many simply defined concepts, making it work is not simple. As the saying* goes, "The devil is in the details." With criminals working overtime to read or destroy databases, it is necessary to work just as hard to prevent them from doing that.

Security versus Privacy

The terms *security* and *privacy* are often confused. Privacy is keeping personal or confidential information out of the wrong hands. It requires four things:

1. An understanding of what information is considered private.
2. An understanding of who "the wrong hands" refers to.
3. A policy that states these clearly.
4. Security to enforce that policy. Privacy is meaningless without security to ensure it.

Types of Security

The usual image that comes to mind when one hears "computer security" is criminals trying to steal credit card information from databases or penetrate a bank system to move funds into their own accounts. Breaches like that make headlines. In the first half of 2018 alone, Hudson's Bay Brands (which owns Saks and Lord & Taylor, among other store chains) had 5 million credit card accounts breached; Ticketfly (owned by EventBrite) had 27 million customers' information exposed; information about 92 million people was stolen from genealogy and DNA testing firm MyHeritage; 150 million user accounts in Under Armour's MyFitnessPal's app were compromised; information on 37 million Panera Bread customers was exposed; a misconfigured cloud server at Amazon Web Services exposed over 100,000 sensitive documents about FedEx customers; a similar problem left a great deal of Los Angeles County information wide open, including notes from 200,000 incidents of child abuse, elder abuse, and more, with names and addresses … the list goes on, and this list was compiled before the Capital One breach discussed in Chapter 3. Cyberattacks, it seems, are everywhere.

Most security breaches are not of this type, though. Security failures fall into four categories:

- Uncontrollable external causes, such as natural disasters and power grid failures.
- Human error.
- Internal malicious attacks (by employees, contractors, etc.).
- External malicious attacks such as those listed just above.

The first two types of security breaches do not send data directly to criminals, but can make it unavailable for intended uses and can result in inadvertent destruction or exposure.

Malicious attacks on enterprise systems differ from those aimed at personal computers. Personal computer attacks may try to trick users into visiting malicious web sites or providing personal information in reply to an email, log keystrokes that may disclose user names and passwords, or retrieve contact information from email address books. These are not useful attacks on enterprise systems. Businesses must guard against such threats, since businesses use personal computers, but attacks on enterprise systems are different. Those attacks may attempt to penetrate a database to retrieve valuable information, or to obtain login information to access such databases by posing as a legitimate user.

Also, attacks on personal computers are usually directed towards thousands, if not millions, at the same time. Criminals often do this by creating programs that spread from computer to computer. The odds of obtaining useful information from one attacked computer may be small, but the large numbers of targets make the effort worthwhile (to the criminal). Attacks on enterprise systems tend to be aimed at a specific enterprise, probing it from every possible angle.

---

* Often attributed to architect Ludwig Mies van der Rohe, but probably not original with him.

STEALING PERUVIAN PRODUCT DESIGNS

The ACAD/Medre.A worm was targeted at users of AutoCAD computer-aided design software in Peru. (It could infect such systems anywhere, but because of where it was sent, 96% of the infected systems were in Peru. Most of the rest were in nearby Ecuador and Colombia.) It used AutoCAD's internal programming language to find AutoCAD designs in infected systems and email them to an address in China. Security experts assume that someone in China wanted to steal a Peruvian firm's confidential product designs and cast a broad net to try to get them.

The Chinese government (through CVERC, the Chinese National Computer Virus Emergency Response Center) and Tencent, the Chinese ISP that hosted the domain to which the drawings were sent, responded quickly and minimized the damage. Autodesk, the supplier of AutoCAD, updated its software, and a free program to remove ACAD/Medre. A is available from security consultancy ESET. There is no longer reason to fear it. Though it may still infect systems whose AutoCAD software has not been updated, any emails it sends will reach a dead end.

CATEGORIES OF SECURITY

Security must be provided at several levels. Database and network security were discussed in Chapters 5 and 6. Application security also matters. Since most database accesses are done via an application, failure to protect application access opens a door to the database. As John Pescatore points out, web applications are prime targets for several reasons:

1. Web applications are on the Internet. Attackers can find them and look for vulnerabilities.
2. Web sites are refreshed frequently to keep users coming back. This may lead to shortcutting test and control procedures.
3. Web applications often mix off-the-shelf software, business- or contractor-developed applications and open-source components. The variety and complexity of these, and the unfamiliarity of system developers with their insides, can hinder software testing.

Applications must be tested for security, not just for correct functioning, before deploying them.

Where you fit in: As a manager, you must be aware of your applications' potential vulnerability and make sure this testing is done.

BUDGETING FOR SECURITY

The business issue is: Does protecting against a breach cost more or less than its expected cost? The *expected cost* of a breach is the cost if a breach occurs times the probability that it will occur:

$$\text{Expected Cost} = \text{Cost} \times \text{Probability}$$

Costs of a breach include all costs of recovering; indemnifying customers, clients and employees from resulting damage; lost business during an outage; and lost future business from those who are affected or who hear about it and take their business elsewhere. The probability of a breach can only be estimated, but historical data can help for some types such as power failures.

Suppose it would cost $5,000 to protect a server room against flooding. The cost of dealing with a flood is estimated to be $100,000, including indirect costs such as lost business while servers are down. If the probability of a flood is estimated at less than 5% over the useful life of this server room, it is statistically best to accept that cost if a flood occurs.

Some security measures have little cost but great benefit. (Businesspeople call these *low-hanging fruit.*) A backup power supply to permit orderly system shutdown if utility power fails is such a measure. Its cost is low, the likelihood of needing it someday is high. Other low-cost, high-value measures include physical control over access to servers and databases, and training employees in the need to protect information and to verify the identity of anyone who requests access to it.

After those come measures that cost more and have less expected benefit because the harm they prevent is smaller, less likely, or both. Figure 12.8 shows increasing protection cost and decreasing protection value. The upper curve is their sum: the total, at any spending level, of the cost of security and the remaining exposure. The ideal spending level is at the minimum of this upper curve. To its left, an additional dollar spent on security is expected to return more than a dollar in benefits. To its right, the organization spends more to prevent a breach than the expected cost of letting the breach happen. (In a real situation the curves will be bumpy, with jumps at each additional security measure or vulnerability, and will in part be estimated.)

Financial numbers may not reflect the true impact of a security breach on a business. The cost of a rare event can be crippling, but insurance to protect against it is often affordable. We know insurance companies make money, but we accept that for protection against unlikely disasters. So it is with computer security. According to statistical calculations, many organizations overspend on security. They do this because those calculations understate the true damage of a breach. In the server room flood example from a few paragraphs back, the organization may decide to purchase insurance. (If the probability of a flood is low, insurance premiums may be less than $5,000.) Utility theory, covered in economics courses, offers one way to quantify the true risk.

Another reason for overspending on security is that security decisions are often made by technical professionals. If there is a breach and they are found not to have taken proper precautions, they face criticism. If they spend on precautions and there is no breach, they can say "we may have prevented one." This motivates them to err on the side of caution.

Countering these is the fact that security has no visible benefit. Ideally, nothing happens, but it still takes funds away from other uses. That can make it difficult to fund security measures.

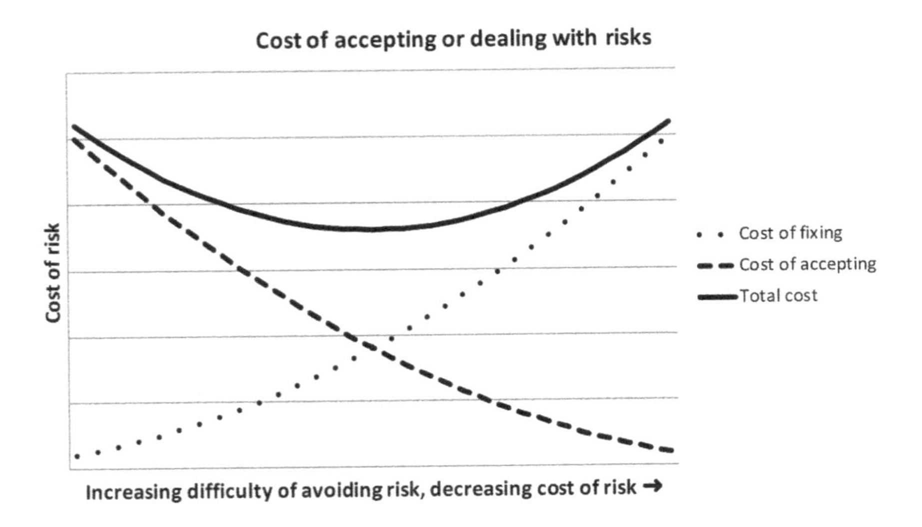

FIGURE 12.8 Security spending curve.

Where you fit in: The important security decisions are business, not technical. As a manager, you'll have to ensure adequate funding for security. Security failures can be damaging, while security success has no visible benefit. You must budget for security *before* something happens. An incident may increase funding for security, but by then it will have done its damage.

NON-TECHNOLOGY SECURITY PRECAUTIONS

Technology security precautions get most of the publicity and security budget, but simpler ones are more important. Most breaches are due to human error. Most outages are due to external causes that can be planned for. All organizations should take these precautions.

Hardware Failures: Data Backup

There are two types of people: those who have lost data to hardware failure, and those who will.

No hardware is failure-proof. This includes the hardware that stores an organization's databases. All databases should be copied regularly to backup media. At least one copy should be kept at a remote location in case the main facility becomes unavailable.

Backup strategy balances the time and expense of frequent backups, so there is always a recent one, with the time it would take to restore a database from an older copy. Most firms create *full backups* from time to time and *incremental backups*, copying changed data, at least daily. They also keep *transaction logs.* Those hold changes since the most recent incremental backup, in addition to their value in reversing transactions that were entered in error. Some organizations maintain database copies at multiple locations, but that can require expensive communication links. Cloud-based backup services are increasingly popular, especially for smaller businesses.

SPECIALIZED TERMINOLOGY

Professionals use two terms in discussing backup. Understanding them will help you make business tradeoffs when this subject comes up:

Recovery Point Objective (RPO): The maximum age of the most recent backup. If the RPO is one hour, the organization will always have a backup that is no more than an hour old. No more than one hour's data can be lost.

Recovery Time Objective (RTO): The maximum time it can take to restore a system from its most recent backup, the maximum time a system can be down due to a problem.

Ideally both are small, but here are tradeoffs between them. Shortening RPO by making frequent incremental backups increases RTO. Making RPO and RTO both small is expensive.

As organizations move data to Software as a Service providers and to the cloud, the need to deal with backups is reduced. Backups for such services are the responsibility of the service provider. In 2020, few organizations have moved all their data to such places. Almost all still need backup.

Where you fit in: As a businessperson, you can safely leave the details of data backup at work to qualified professionals. It is important for you to make sure that this area isn't overlooked.

Physical Outages: Power Protection

Power failures are frequent, as anyone who has returned to a microwave blinking "12:00" can confirm. The simplest protective measure is an *uninterruptible power supply* (UPS). It contains a battery or flywheel that takes over instantly if external power fails. A UPS can run systems for a few minutes at most, but that's longer than nearly all outages. If one lasts longer than seconds, a UPS provides time to shut equipment down in an orderly fashion or switch to alternate power.

A standby generator can enable continued normal or essential operations. The running time of a generator is limited only by the size of its fuel tank.

There are other protections against power failures. An organization can bring in power from more than one utility substation, reducing vulnerability to substation failure. It can bury cables instead of using overhead wiring, reducing the likelihood of damage due to weather or accident.

Physical Outages: Backup Sites

Disasters strike in unexpected ways. Fire, flood, earthquake, and tornado insurance can pay to rebuild a computer center that was damaged or destroyed, but will not pay for loss of business during the months that can take. Companies should plan alternate processing locations for use if their main site becomes unavailable.

Database backup is part of this picture, but only part of it. Backup ensures that a database can be restored—but where? Facilities must be planned for, to be ready when needed.

Backup sites include *hot sites* and *cold sites.*

Hot sites are fully equipped data centers that require only a copy of an organization's database to start running. They can be online in seconds to hours. An organization with a hot site should plan its backup strategy to include that site. Switchover time is then near the short end of this range.

Cold sites are prewired for power and communications services, equipped with air conditioning to remove computer-generated heat, but are otherwise empty or nearly so. If a company's main data center becomes unusable, it moves similar equipment to the cold site, installs its software, restores its databases, and begins to operate. A cold site is less expensive than a hot site, but time to restart operations is measured in hours or days.

Backup site operators assume their customers won't all have failures at the same time, so they spread the cost of their site over many customers. This usually works, but area-wide disasters, such as flooding after a hurricane, can strain cold sites in the region past the breaking point.

Physical Access Control

Potential troublemakers must be kept away from servers and other equipment. It's easier for them to enter a busy area than a secured space used by people who know each other on sight. Once inside, they can damage computers or steal data in ways that they can't over a network. Also, having more people in an area increases the likelihood of mishaps such as tripping over a cable. Access cards, keypad codes (changed often), and biometric scans are effective control methods.

Physical access control should minimize the potential for tailgating, described in the next section.

Social Engineering

Social engineering is persuading an authorized employee to allow inappropriate access to an information resource.

Social engineers employ many tricks. One is *tailgating* or *piggybacking.* A criminal starts up a conversation with someone in the line to enter a secure area, shows a box of donuts and says "I'm bringing these to my project team—want one?", and follows the authorized employee through a secure door without using a separate ID card.

Much social engineering involves the telephone. Social engineers excel at explaining why they need a network access code, database password, or administrator access to a server—right now!

All employees should be aware of the dangers of social engineering. Taking time to think is the social engineer's worst enemy. No matter how trustworthy someone sounds on the phone, no matter how well a person (perhaps briefed by a disgruntled ex-employee) knows company jargon and procedures, no confidential information should be given out and no requests should be honored until the caller is positively identified. A good response to an urgent-sounding phone call is "Spell your name, please. I'll call you back at the number listed for that name in the employee directory. If you really are who you say you are, you know that's the only way you'll get this." A caller who hangs up at that point was a social engineer.

Where you fit in: As a manager, you will be responsible for making sure the people who work for you are aware of the dangers of social engineering and don't fall for such schemes.

Technology-Based Security Precautions

Enterprises must protect against attacks on their enterprise systems and also against attacks on the personal computers that their employees, suppliers, and customers use.

A person who tries to break into computer systems is commonly called a *hacker.* Hackers were originally programmers who came up with *hacks*, ingenious ways to solve tough problems. Many people resent the way this term has taken on a negative meaning. Some have tried to get *cracker* accepted instead, but with little success. The use of *hacker* to mean a criminal is well-established, though you should expect to find people who object to this usage. A *white-hat hacker* is someone who uses similar methods to test organizations' defenses against real criminals.

Hackers may attack information systems with software that damages it or accesses unauthorized information when it runs. They use subterfuges to get their programs into computers, such as attaching them to emails with content such as "Review this customer proposal" or "Great photo of Scott at a party!" When the attachment is opened, the program runs and the computer is compromised. Such programs may propagate themselves further by finding email address lists used by popular email programs such as Microsoft Outlook.

Security experts classify software threats into categories such as *virus*, *worm*, and *trojan* based on how they work. Businesspeople need not be familiar with the differences. The umbrella term for all types is *malware.* Informally, "virus" is often used as a catch-all term.

Antivirus programs protect personal computers against many threats. Personal computer users* should install one and keep it up to date. Suppliers of such programs learn of new threats within days, often hours, and update their programs to deal with them. Many enterprises install antivirus software on all their personal computers and require it on any that connect to internal networks.

Such programs are of little use against many attacks that enterprise systems are subjected to, however. They are designed to detect broad attacks, not targeted ones.

A wide range of devices, programs, and services is available to keep networks and systems secure. All organizations should use them: not because technology-based security breaches are frequent, but because precautions keep them from becoming frequent. An organization that doesn't take them will quickly find itself the target of mischief-makers, from the greenest *script bunnies* up to more experienced criminals. It will have nobody but itself to blame.

Where you fit in: As a manager, you can leave the details of technology-based security precautions to the pros (as with data backup). Your role will be to ensure that they have the necessary management support and resources for the job they are asked to do.

---

* You may hear that Mac OS and Linux "don't get viruses." While their UNIX base may be more secure than Windows, malware for Mac OS/Linux is rare primarily because their market share makes them less attractive targets. Some malware creators, knowing that these systems are often unprotected, target them.

The Information Security Policy

Organizations have policies so employees will operate consistently according to management wishes. Information security is no exception. Security policies help ensure that people do not, through ignorance or neglect, threaten the security of information resources.

No policy can cover every situation, so part of an information security policy deals with general principles. It covers the need to protect corporate data; explains that protecting it is everyone's responsibility, so employees must avoid actions that could jeopardize it; and states that the policy applies to everyone with access to company information systems from the CEO down. It states who is responsible for making sure that the policy is followed, and the penalties for a violation.

The rest of the policy describes actions that the organization and its employees must take in support of these principles. Such actions can include:

- Using only the employer's standard personal computing platform for company business. (There must be a procedure for obtaining exceptions, but exceptions should be rare.)
- Regular remote "cleansing" of all personal computers by the MIS department.
- Prohibiting applications that are not installed by that department, since software obtained from unknown sources can contain malware. (An exception procedure is needed here too.)
- Requiring all databases to enforce access control on a "need to know" basis.
- Requiring annual, or more frequent, review of access permissions, so that employees who change jobs don't keep their old permissions in addition to acquiring those of their new job.
- Prohibiting downloading business data to a removable storage device.
- Requiring a *clean screen:* A computer screen may not show any sensitive information when it is not being actively used, even briefly.
- Requiring encryption of any remote organizational information system access.
- Requiring employees to complete information security awareness training annually.

These are examples, not a complete list. Figure 12.9 shows the opening of an IS security policy.

An organization may have other policies that focus on specific information security areas within the overall policy guidelines. For example, an organization may have a policy that describes how to define the sensitivity of information and thus the need to control access, on acceptable use of the Internet, or on disposal of equipment that may have held sensitive information.

Implementation of an organizational information security policy is not part of the policy itself. Implementation must be left to professionals in technology, in training, and in other fields. The policy guides their actions, too.

Where you fit in: As a manager, you will be involved in setting security policies. You will also have to set an example for those who work for you. If you disrespect information security, your subordinates will conclude that information security doesn't matter. If you are seen to care about information security, those who watch you will care too.

BYOD: Complicating the Picture

The security picture is complicated by the *BYOD* trend: Bring Your Own Devices.

Organizations know employees have phones and tablets. Taking them into account is a necessity. The result, however, is that company data is on a variety of individually controlled devices and is vulnerable to anyone who steals, finds, or penetrates one. Policies are needed to prevent problems from arising. These can include training, requiring secure login to networks, encrypting business data on devices, or using cloud-based applications that don't store data on clients.

Information Security Policy

I. POLICY

A. Information in all its forms—written, spoken, recorded electronically or printed—will be protected from accidental or intentional unauthorized modification, destruction or disclosure throughout its life cycle …

B. Additional policies, standards and procedures will be developed detailing the implementation of this policy and set of standards, and addressing any additional information systems functionality in each entity/department …

II. SCOPE

A. Information security includes protection of the confidentiality, integrity and availability of information.

B. This policy applies to all entities and workers, and other involved persons.

C. This policy and all standards apply to all protected health information and other classes of protected information in any form …

III. RISK MANAGEMENT

A. Periodic analysis of information systems and networks will be performed …

B. Measures to reduce the amount and scope of vulnerabilities will be implemented …

IV. INFORMATION SECURITY DEFINITIONS

V. INFORMATION SECURITY RESPONSIBILITIES

(Information Security Officer, Information Owner, Information Custodian, User, User Management)

VI. INFORMATION CLASSIFICATION

(Protected Health Information/PHI, Confidential Information, Internal Information, Public Information)

VII. COMPUTER AND INFORMATION CONTROL

Systems and information are assets of this organization and are to be protected from misuse, unauthorized manipulation, and destruction …

FIGURE 12.9 Opening of sample security policy.

KEY POINT RECAP

An information system that doesn't fit the needs of the company isn't a good information system. *Needs differ from one company to another, even within the same industry.*

Assessing needs is a business issue, not a technical one. As a businessperson, you will be in a better position than your company's technical staff to do that. You must therefore understand how those needs relate to information systems.

The information systems department can be organized in a variety of ways. *Alternatives include the centralized, decentralized, hybrid, and matrix structures.*

Businesspeople like yourself are ultimately responsible for the way your employer chooses to organize its functions. That means recognizing the advantages and disadvantages of different approaches. When you are faced with an existing information systems organizational structure, understanding how it works will be essential to working with it to get your job done.

Information systems security is never complete.

The most important information systems security precautions are not technical. They must be put in place and managed by businesspeople such as yourself. While technical measures are also important, they cannot be put in place without strong management support and funding from the user side of the organization.

KEY TERMS

Backup: Copy of data that can be used to restore a file or database in case the main copy is damaged or becomes unavailable. See *full backup, incremental backup.*

Backup site: Alternate processing site to use in case the main location of an organization's computers becomes unusable. See *hot site, cold site.*

Bring Your Own Devices (BYOD): Organizational information systems environment that integrates personally owned devices such as smartphones and tablets.

Centralized IS department structure: Organizational structure in which IS-related human resources work in the same department.

Centrally planned: *Linking strategy* in which top management supports and coordinates an organization's information systems.

Chief Information Officer (CIO): Top manager responsible for information systems and information technology.

Clean screen policy: Requirement that a computer screen not show sensitive information when it is not in use.

Cold site: *Backup site* ready for computers to be installed but without the computers themselves.

Cracker: See *hacker*, sense (b). Not widely used.

Decentralized IS department structure: Organizational structure in which all IS-related human resources work in the departments whose activities they support.

Expected cost: The cost of something (such as a breach) multiplied by the probability of its occurrence.

Free Market: *Linking strategy* in which departments are free to choose the information systems they will use.

Full backup: Backup of a complete file or database. Compare *incremental backup.*

Hacker: (a) Originally, a clever programmer who solves tough problems in ingenious ways. (b) Today, a person who attempts to penetrate information systems without authorization.

Hot site: *Backup site* complete with computers and other equipment, ready for use.

Hybrid IS department structure: Organizational structure in which some IS-related staff work in a central department, while others work in departments whose activities they support.

Incremental backup: Backup of parts of a file or database that have changed since the previous backup. Compare *full backup.*

Leading Edge: *Linking strategy* in which an organization tries to pioneer with new technologies.

Linking strategy: The overall management approach an organization takes to information systems and technology.

Malware: Software sent to a computer in an attempt to penetrate it without authorization.

Matrix organization: Organizational structure in which each employee has two managers, one in his or her profession and another in the department whose activities the employee supports.

Monopoly: *Linking strategy* in which an internal information systems department is expected to provide efficient systems to support business operations.

Necessary Evil: *Linking strategy* in which information systems are tolerated as a necessity.

Piggybacking: See *tailgating.*

Privacy: Keeping personal or confidential information away from unauthorized users.

Recovery Point Objective (RPO): Maximum time that can elapse between backups.

Recovery Time Objective (RTO): Maximum time it can take to restore a database from its backups.

Scarce Resource: *Linking strategy* in which information systems are expected to operate at minimum cost to the organization.

Script bunny: Inexperienced malware creator who modifies and uses programs (scripts) created by others. Often used as an insult.

Security: Protection of information against unauthorized access, modification, or destruction.

Social engineering: Persuading an employee to grant access to protected information assets.

Tailgating: The *social engineering* practice of persuading an authorized person to let a potential criminal through a secure access point by following the authorized person through the gate.

Transaction log: Record of transactions that can restore changes to a database since its last backup.

Trojan: Type of malware that appears to perform a useful function but instead harms a system.

Uninterruptible Power Supply (UPS): Computer power supply that is independent of utility power and can operate equipment for a short period of time (typically a few minutes).

Virus: Type of malware that attaches itself to another program and runs when that program is executed.
White-hat hacker: A person who uses criminal methods to test defenses against true criminals.
Worm: Type of malware that operates as a stand-alone program that must be executed directly.

REVIEW QUESTIONS

1. What are two key factors that should influence a firm's approach to managing its information systems?
2. What are the four quadrants in the grid that is based on these two factors?
3. Identify, and describe briefly, the six basic linking strategies.
4. How are the position of an organization in this grid and its linking strategy related?
5. What is a *CIO*, and for what is this person responsible?
6. Why does the information systems function often directly under the CEO today even though it wasn't always in the past?
7. What are some of the functions that the information systems department performs?
8. Describe three different approaches to organizing a firm's information systems staff.
9. What is the difference between *security* and *privacy*? Which cannot exist without the other?
10. What are the four main categories of security breaches?
11. Why can it be difficult to obtain adequate funding for security measures?
12. Define *RPO* and *RPO*.
13. What are two types of sites at which a company can resume data center operations after a natural disaster?
14. Define social engineering.
15. What is *malware*?
16. Why is an information security policy necessary?

DISCUSSION QUESTIONS

1. Consider three organizations: the two example firms at the beginning of this chapter (Amazon and a local heating oil supplier) and the school at which you are studying.
   a. Put each of these three in the correct quadrant of the linking strategy grid.
   b. Select an appropriate linking strategy for each of them. Justify your choice.
2. You work for a car dealer. Customers choose it because it is the only one for a popular make in its area, has a good reputation for service, and its prices are competitive. Its owner and CEO spends a great deal of time testing its web site, suggesting changes to the user interface of its sales support system, and otherwise getting involved with its information systems. What Linking Strategy does this reflect? Is it a good one for this dealership? What would you say to this CEO about this, if you were asked? (Assume he is competent, his testing is useful, and his suggestions are usually good.)
3. You work for a stock brokerage. Customers choose it because it gives good investment advice, has a good online system for managing investments, and its fees are competitive. Its owner and CEO spends a great deal of time testing its web site, suggesting changes to the user interface of its sales support system, and otherwise getting involved with its information systems. What Linking Strategy does this reflect? Is it a good one for this brokerage? What would you say to this CEO about this, if you were asked? (Assume he is competent, his testing is useful, and his suggestions are usually good.)
4. Your university probably has a centralized IS department structure. Should it keep it, or move to a decentralized, matrix, or hybrid structure? Justify your recommendation.

5. You are a programmer who wants to become a database administrator. You have a choice of jobs in organizations with centralized, decentralized, and matrix IS department structures. The jobs are equally attractive in every other way. Which do you choose, and why?
6. As a market analyst in a consumer goods company, you have an idea for using its CRM data to predict demand for new products. You start with your boss, who thinks your idea makes sense and tells you to move it forward but does not say how. For each of the four IS organizational structures, how would you proceed?
7. Search popular job sites for entry-level positions for business graduates with a focus on information systems. Create a table comparing five jobs that you found, including salary information if available. Do you think you would be good at these jobs? Do you think you would enjoy them? Explain your answers.
8. Consider the four privacy requirements in Section 12.4. List one item of information in some business database that should be available to the general public and one that should not be. For this second item, state two categories of people who should be allowed to access it.
9. Search the Internet for systems that control physical access to areas inside a building. Create a chart comparing three such systems that use three different technologies.
10. Carry out a class survey: How many of your classmates use antivirus software on their computers? Of these, how many update it regularly, whenever they receive a notice of new malware it can guard against? Do the answers depend on the platform: Windows, Macintosh, or Linux? Discuss the findings.
11. Find the information security policy for your school, or do a web search to find that of another organization. (Don't use a general policy or template that must be adapted to a specific organization.) Discuss both its general principles and its specific requirements.

KHOURY CANDY DISTRIBUTORS CHOOSES A NEW CIO

Chris Evans, who had been KCD's chief information officer for four years, moved to that position from the operations side of the business: sales rep, business systems analyst, manager of systems analysis, and then CIO. He never hid his preference for more customer contact than a CIO usually has. In fact, he often credited his success as CIO to time spent at distribution centers and talking to customers. With Jennifer Khoury to become Chief Operating Officer, the Atlanta distribution center needed a new manager. Chris applied for the position and was selected. It was a win-win: Atlanta got a manager who knew the business; Chris got the career move he wanted; and KCD kept a valued employee.

This created a vacancy in the headquarters top management team. Lakshmi and Armand took themselves out of the running: Lakshmi because she'd just been promoted to her present position, Armand because he didn't think his background was broad enough. For various reasons, none of Chris's other direct reports was considered suitable either. KCD thus had to look outside. They found four possible candidates. They were identified by letters to prevent any reaction, pro or con, to their names:

A: BS in computer science from an Ivy League university; career in MIS; currently CIO of a nonprofit; would need to relocate; now earning $110,000 per year; seeks career advancement.

B: BS in music and MBA with concentration in information systems from Standish; career in MIS, much of it in database administration; currently IS project manager at a large manufacturer; could commute but might want to move closer; now earning $90,000 per year; seeks career advancement.

C: BS and MS in computer science from California public universities; career in technical positions; currently lead programmer at a software vendor; lives locally; now earning $100,000 per year, seeks to move to a user organization.

D: BS in mathematics and MS in computer science from State (Jason Khoury's undergraduate school); career in MIS at educational institutions; currently vice-chancellor for IS (equivalent to CIO) at a large university; would need to relocate; now earning $100,000 per year; seeks to return to Springfield for family reasons.

Chris Evans chaired the initial meeting to decide how to proceed. One option was to interview all four by videoconference before deciding who to invite to Springfield, keeping in mind that two of the candidates could drive but two would have to fly in. He first passed out information sheets about the candidates. Then, he asked "To see if we can get away with doing less work, does anyone see any reason to cut any of the four now?"

Lakshmi raised her hand. Seeing that nobody else asked to speak, Chris called on her. "I have misgivings about Candidate C," she said. "C doesn't seem to have any management experience at all. I get someone wanting to come to the user side of the industry, but I don't think CIO is where they should start, and I don't want to train a boss who's never managed anyone before."

"Anyone else have any thoughts about dropping C?" Chris asked.

Brian Greenwood spoke up. "I tend to agree. I've gone over C's full information sheet, there's some managing volunteer organizations, but a charity isn't a business. I wouldn't mind seeing C here as a project leader. Maybe we should hold on to this résumé in case we get an opening like that since C's local, but not in the top spot."

Chris, turning to the group, said "And the others? Anyone feel we should rule out any of them?"

After a brief silence, Chris continued with "At least the others all have management experience on the user side, though A and D have more of it than B does. Suppose we do a weighted evaluation of all three. What are some factors we should look at?

After some discussion, the group came up with the factors listed in Figure 12.10.

"We can't evaluate the candidates on all of these yet, of course," said Chris. "For one thing, we don't have references yet. Still, we should be able to get a consensus on how important each one is. That will let us do a sensitivity analysis for how important getting more information is."

The group reached a quick consensus that 2, 3, 4, and 7 were the most important; that education was less important at the CIO level than for more junior positions; that differences in salary and relocation cost were less important than the need to get the right person; and that candy distribution experience (item 8) would be hard to find. They then agreed that, at this point, B seemed to have the

Candidate decision factors

1. Education
2. Amount of experience
3. Quality of experience
4. References
5. Salary requirements
6. Relocation expense
7. Knowledge of business
8. Knowledge of our business
9.

FIGURE 12.10 CIO candidate decision factors.

edge in business knowledge while D probably had the most relevant management experience, with A a close second in that regard. Finally, they agreed that references would be crucial, and asked the HR representative to check those for all three.

"Now that that's settled," Chris said, "there's one other item I wanted to cover today. With that, he reached under the table and pulled out a large white box.

"Surprise!" called out the KCD staff members as Chris opened the box. It held a white-frosted cake, with a graduation cap and diploma in dark blue icing under an excellent rendition of the Standish College seal.

"We know it's a few weeks early, but we're not worried about your making it," he explained. "Besides, with your project report due next week, you might not be coming back."

"How can you say that? How could we not come back?" asked Isabella, astonished. "We have to see how some things, like routing software and your CIO search, come out!"

"Well, we were hoping you'd say something like that," said Chris. "Actually"—he paused briefly here—"our business planners were hoping for a bit more. Do you two have jobs yet?"

"Sorry, but I do," answered Isabella with visible regret. "I interviewed with a few companies on campus last term, before we met Jason, and I accepted a job. I'll start their training program in about six weeks."

"And Jake, how about you?" he continued.

"I'd been planning to stay at Standish for a bit and get my MBA. I could use more business education to go with the technical stuff I've been focusing on. When I took Professor Potter's business strategy course, I realized that more electives in marketing and organizational behavior would have been at least as useful as the extra programming courses I took. If I get an MBA, I'll have the best of both worlds."

"Have you considered working for a while and then going back for your MBA?" Chris persisted.

"Now that you mention it, the MBA program people said most of their students have a few years of work experience, and it helps them academically. It might not be a bad idea. I'll talk to those folks next week!"

Questions

1. While three of the candidates seem qualified to replace Chris, the summaries in the case suggest that none is outstanding. If that preliminary feeling is correct, there might be something in the job posting that makes this opening less attractive than it could be. You haven't seen that posting, so you can't say for sure. List four things it could be, and say what KCD can do differently if they decide they want to reject all four current candidates and start a new search.
2. Come up with weights adding up to 100% for the eight factors in Figure 12.10. They don't have to agree with anything in the case. Using the Delphi process (you may need to read up on it), reach a class consensus on them. Do you think this approach to choosing weights would be useful in practice? Why or why not?
3. Consider getting a few years of business experience before going back to school for an MBA. Give two reasons why this would be a good idea, two why it wouldn't. What do you recommend for Jake? Why?

CASE 1: UNDERWATER DATA CENTERS: SHOULD YOU WORRY?

Security was defined as "ensuring that information resources are available for authorized uses, and only for those." A data center under water is not available at all. That makes it a security problem.

INAP (Internap until late 2016) is an infrastructure colocation* provider with facilities in the U.S., Europe, Asia, and Australia. Events have led them to pay attention to this issue. "No one

* *Colocation* is using a data center for several firms to obtain high-speed communication links, specialized services, and other economies of scale that small companies can't afford on their own.

expected Sandy to become as catastrophic as it was," says senior vice president Steve Orchard. "With Hurricane Irene the previous year, we're seeing a trend that's a little alarming."

INAP is strengthening at-risk facilities, including the building at 75 Broad St. in lower Manhattan that flooded after Sandy hit. During the storm, fuel pumps shut down and INAP switched to a 1,200-gallon reserve fuel tank on a higher floor to keep servers running. Since then INAP also built a new data center in Secaucus, N.J., outside the flood plain. "We take climate change very seriously, and it does factor into our new site selection," Orchard says.

Another approach to dealing with flooding proactively is to have multiple data centers and not depend on any single one. This is based on reasoning that natural disasters are unlikely to affect widely separated locations at the same time. Entergy, a $10 billion electrical power company with 2.8 million customers, is taking this approach. Before Hurricane Katrina in 2005, they had two data centers: one in New Orleans and one just across the Mississippi River in Gretna, La. It was not hard to imagine a single natural disaster affecting both of them.

By 2010, Entergy had completed a brand-new $30 million data center in Jackson, Miss., 150 miles (240 km) inland. The company balances the load on several systems between the two, says CIO Jill Israel. "Moving applications from New Orleans involved quite a choreography plan. Subsequent to Katrina we've had [major] storm events, including ice storms in Little Rock, but I no longer have to hold my breath," she says.

The company holds hurricane and storm drills every year "to get better and better at responding," Israel says. "One of the things we quickly recognized was how effective a dispersed workforce can be. Our employees can do a lot more things from remote locations and that has served us very, very well."

Heroic efforts may save the day. Power cuts due to Superstorm Sandy prevented Peer 1 Hosting's rooftop emergency generator, which was unharmed, from accessing the 20,000-gallon fuel tank in the building's flooded basement. Its "day tank" on the roof had limited capacity. Fuel trucks could reach the entrance to the building, but without power, how to get the fuel to the roof? The answer was a "bucket brigade" of Peer 1 employees, customer employees, and a few hired day laborers who passed fuel from hand to hand up the stairs. "Over the next night into the morning, we were able to successfully continue to coordinate fuel trucks coming and manually move hundreds of gallons of diesel to the roof and keep the thing online," says Anthony Casalena, founder and CEO of Squarespace, one of Peer 1's customers.

Some things can't be moved. Communication cables must go to where their users are. Verizon is replacing 150 tons of copper cable, most of it under New York City streets, with fiber optic cable. "If you take a fiber optic cable and lay it in your bathtub it probably will still work; fiber is submersible," says Chris Kimm, vice president of global customer assurance for Verizon Enterprise Solutions.

And some are doing nothing. "Until we have a major data outage most clients are not calculating for risk or change; they're turning a blind eye to it," says Rakesh Kumar, a Gartner vice president who specializes in data centers and infrastructure. He feels this is wrong: "What was perceived as a safe area before may not be now." And it's not just flooding: "In London and Germany, winters seem to be getting slightly worse; we've had … bits of electrical equipment freezing up."

Questions

1. As an entry-level new hire, you note that your employer's servers are all at its headquarters: the ground floor of a historic building in Charleston, S. C., on E. Battery St. near Atlantic and Water Streets. (If you infer from the street names that it's near the ocean, you're right.) It has no protection against flooding or other natural disasters. Write a memo to the company CIO pointing out the risks and suggesting at least one course of action. Remember your position in the company as you write.

   This question was written before Hurricane Dorian hit Charleston in September 2019. This data center would probably have been destroyed. Assume you were writing before Dorian.

2. Most water damage to computers is not due to flooding, but to plumbing failures on higher floors. (Water may travel a long distance in a ceiling before it finds an opening.) Given that, do articles about floods contribute to misplaced corporate priorities? Why or why not?
3. A data center operator can't count on customer help in a disaster. People like Casalena wouldn't have helped Peer 1 Hosting if that firm hadn't sent an email to customers about its situation. What does this suggest to you about handling a natural disaster, should one occur?

CASE 2: SNOOPS ON THE INSIDE

Cedars-Sinai Medical Center, Los Angeles, California, fired six employees for accessing Kim Kardashian's medical records shortly after she gave birth to her daughter North West.

Norfolk General Hospital in Simcoe, Ontario, fired a nurse who was found to have improperly accessed the protected health records of about 1,300 patients over a nine-year period.

University Hospitals in Cleveland, Ohio, notified nearly 700 patients after an employee was caught looking at confidential medical records—and was found to have looked at them for over three years. The employee was fired.

Five employees were fired by Nanaimo Regional General Hospital, on Vancouver Island in British Columbia, during 2013–2019 for snooping through patient records. B.C. law does not specifically prohibit snooping, though it does prohibit disclosing information obtained that way.

The common thread is clear. There is a great deal of personal temptation to see someone else's medical records: perhaps a celebrity such as Ms. Kardashian, perhaps an acquaintance, perhaps anyone. With about 20 million healthcare workers in the U.S. and Canada, it is not surprising that some yield to temptation. While some might say it's impressive that so few do, any breach is cause for concern.

According to Experian, many healthcare organizations have invested in security technologies and most have developed data breach response plans. Cyber insurance policies are increasing in popularity. Yet, "Although businesses will increase focus on security protocols against external hackers ... many will miss the mark on protecting against insider threat." No type of business, especially but not only in healthcare, can afford to ignore it. "Security can no longer be viewed as just an IT issue. Scrutiny of corporate leadership's management of security may continue to increase in the form of legal and regulatory action after a major incident."

In plain English: it's a business problem, not a technology problem.

How were these breaches discovered?

The Kardashian snoopers were caught after media reports included confidential information about her baby that she hadn't told anyone.

The Norfolk General employee was caught after a patient heard confidential information about herself from friends and neighbors and called the hospital, which launched an investigation.

The University Hospitals employee and the Nanaimo Regional General Hospital employees were caught in the act.

None of these is a reliable method of catching offenders in the future. Most patients are not of interest to tabloids. Most friends and neighbors don't repeat what they hear. Most snoops aren't caught—or aren't caught quickly.

Mary Chaput of Clearwater Compliance, a HIPAA advisory firm in Brentwood, Tenn., advises:

- **Conduct a security risk analysis.** This step shows "due diligence" in protecting confidential patient data, and is required by HIPAA.
- **Communicate your no-snooping policy clearly to all employees.** Give all new hires both a written and a verbal orientation to a zero-tolerance policy on snooping. This policy should also extend to business associates, including accountants, lawyers, and IT professionals.

- **Give employees only minimum necessary access to protected health information.** A receptionist doesn't need access to clinical data.
- **Password-protect medical files depending on need to know.** Remind employees frequently that sharing passwords and user IDs is prohibited.
- **Document a formal process for initiating and terminating access.** Shut down access immediately when an employee leaves.
- **Communicate and enforce disciplinary actions for snooping.** Inform employees up front what the consequences will be, such as suspension or termination of employment.
- **Conduct background checks** before new employees start. Running checks shows reasonable due diligence and may prevent costly fines and a tarnished reputation.

Healthcare organizations can't ignore electronic precautions. The 2019 Cybersecurity Survey by the Healthcare and Information Management Systems Society found that 82% of hospitals had a "significant" security incident in the previous 12 months. The most common perpetrator was an online scam artist, with most of those attacking via email, but negligent insiders were in a close second place. Fortunately, healthcare organizations are waking up. From 2018 to 2019, 70 of 135 organizations had increased their cybersecurity spending, while only 4 had decreased it.

Doug Brown, president of Black Box Market Research, points out that "Healthcare organizations are hyper-focused on patient care." Perhaps people want it that way, but not at the expense of security. They must put someone in charge of cybersecurity, a person who will not be distracted by other issues no matter how important they are to the mission of the hospital overall.

Not in healthcare? That doesn't mean you can ignore the threat of insider security breaches or the need to be proactive about security. The details may differ, but the principles don't.

Questions

1. What sort of confidential information might unauthorized employees be motivated to snoop into, and why, in these types of organizations:
   a. An automobile dealership
   b. A university
   c. A city government
   d. A clothing store
2. You are head of a hospital that discovered a situation such as those of Norfolk General, University, and Nanaimo Regional General hospitals. You terminate the offending employee and apologize, in writing, to patients whose records were improperly accessed. A reporter for local TV calls with "That's fine for this time, but what are you doing to make sure it doesn't happen again?" and says she would like to play your answer, not over 30 seconds long, on the 7 o'clock news. She will call back in half an hour to record it. Write your answer. Read it out loud, without rushing, to make sure it takes under 30 seconds. (That's probably 100 words or so.) If it's too long, shorten it and time it again.
3. Cedars-Sinai Medical Center, because of its location and reputation, has many patients from the entertainment field. Like Ms. Kardashian, they are of more interest to the press than most people. That puts employees under great temptation to access medical records, because of curiosity about celebrities and opportunities to sell information. Should Cedars-Sinai do anything differently than Ms. Chaput's general recommendations?

BIBLIOGRAPHY

Black Box Market Research, "State of the healthcare cybersecurity Industry 2018 user survey results," May 2018, blackbookmarketresearch.com/uploads/pdf/2018%20Black%20Book%20State%20of%20the%20Cybersecurity%20Industry%20&%20User%20Survey%20Results.pdf, accessed September 23, 2019.

Chaput, M., "Don't let your practice get stung by 'snooping'," *Medical Practice Insider*, July 22, 2014, https://web.archive.org/web/20161225105340/www.medicalpracticeinsider.com/best-practices/dont-let-your-practice-get-stung-snooping, accessed September 23, 2019.

Chickowski, E., "The biggest cybersecurity breaches of 2018 (so far)," *Tech Digest*, July 2018, twimgs.com/custom_content/the-biggest-cybersecurity-breaches-of-2018-so-far.pdf, accessed September 25, 2019.

Editorial staff, "Crack down on snooping into health records," *Times-Colonist, Victoria, B.C., Canada*, April 13, 2019, www.timescolonist.com/opinion/editorials/editorial-crack-down-on-snooping-into-health-records-1.23789117, accessed September 23, 2019.

Emmett, A., "Data centers under water: What, me worry?" *Computerworld*, May 2, 2013, www.computerworld.com/s/article/9238763/Data_centers_under_water_What_me_worry_, accessed September 23, 2019.

ESET Press Center, "ESET Uncovers ACAD/Medre.A Worm: Tens of thousands of AutoCAD design files leaked in suspected industrial espionage," June 21, 2012, https://web.archive.org/web/20121101022716/http://www.eset.com/us/presscenter/press-releases/article/eset-uncovers-acadmedrea-worm-tens-of-thousands-of-autocad-design-files, accessed September 23, 2019.

Healthcare and Information Management Systems Society, *2019 HIMSS cybersecurity survey*, 2019, www.himss.org/sites/himssorg/files/u132196/2019_HIMSS_Cybersecurity_Survey_Final_Report.pdf, accessed September 23, 2019.

INAP web site, www.inap.com, accessed September 23, 2019.

Kumar, M., "A Virus Specialized for AutoCAD, a perfect cyber espionage tool," *The Hacker News*, June 23, 2012, thehackernews.com/2012/06/virus-specialized-for-autocad-perfect.html, accessed September 23, 2019.

McCann, E., "Nurse sacked for snooping patient files," *Healthcare IT News*, August 14, 2013, www.healthcareitnews.com/news/nurse-sacked-snooping-patient-files, accessed September 23, 2019.

McCann, E., "Six fired for keeping up with Kardashian," *Healthcare IT News*, July 15, 2014, www.healthcareitnews.com/news/kardashian-hipaa-breach-catastrophe, accessed September 23, 2019.

McCann, E., "Employee sacked after snooping patient EMR records," *Healthcare IT News*, December 2, 2014, www.healthcareitnews.com/news/employee-sacked-after-snooping-patient-emr-records, accessed September 23, 2019.

McFarlan, W. and J.L. McKenney, "The information archipelago—Governing the New World," *Harvard Business Review*, vol. 61, no. 9 (July–August 1983), p. 91.

Millard, M., "Healthcare to be 'plagued' by data breaches in 2015," *Healthcare IT News*, December 12, 2014, www.healthcareitnews.com/news/healthcare-plagued-breaches-2015.

Pescatore, J., "Application security: Tools for getting management support and funding," (A SANS Whitepaper), September 2013, www.sans.org/reading-room/analysts-program/whitehat-appsec-2013 (requires free registration), accessed September 23, 2019.

Thibodeau, P., "Huge customer effort keeps flooded NYC data center running," *Computerworld*, November 1, 2012, www.computerworld.com/s/article/9233136/Huge_customer_effort_keeps_flooded_NYC_data_center_running_, accessed September 23, 2019.

U.S. Department of Labor, Bureau of Labor Statistics, "Employment by detailed occupation," September 4, 2019, www.bls.gov/emp/tables/emp-by-detailed-occupation.htm, accessed September 4, 2019.

Index